AFFINITY CHROMATOGRAPHY

AFFINITY CHROMATOGRAPHY

Biospecific Sorption — The First Extensive Compendium on Affinity Chromatography as applied to Biochemistry and Immunochemistry

Proceedings of an International Symposium held at Vienna, Austria, 20-24 September 1977

Edited by

O. HOFFMANN-OSTENHOF M. BREITENBACH

F. KOLLER D. KRAFT O. SCHEINER

University of Vienna, Austria

PERGAMON PRESS

OXFORD · NEW YORK · TORONTO · SYDNEY · PARIS · FRANKFURT

U.K.	Pergamon Press Ltd., Headington Hill Hall, Oxford OX3 0BW, England
U.S.A.	Pergamon Press Inc., Maxwell House, Fairview Park, Elmsford, New York 10523, U.S.A.
CANADA	Pergamon of Canada Ltd., 75 The East Mall, Toronto, Ontario, Canada
AUSTRALIA	Pergamon Press (Aust.) Pty. Ltd., 19a Boundary Street, Rushcutters Bay, N.S.W. 2011, Australia
FRANCE	Pergamon Press SARL, 24 rue des Ecoles, 75240 Paris, Cedex 05, France
FEDERAL REPUBLIC OF GERMANY	Pergamon Press GmbH, 6242 Kronberg-Taunus, Pferdstrasse 1, Federal Republic of Germany

First edition 1978

British Library Cataloguing in Publication Data

Affinity chromatography.
1. Affinity chromatography - Congresses
I. Hoffmann-Ostenhof, O II. Österreichische
Biochemische Gesellschaft
574.1'9285 QP519.9.A35 78-40289

ISBN 0-08-022632-9

In order to make this volume available as economically and as rapidly as possible the authors' typescripts have been reproduced in their original forms. This method unfortunately has its typographical limitations but it is hoped that they in no way distract the reader.

*Printed in Great Britain by William Clowes & Sons Limited
London, Beccles and Colchester*

CONTENTS

PART IV

MATRIX-BOUND ANTIGENS AND ANTIBODIES

LIST OF PARTICIPANTS

ALLEN, G.J., G.D. Searle & Co. Ltd., Research Div.,Lane End Road, High Wycombe, Bucks., HP12 4HL, United Kingdom.

AEBISCHER, M.L., Centro Nazionale Transfusione Sangue, Via Ramazzini 15, Roma, Italy.

ALLENMARK, S., Clinical Research Centre, Linköping University Hospital, S-581 85 Linköping, Sweden.

ANDERSON, U., Clinical Research Centre, Linköping University Hospital, S-581 85 Linköping, Sweden.

BACHNER, J., Institut für Medizinische Chemie, Währingerstraße 10, A-1090 Vienna, Austria.

BARNA-VETRO, I., Human Institute for Serobacteriological Research and Production, Szállás u.5, H-1107 Budapest X, Hungary.

BARROS, M.F., Cell Biology Laboratory, Gulbenkian Institute, Oeiras Portugal.

BAUSCH, J., c/o F. Hoffmann-La Roche & Co. AG, Grenzacherstr. 124, CH-4002 Basel, Switzerland.

BIEVA, C., Laboratoire national de référence pour l'histocompatibilité, Hôpital Saint-Pierre, rue Haute 322, B-1000 Bruxelles, Belgium.

BIRKE, F., Institut für Experimentelle Chirurgie, Abteilung f. Exper. Chirurgie u. Immunologie, D-74 Tübingen, German Federal Republic.

BINDER, B., Physiologisches Institut der Universität Wien, Schwarzspanierstraße 17, A-1090 Vienna, Austria.

BLUM, W., Vogtshaldenstr. 47, D-74 Tübingen, German Federal Republic.

BOKHOUT, B.A., Centraal Diergeneeskunding Inst., Afdeling Rotterdam, Prof. Poelslaan 35, Postbus 6007, Rotterdam 7, The Netherlands.

BOLLMANN, O., Institut f. Allgemeine Biochemie, Universität Wien, Währingerstraße 38, A-1090 Vienna, Austria.

BOSCHETTI, E., IBF, 35 quai du Moulin de Cage, F-92231 Genneviliers, France.

BRANDT, J., AB Kabi, S-112 87 Stockholm, Sweden.

BREITENBACH, G., Institut f. Allgemeine Biochemie, Universität Wien, Währingerstraße 38, A-1090 Vienna, Austria.

BREITENBACH, M., Institut f. Allgemeine Biochemie, Universität Wien, Währingerstraße 38, A-1090 Vienna, Austria.

BROCAS, H., I.R.I.B.H.N., Faculté de Médicine, 115, Bld. de Waterloo, Rue Evers 20, B-1000 Bruxelles, Belgium.

BRUNNER, G., Max Planck Institut für Immunbiologie, Stubeweg 51, D-78 Freiburg, German Federal Republic.

BÜNEMANN, H., Universität Bielefeld, Biologie III M, Morgenbreede 45, D-48 Bielefeld, German Federal Republic.

CACACE, M.G., Laboratory of Molecular Embryology, Via Toiano 2, I-80072 Arco Felice (Napoli), Italy.

CASATI, P., Lepetit S.p.A., Via M. Gorky 30/32, Cinisello Balsamo (Milano), Italy.

CLAEYSSENS, M., Labor. voor Algemene en Biologische Scheikunde, Rijksuniversiteit Gent, K.L. Ledeganckstraat 35, B-9000 Ghent, Belgium.
CLERICI, L., Biology Group - D.G. XII, Joint Research Center, I-21020 Ispra, Italy.
CLIFFORD, H.K., Shell Research Ltd., Biosciences Laboratory, Sittingbourne Research Centre, Broad Oak Road, Sittingbourne, (Kent), United Kingdom.
COOKE, K.B., Dept. of Chemical Path. Page Street Wing, 17 Page Street, London SW1P 2AR, United Kingdom.
CORCORAN, E., Imperial Chemical Industries Ltd., Agricultural Division, Research & Development Dept., Biochemical Group, Billingham, Cleveland, United Kingdom.
CORCUFF, P., L'Oreal Laboratoire de Recherche, L'avenue de Saint Germain, F-93600 Aulnay-Sous-Bois, France.
CORGIER, M., IBF, 35 Quai du Moulin de Cage, F-92231 Gennevilliers, France.
COURTENAY, J.S., Wellcome Research Laboratories, Langley Court, Beckenham, United Kingdom.
DEAN, P.D.G., Department of Biochemistry, University of Liverpool, Liverpool L69 3BK, United Kingdom.
DECKX, R., Centraal Laboratorium, Appelmanstraat 29, B-2000 Antwerpen, Belgium.
DIMENT, J.A., Wellcome Reagents Ltd., Beckenham, Kent, BRS3 3BS, United Kingdom.
DUMONT, J.E., Ave. Louis 199 A, B-1050 Bruxelles, Belgium.
DWORSKY, P., Institut für Allgemeine Biochemie, Universität Wien, Währingerstraße 38, A-1090 Vienna, Austria.
EGLY, J.M., Centre de Neurochimie du CNRS and Unité 44 de l'INSERM, 11, rue Human, F-67085 Strasbourg Cedex, France.
EKETORP, R., AB Kabi, Fack, S-11287 Stockholm, Sweden.
EMÖD, I., Inst. Pasteur Chimie des Proteins, 28, Rue du Docteur Roux, F-75015 Paris, France.
ERARSLAN, A., İstanbul Medical Faculty, Dept. of Microbiology, Capa-Istanbul, Turkey.
ERSSON, B., Institute of Biochemistry, University of Uppsala, Box 576, S-751 23 Uppsala, Sweden.
FABRE, J., Nuffield Dept. Surg., The Radcliffe Infirmary, Oxford, OX2 6HE, United Kingdom.
FAUCHERE , J.L., Institut f. Molekularbiologie und Biophysik, ETH-Hönggerberg, CH-8093 Zürich, Switzerland.
FLEISSNER, A., Psychiatrische u. Nervenklinik, Universität Hamburg, Neurochem. Abteilung, Martinistr. 52, German Federal Republic.
FÖRSTER, O., Institut f. Allgem. u. Exp. Pathologie, Universität Wien, Währingerstraße 13, A-1090 Wien, Austria.
FRIED, W., Industriestr. 72, A-1220 Vienna, Austria.
GABEL, D., Dept. of Chemistry, University of Bremen NW2, Postfach 330 440, D-28 Bremen 33, German Federal Republic.
GEMEINER, M., Tierärztliche Universität, Institut f. Biochemie, Linke Bahngasse 11, A-1030 Vienna, Austria.
GERFAUX, G., France
GLATTHAAR, B.E., Diagnostische Forschungsabteilung, F. Hoffmann-La Roche & Co AG, CH-4002 Basel, Switzerland.
GRANT, D., Dept. of Surgery, St. Georges Hospital Medical School, Cranmer Terrace, Tooting, London SW17 ORE, United Kingdom.
GREIMEL, A., Schwechaterstraße 22, A-2325 Himberg, Austria.
GSCHWENDER, H.H., Postfach 1109, MPI f. Zellbiologie, D-2940 Wilhelmshaven, German Federal Republic.

GSCHWIND, H.P., Biozentrum d. Universität Basel, Klingelbergstr. 70, CH-4056 Basel, Switzerland.
GÜRLER, B., Istanbul Medical Faculty, Dept. of Microbiology, Capa-Istanbul, Turkey.
HALPERIN, G., The Weizmann Institute of Science, Dept. of Chemical Immunology, Rehovot, Israel.
HAIGIS, E., Kinderspital Zürich, Steinwiesstr. 75, CH-8032 Zürich, Switzerland.
HAMMARSTRÖM, S., Wenner-Gren-Institute, University of Stockholm, Dept. of Immunology, FAC, S-104 05 Stockholm 50, Sweden.
HAMMERSCHMID, F., Sandoz Forschungsinstitut, Brunnerstr. 59, A-1235 Vienna, Austria.
HANSEN, S.I., Frederiksborg Amts Centralsygehus, Centrallaboratoriet, DK-3400 Hillerød, Denmark.
HARTIG, A., Institut f. Allgemeine Biochemie, Universität Wien, Währingerstraße 38, A-1090 Vienna, Austria.
HELLWIG, E., Bundesanstalt f. Lebensmitteluntersuchung und Forschung, Kinderspitalgasse 15, A-1090 Vienna, Austria.
HERMON-TAYLOR, J., Department of Surgery, St. Georges's Hospital Medical School, Cranmer Terrace, London, SW17 ORE, United Kingdom.
HORÁNYI, M., Daróczi ut. 24, H-113 Budapest XI, Hungary.
HINTEREGGER, Ch., Institut für Biochem. Technol. und Mikrobiol., Technische Universität Wien, Getreidemarkt 9, A-1060 Vienna, Austria.
HITSCHMANN, A., Institut f. Angew. Mikrobiologie, Universität f. Bodenkultur, Peter Jordan Str. 82, A-1190 Vienna, Austria.
HOFFMANN-OSTENHOF, O., Institut f. Allgemeine Biochemie, Universität Wien, Währingerstraße 38, A-1090 Vienna, Austria.
HUBERT, P., Laboratoire de Chimie Physique Macromoleculaire, ENSIC, 1 rue Grandville, 5400 Nancy, France.
JARISCHKO, E., 2. Universitäts-Hautklinik (Immunologie), Alserstraße 4, A-1090 Vienna, Austria.
JOHNSON, R., G.D. Searle & Co. Ltd., Research Division, P.O. Box 53, Lane End Road, High Wycombe, Bucks. HP12 4HL, United Kingdom.
JÖNSSON, A., Dept. of Tumor Immunology, Wallenberglaboratory, University of Lund, Fack., S-220 07 Lund, Sweden.
KAMBERGER, W., Inst. f. Mikrobiologie u. Biochemie, Lehrstuhl f. Mikrobiologie, Universität Erlangen-Nürnberg, Friedrichstr. 33, D-8520 Erlangen, German Federal Republic.
KATINGER, H., Institut f. Angew. Mikrobiologie, Universität f. Bodenkultur, Peter Jordanstraße 82, A-1190 Vienna, Austria.
KIERULF, P., Central Laboratory, Ulleval Hospital, N-Oslo 1, Norway.
KOLLER, E., Physiologisches Institut, Schwarzspanierstraße 17, A-1090 Vienna, Austria.
KOLLER, F., Institut f. Allgemeine Biochemie, Universität Wien, Währingerstraße 38, A-1090 Vienna, Austria.
KOKOSCHKA, E.M., 2. Universitäts-Hautklinik (Immunologie), Alserstraße 4, A-1090 Vienna, Austria.
KRAFT, D., Institut f. Allgem. u. Exp. Pathologie, Universität Wien, Währingerstraße 13, A-1090 Vienna, Austria.
KRISTIANSEN, T., Biochemical Institute, Biochemical Center, P.O. Box 576, S-751 Uppsala 23, Sweden.
KRUSE, V., Institut f. Physiologie, Technische Universität München, D-8050 Freising, German Federal Republic.
LANG, R.E., Universität Ulm, Abteilung f. Innere Medizin, Steinhövelstr. 9, D-7900 Ulm/Donau, German Federal Republic.

LANNÉR, M., Dept. of Parasitology, National Bacteriological Labor., S-105 21 Stockholm, Sweden.
LAKATOS, S., Institute of Biochemistry, P.O. Box 7, H-1502 Budapest, Hungary.
LAWNY, F., Institut de Technologie des Surfaces Actives, U.T.C., B.P. 233, F-60206
LEMIUS, J., Pharmacia Fine Chemicals, Uppsala, Sweden.
LENAUER, K., Inst. f. Biochem. Technol. und Mikrobiol., Technische Universität, Getreidemarkt 9, A-1060 Vienna, Austria.
LIN, J.T., Hedderichstr. 43a, D-6 Frankfurt/Main 70, German Federal Republic.
LINNAU, Y., Immuno AG, Industriestraße 72, A-1220 Vienna, Austria.
LÖFFELHARDT, W., Institut f. Allgemeine Biochemie, Universität Wien, Währingerstraße 38, A-1090 Vienna, Austria.
LOWE, Ch., University of Southampton, Department of Physiology and Biochemistry, Medical and Biological Science Building, Bassett Crescent East, Southampton SO9 3TU, United Kingdom.
LUKACS, N., Inst. f. Chem. Pflanzenphysiologie, Correntstraße 41, D-74 Tübingen, German Federal Republic.
LUKAS, E.V., Elisabethstraße 60, A-2500 Baden, Austria.
MAGEE, A.I., Department of Surgery, St. George's Hospital, Medical School, Cranmer Terrace, London SW17 ORE, United Kingdom.
MARQUET-ESPINOSA, A., Service de Microbiologique, Institut de Botanique, Univ. de Liège, B-4000 Liège, Belgium.
MATTAUSCH, M., Abt. f. Experimentelle Chirurgie, I. Chirurgische Univ. Klinik, Alserstraße 4, A-1090 Vienna, Austria.
MATTHES, S., Schering, Dept. Allgem. Biochemie, Müllerstraße 170-178, D-1000 Berlin 65, German Federal Republic.
MAUCH, H., Med. Univ. Klinik u. Poliklinik, Lehrstuhl Innere Medizin I, Immunologisches Labor, Univ. d. Saarlandes, D-665 Homburg/Saar, German Federal Republic.
MEILLER, M., Centre de Recherches de la Croiy-de-Berny, 182-184 Ave. Aristide-Briand, F-92160 Antony, France.
MESTER, J., Université de Paris Sud, Dépt. de Chimie Biologique, Fac. de Méd. de Bicêtre, 78, Av. Du Gén. Leclerc, F-94 Bicêtre, France.
MEYRATH, J., Institut f. Angewandte Mikrobiologie, Universität f. Bodenkultur, Peter Jordan-Str. 82, A-1190 Wien, Austria.
MOHR, P., Zentralinstitut f. Molekularbiologie d. Akademie d. Wissenschaften der DDR, DDR-115 Berlin-Buch, German Democratic Republic.
MOLINARI, E., Industriestraße 72, A-1220 Vienna, Austria.
MOSBACH, K., The Lund Inst. of Technology, University of Lund, Biochemistry II, P.O. Box 740, S-220 07 Lund 7, Sweden.
MOUCHES, C., Univ. Bordeaux II, Lab. du Biologie Cell. et Molec. de Bordeaux, Pont-de-la-Maye, France.
NAKAMURA, S., Yamaguchi University, Yoshida, 753 Yamaguchi, Japan.
NELBÖCK, M., Boehringer Mannheim, Werk Tutzing, Bahnhofstraße 9-15, D-8132 Tutzing, German Federal Republic.
O'CARRA, P., Department of Biochemistry, University College, Galway, Ireland.
OGLE, C., Department of Biochemistry, College of Medicine, 231 Bethesda, Cincinnati, Ohio 45267, USA.
OGLE, J., Department of Surgery, College of Medicine, University of Cincinnati, Eden and Bethesda Ave., Cincinnnati, Ohio 45267, USA.
OGUNYEMI, O., Institut für Allgemeine Biochemie, Universität Wien, Währingerstraße 38, A-1090 Vienna, Austria.

PAHLMANN, S., Institute of Biochemistry, University of Uppsala, Box 576, S-751 23 Uppsala, Sweden.

PAHUD, J.-J., Nestec. Lab., 5 RO8, CH-1814 La Tour de Peilz, Switzerland.

PAOLUCCI, F., Centre de Recherches Clin-Midy, F-34082 Montpellier-cedex, France.

PARIKH, I., Molecular Biology Dept., The Wellcome Res. Laboratories, Burroughs Wellcome Co., 3030 Cornwallis Road, Research Triangle Park, N.C. 27709, U.S.A.

PELLE, T., Institute of Microbiology, University Medical School, H-6720 Szeged, Dóm tér, Hungary.

PETERLIK, M., Institut f. Allgem. und Experiment. Pathologie, Universität Wien, Währingerstraße 13, A-1090 Vienna, Austria.

PIQUET, Y., Centré de Transfusion Sanguine et d'Hématologie, Place Amélie-Raba-Léon, F-3303 Bordeaux, France.

PITTNER, F., Institut für Allgemeine Biochemie, Universität Wien, Währingerstraße 38, A-1090 Vienna, Austria.

POHL, A., 1. Medizinische Univ.-Klinik, Lazarettgasse 14, A-1090 Vienna, Austria.

PORTMANN, T., Schweiz. Serum u. Impfinstitut, Postfach 2707, CH-3001 Bern, Switzerland.

POTUZAK, H., University of Liverpool, Department of Biochemistry, Liverpool L69 3BX, United Kingdom.

PLUZEK, K.-J., Dakopatts A/S, Guldborgvej 22, DK-2000 Kobenhavn F, Denmark.

RAPER, J., Wellcome Res. Lab., Langley Court, Beckenham, United Kingdom.

RAVAZZOLO, R., I.S.M.I.-Ematologia, Via Benedetto XV,6, I-16132 Genova, Italy.

RÖDER, A.H., Bahnhofstraße 41, D-8124 Seeshaupt, German Federal Republic.

ROSSI, M., Instituto Internazionale di Genetica e Biofisica, Via G. Marconi 10, I-80125 Napoli, Italy.

RUIS, H., Institut für Allgemeine Biochemie, Universität Wien, Währingerstraße 38, A-1090 Vienna, Austria.

SCHEINER, O., Institut f. Allgemeine u. Experimentelle Pathologie, Universität Wien, Währingerstraße 13, A-1090 Vienna, Austria.

SCHERBERICH, J.E., Universitäts-Klinik Z.I.M., Immunol. Biochem. Laboratorium, Dept. Nephrologie, Th. Stern Kai 7, D-6000 Frankfurt/Main, German Federal Republic.

SCHWAGER, J., Hygiene-Institut der Universität Graz, Universitätsplatz 4, A-8010 Graz, Austria.

SCHONNE, E., Université Catholique de Louvain-La-Neuve, Place Croix du Sud 5, B-1348 Louvain-La-Neuve, Belgium.

STEMBERGER, H., Institut f. Spez. Prophylaxe u. Tropenmedizin, Kinderspitalgasse 15, A-1090 Vienna, Austria.

SEELICH, Th., c/o Immuno-AG, Industriestraße 72, A-1220 Vienna, Austria.

SEBILLE, B., Univ. Paris-Val de Marne, Av. du Général de Gaulle, F-94 Créteil, France.

SHALTIEL, S., The Weizmann Institute of Science, Department of Chem. Immunology, Rehovot, Israel.

SLOR, H., Tel Aviv-University, Sackler School of Medicine, Dept. Human Genetics, Tel Aviv, Israel.

SLUYTERMAN, L.A.Ä., Philips Research Labor, Eindhoven, The Netherlands.

SPIRO-KERN, A., Zoolog. Institut der Universität Zürich, Künstlerg. 16, CH-8006 Zürich, Switzerland.

SUSANI, M., Institut für Allgemeine Biochemie, Universität Wien, Währingerstraße 38, A-1090 Vienna, Austria.
SVENSON, A., Swedish Water and Air Pollution Research Institute, P.O. Box 21060, S-10031 Stockholm, Sweden.
SWART, W., Inst. for Experimental Gerontology, T.N.O., 151 Lange Kleinweg, Rijkswijk z.h., The Netherlands.
TAKEO, K., Yamaguchi University School of Medicine, Department of Biochemistry, Ube 755, Japan.
TAYLOR, J., The Milton S.Hershey Medical Center, The Pennsylvania State University, Hershey, Pennsylvania 17033, U.S.A.
TAYOT, J.L., Institut Merieux, Marcy L'Etoile, F-69260 Charbonnieres, France.
THWAITS, B., Western Province Blood Transfusion Service, P.O. Box 3788, Cape Town 800, South Africa.
TOTH, G., Institute of Isotopes, H-1525 Budapest, P.O.B. 77, Hungary.
VIJAYALAKSHMI, M.-W., Institut de Téchnologie, Surfaces Actives, B.P. 233, F-60206 Compiègne, France.
VISSER, J., Department of Genetics, Agric. University, Gen. Foulkesweg 53, Wageningen, The Netherlands.
VRETBLAD, P., Pharmacia Fine Chemicals AB, Box 175, S-751 04 Uppsala, Sweden.
WAGENMAKER, G., REP-Radiobiological Inst. für Experimental Gerontology, 151 Lange Kleinweg/Rijkswijk, The Netherlands.
WAGNER, K., Institut für Angewandte Mikrobiologie, Universität f. Bodenkultur, Peter Jordan Straße 82, A-1190 Vienna, Austria.
WAIBEL, R., Institut f. Lebensmittelwiss., ETH Zürich, Universitässtraße 2, CH-8092 Zürich, Switzerland.
WATSON, D.A., Department of Biochemistry, P.O. Box 174, University of Liverpool, Liverpool L69 3BS, United Kingdom.
WEBSTER, M.F., G.P.Searle Co. Ltd., Research Division, Lane End Road, High Wycombe, Bucks, HP12 4HL, United Kingdom.
WEINS, D., Hans Geiger Weg 26, D-74 Tübingen, German Federal Republic.
WENINGER, P., Institut f. Biologie, Forschungszentrum Seibersdorf, A-2444 Seibersdorf, Austria.
WESTON, P.D., Wellcome Reagents Ltd., Beckenham BR3 3BS, United Kingdom.
WICK, G.K., Institut f. Allgemeine u. Experimentelle Pathologie, Universität Innsbruck, Fritz Pregelstraße 3, A-6020 Innsbruck, Austria.
WIEDERMANN, G., Institut f. spez. Prophylaxe u. Tropenmedizin der Universität Wien, Kinderspitalgasse 15, A-1090 Wien, Austria.
WOLF, H., Institut f. Allgemeineine u. Experimentelle Pathologie, Fritz Pregelstraße 3, A-6020 Innsbruck, Austria.
YANG, C.-y., Hermann-Rein-Str. 3, D-34 Göttingen, German Federal Republic.
YEATES, R., Schweizerisches Tropeninstitut, Socinstraße 57, CH-4051 Basel, Switzerland.
ZEEGERS, J., Inst. voor Virologie, Yelelaan 1, Utrecht, The Netherlands.
ZIEGLER, D., Institut f. Biochemie II, Universität Heidelberg, Im Neuenheimer Feld 328, D-6900 Heidelberg 1, German Federal Republic.

INTRODUCTORY REMARKS

Otto Hoffmann-Ostenhof

Chairman of the Organizing Committee

Ladies and Gentlemen, dear colleagues,

It is for me a most agreeable duty to welcome you in Vienna on behalf of the Organizing Committee of this Symposium and also on behalf of the Österreichische Biochemische Gesellschaft, under whose aegis this meeting is organized.

Although affinity chromatography has been dealt with at various local meetings and was also, together with other chromatographic methods, a topic discussed at several international conferences, this symposium is, as far as we know, the first international symposium exclusively devoted to affinity chromatography.

The idea of using biospecific ligands coupled to an insoluble support for the purification of proteins is much older than most of us are aware of. It was already in 1910 that the pharmacologist Starkenstein of Prague tried to purify amylase by adsorbing it on insoluble starch (1). As at that time no methods for the determination of protein purity were available, we do not know the degree of purification that was obtained in these experiments. It must be mentioned that Starkenstein, who was also a very active researcher in other fields of biochemistry, as well as in pharmacology and physiology, perished in a German concentration camp during world war II.

It took around forty years until Campbell, Luescher and Lerman in 1951 started similar experiments by using cellulose as solid support for ligand attachment (2). They bound an antigen to it and used the product in a column for chromatographic isolation of antibodies from the serum. This method was to become a standard immunological procedure.

Two years later Lerman purified to so-called tyrosinase (monophenol oxygenase) by adsorption on p-azophenol cellulose (3). In the early sixties it was McCormick who tried affinity chromatography for the purification of flavokinase (4) and also for that of avidin (5). However, most of these attempts to develop a purification method for proteins other than antibodies on the basis of biospecific adsorption did not meet with more than limited success.

The real breakthrough which initiated the rapid development of affinity chromatography occurred in the later sixties. The introduction of agarose as an insoluble support by Hjertén (6), and the use of cyanogen bromide for the activation of agarose, dextran and cellulose, suggested by Porath, Axén and Ernbäck (7), and the studies of Cuatrecasas (8) and others on the necessity to separate the ligand from the matrix backbone by a "spacer" in order to minimize steric interference, were milestones in the history of our method.

It is of no use to describe the further development of affinity chromatography before this auditorium. We are in the agreeable situation to have representatives from most of the leading schools of our field with us, and affinity chromatography is still a method young enough that we can expect that their reports will give us a complete picture of the development and the actual state of our method.

It should, however, be mentioned that the technique of affinity chromatography has now found application in almost all major fields of biochemistry and related branches of science. Besides reports on the purification of antibodies and enzymes, we shall also hear of its application for the purification of hormone receptors, of specific messenger ribonucleic acids and of cell populations.

It remains for me to thank all the authorities and industrial companies who helped to make this symposium possible. Our special thanks are due to the Austrian Federal Ministery of Science and Research, to the City Administration of Vienna, and last not least to the University of Agriculture of Vienna who kindly put this beautiful lecture hall at our disposition. I also thank all colleagues from abroad for having come to Vienna and wish them an enjoyable time in our city. May this symposium be fruitful to all participants.

LITERATURE

(1) E. Starkenstein, Biochem. Z. 24, 210 (1910)
(2) D.H. Campbell, E. Luescher, L.S. Lerman, Proc. Nat. Acad. Sci. USA 37, 575 (1951)
(3) L.S. Lerman, Proc. Nat. Acad. Sci. USA 39, 232 (1953)
(4) D.B. McCormick, J. Biol. Chem. 239, 3093 (1964)
(5) D.B. McCormick, Anal. Biochem. 13, 194 (1965)
(6) S. Hjertén, Arch. Biochem. Biophys. 99, 446 (1962)
(7) J. Porath, R. Axén, S. Ernbäck, Nature 214, 1302 (1967), Nature 215, 1491 (1967).
(8) P. Cuatrecasas, Biochem. Biophys. Res. Comm. 33, 235 (1968)

PART I

GEL MATRICES, COUPLING METHODS AND CHARGE TRANSFER CHROMATOGRAPHY

GEL MATRICES, COUPLING METHODS AND CHARGE TRANSFER CHROMATOGRAPHY

Jean Marc Egly* and Jerker Porath**

**Centre de Neurochimie du CNRS and Unité 44 de l'INSERM, 11, Rue Humann, 67085 Strasbourg Cedex, France*

***Institute of Biochemistry, University of Uppsala, Box 576, S-751 23 Uppsala, Sweden*

CONTENTS

1. INTRODUCTION

A variety of methods for the preparation of immobilized active substances have been reported in recent years (1-5). The immobilization of a substance can be accomplished by different principles: entrapment in a polymeric network, physical adsorption, extensive cross-linking, and covalent coupling to an insoluble carrier. This last technique has been extensively used in connection with affinity chromatography which is based on the exceptional stereospecificity of biologically active substances to bind reversibly other substances generally called ligands or affinants. These ligands can be fixed to an insoluble support or matrix, either directly or with the mediation of a spacer arm. Then, under appropriate experimental conditions all compounds which have no affinity for the fixed ligands pass through, whereas the substances having affinity for the ligand are adsorbed on the column. The bound substances can be released from the stereospecific complex by bioelution using a soluble biospecific counterligand which can compete with the immobilized ligand. Another possibility is non-biospecific elution whereby desorption is effected by changing physico chemical parameters (ionic strength, pH, temperature) with so-called deforming buffers.

Affinity chromatography, hydrophobic interaction chromatography or charge transfer chromatography (EDA-chromatography) are depending on the choice of the matrix and the choice of the coupling method. These two important parameters are the subjects of this lecture. We shall also present here our first observations and comments concerning charge transfer chromatography about the separation of aromatic amino acids and oligopeptides, or nucleosides, nucleotides and oligonucleotides.

2. MATRICES, COUPLING METHODS

2.1. Matrices

A great deal of discussion has been generated on the applicability of various inert materials as granular support matrices in affinity chromatography. There is general agreement that in selecting a support one has to consider the following factors:

1) Insolubility
2) Permeability
3) Rigidity and physical form
4) Hydrophilic character
5) Chemical stability
6) Chemical reactivity allowing ligands and spacers to be introduced
7) Resistance toward microbial and enzymatic attack
8) Cost.

Virtually any water-insoluble substance that has a high surface density groups capable of reaction with ligand forming substances and satisfies all the other criteria listed can serve as a support. However, according to the opinions given in the literature no ideal matrix has been introduced yet. Many of the matrix substances in use suffer from drawbacks in one way or another. For example, it is almost impossible to find a matrix material which does not itself interact with biomolecules by ionic attraction, hydrogen bonding or hydrophobic interaction. Occasionally these interactions fortuitously improve adsorption selectivity but it is our opinion that they should be minimized to allow better control of the production and performance of the derived adsorbents. The elimination of disturbing secondary interactions is not always understood or appreciated by the workers in the field and is particularly important when extraordinary high selectivity is required, as is the case with bioaffinity adsorption methods both in columns and in batchwise extractions.

For a given application one must make some compromise as to the demands made on the matrix. Bead size and permeability determine the rate of attainment of equilibrium. Therefore, in batchwise operations the gel particles should be steady enough to withstand agitation without disintegration. In column procedures rigidity is important, since the degree of bead deformation and compaction determines the maximum usable or attainable flow rate.

Experience has revealed that the hydrophilic character of the matrix is an absolute prerequisite for expansion of the network in aqueous media, which permits permeation of water-soluble macromolecular substances into the gel phase and affords a microenvironment that reasonably approximates biological conditions. This reduces or eliminates detrimental effects on the solutes, such as denaturation or uncontrolled hydrophobic interaction.

Chemical stability is required within the range of working conditions: the gel must withstand the conditions to be used for adsorption, desorption and regeneration. This may or may not require stability at extreme pH at room temperature or resistance to reducing or oxidizing agents, organic solvents etc. Biopolymeric gels can be susceptible to microbial attack, although this risk is often overestimated.

2.1.1. Porous glass

Glass offers distinct advantages in several respects: it is resistant to acids and organic solvents as well as to microbial attack. It has exceptional rigidity, which permits outstanding flow properties at elevated pressures and the linkage between the ligand and the arm is chemically well defined and stable. In spite of these appealing features glass is not an ideal support in its present forms. It is expensive, shows a limited binding capacity (compared to agarose for example) and it often shows a considerable nonspecific adsorption due to SiOH-groups. These properties, especially the last one, account for the fact that only a few publications on affinity chromatography have appeared in which porous glass has been used as the matrix.

However, porous glass biospecifics have been used for different purposes. Glycyl-D-phenylalanine immobilized by attachment with glutaraldehyde to aminoalkylsilyl glass was used for the purification of carboxypeptidase A (6). Catecholamines like epinephrine bound covalently to arylamine glass have been found to have biological activities in several systems (7). Another interesting application should be mentioned: mitochondria attached on alkylsilylated glass can be maintained in a viable condition to permit the study of some respiratory properties (8).

Its potential use in large scale application has prompted attempts to diminish or circumvent the disadvantages mentioned. For example, amplification of binding capacity was achieved by treating colloidal silica or porous glass with organic polymers like polyethyleneimine or polyacrylic acid (9).

2.1.2. Polyacrylamide

The three-dimensional network of polyacrylic acid amide is synthesized by copolymerization of acrylamide and N,N'-methylene bisacrylamide. The acrylamide thus formed consists of hydrocarbon backbones carrying carboxamide side chains which confer a hydrophilic character upon the gel. Polyacrylamide is chemically stable in the pH-range 3-10 and at temperatures below 35°C, but does not stand some organic solvents. A most attractive feature is the high density of the chemically reactive carboxamide groups, but this advantage is diminished by the fact that during derivatization (*e.g.*, azide formation) the gel will shrink with a consequent reduction of porosity. Cuatrecasas (10) has illustrated this phenomenon during the purification of β-galactosidase on the agarose or polyacrylamide matrix containing the same inhibitor (p-aminophenyl-β-D-thiogalactopyranoside). With the polyacrylamide the inhibitor does not bind β-galactosidase whereas the enzyme is adsorbed to the agarose gel (probably in large part by hydrophobic interaction) (11).

In order to minimize the reduction of porosity during activation different new composite gels were introduced; for example, copolymers of acrylamide and agarose (Sephacryl) would seem to have the advantages of each polymer, and the availability of both amide and hydroxyl groups affords reactivity with more than one type of ligand function (12). In the preparation of some effective immunoadsorbents the non-specific adsorption seems to be minimized (13) even though the polyacrylamide itself might favor hydrophobic interactions.

2.1.3. Polysaccharides

The most widely employed supports in chromatography are cellulose, dextran and agarose.

2.1.3.1. Cellulose

From a purely economic point of view, cellulose is the most attractive material for immobilization of the ligand. Its chemistry is well understood. It was the first solid support used for ligand attachment by Campbell *et al.* in 1951 (13). They bound an antigen to the cellulose and used a column of this derivative for chromatographic isolation of antibodies from serum. In spite of these facts, cellulose has certain disadvantages that make it less suitable for affinity chromatography. Charged groups are present prior and after derivatization or activation. Another disadvantage, which one might miss at first sight, is the physical structure of the cellulose. Small and sometimes large ligands can be introduced into the cellulose matrix, but permeation of proteins and nucleic acids is strongly restricted by sterical hindrance. Therefore, a high degree of substitution will not guarantee a high capacity for adsorption, especially if both the ligand and the solute are proteins.

2.1.3.2. Dextran

Sephadex is synthesized from partially hydrolyzed dextran, a glucan obtained by biopolymerization of the glucose unit of sucrose. By controlling the degree of cross-linking one can prepare gels with different permeation limits determined mainly by the molecular size of the solute and the matrix density of the gel (14).

Sephadex is used as a starting material for a number of different adsorbents, but it is not well suited for affinity systems in which both the ligand and the solute are large molecules. However, cross-linked dextran is an excellent starting material for preparing hydrophilic adsorbents with small ligands.

Sephadex is also well suited for the preparation of immunoadsorbents, in which case permeation and capacity play minor roles (15).

2.1.3.3. Agarose

Agar and agarose, which consists of monomer units of D-galactose and 3,6 anhydro-L-galactose, were introduced as supports for electrophoresis and as molecular sieves by Polson (16) and Hjertén (17). The characteristics of gels based on agar and agarose have been explored by Porath and coworkers (18-20). In a series of papers they have described how to further improve agar and agarose to approach the goal of an ideal support for the production of adsorbents for biopolymers. By cross-linking agar/agarose under reducing alkaline conditions with epihalohydrins, αβ-halogenated alcohols, bisoxiranes or divinylsulfone most of the undesirable properties of the agar gels can be eliminated. For example, the cross-linked gel can be autoclaved in 1-2 M NaOH without destruction of the matrix, and this treatment removes most of the sulfate groups. The gel is stable down to pH 3 and can be treated with chaotropic ions such as KSCN (3 M) and with organic solvents. The flow rate is greatly improved and the leakage of carbohydrate virtually eliminated. The structure of this polysaccharide gel favors some nonspecific interactions which can be exploited to advantage. For example, the adsorption of some ribonucleic acid on unsubstituted Sepharose has been attributed to hydrophobic interaction, but hydrogen bonding may also play a role (21). The present agarose products suffer from the disadvantage of being rather expensive.

2.1.4. Other matrices

Collagen has been proposed as a matrix by Vieth *et al.* (22). The biomaterial collagen offers a number of advantages as an inexpensive carrier. It is an abundant constituent of higher vertebrates. Its amino acid side chains offer a number of sites for ligand attachment, and collagen polypeptide chain tends to stabilize the conjugated enzyme. However, its range of application is limited by the fact that being a protein it is subject to structural alterations with variation of pH, temperature, salt concentration and organic solvent content.

Other possible matrices such as cross-linked pectin and polyvinyl alcohol may be mentioned.

2.2. Coupling Methods

The coupling method to be used in affinity chromatography, hydrophobic adsorption or charge transfer adsorption chromatography is dictated by the choice of ligand, the desired length of the spacer arm and the importance of avoiding creation of undesired adsorption centers.

The selection of ligand is determined by two requirements:

1) The ligand should have chemically modifiable groups (NH_2, OH, CO, COOH etc) which permit immobilization without much effect on the biomolecule-ligand interaction.

2) The association constant should preferably be of a magnitude that allows linear adsorption, but failing this one ought to be able to effect desorption by changes in conditions that will not damage either the ligand or the solute under study. Knowledge of the kinetics of adsorption and desorption in free solution may help in the design of the chromatographic experiments.

Another important problem is the optimization of the procedures used for ligand attachment. Very few systematic studies have been made so far (23-24) which makes it somewhat difficult to compare merits and shortcomings of the published methods.

The ligand density is of great importance. A high ligand density might not be desirable in the purification of proteins by bioaffinity chromatography, since extensive multipoint attachment can make desorption difficult or impossible. Too low ligand densities, on the other hand, will unduly diminish the capacity and increase the contribution of unspecific adsorption. For affinity chromatography of low molecular weight solutes a high ligand density is usually required.

The attachment of a ligand substance is usually made in two steps: activation and coupling. When activation is done with a bifunctional reagent one cannot avoid some additional cross-linking of the matrix in the process.

We shall only present a few coupling methods that we consider especially useful or promising.

2.2.1. Cyanogen bromide coupling

The use of CNBr for activation of dextran and cellulose (25) and agarose (26) initiated the rapid development of bioaffinity chromatography (27). In a recent review Porath and Axén (5) have compiled descriptions and discussions of the immobilization technique as applied to hydroxylic supports. The reaction follows the scheme:

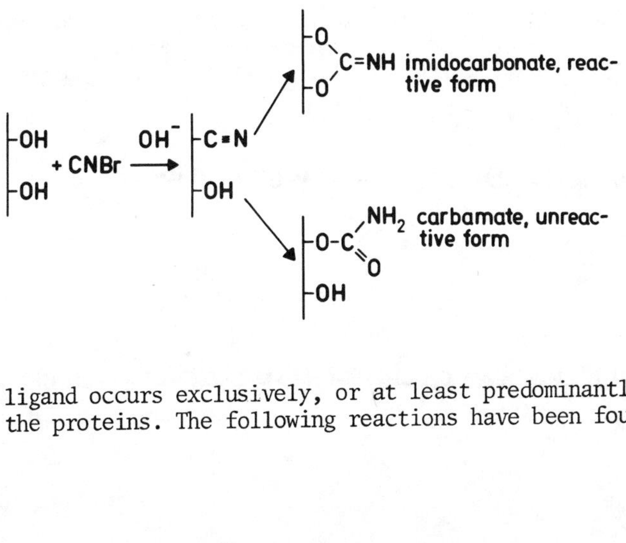

The coupling of ligand occurs exclusively, or at least predominantly, via the free amino groups of the proteins. The following reactions have been found to occur:

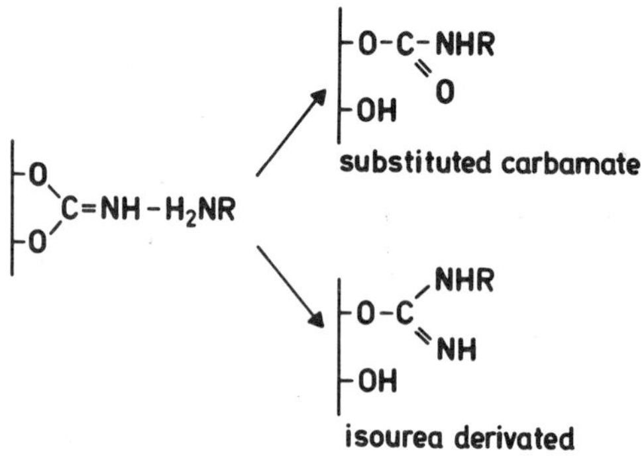

The amine reacts in the unprotonated state. The following alkalinities are appropriate for coupling: aliphatic amines, pH ∿10; amino acids, pH ∿9; aromatic amines, pH 7-8. At pH 6 ribonucleotides are fixed, probably by another mechanism: it has been suggested that terminal nucleic acid phosphate reacts with cyanate or imidocarbonate followed by an intramolecular rearrangement to give the following product (28-29):

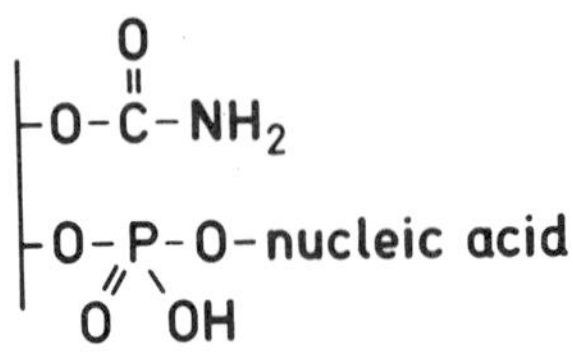

The instability of the isourea linkage in the presence of primary amines and ammonia is a weakness of the method when used for single attachment since the O-C linkage can hydrolyze or aminolyze slowly (30).

2.2.2. Oxirane

Oxirane groups can be introduced into many hydroxyl polymers and epoxides are particularly useful reagents for coupling low molecular weight ligands. The general activation procedure requires bifunctional reagents such as bisepoxides or epihalohydrines (31).

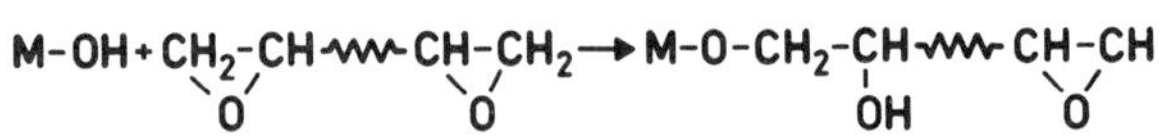

M=Matrix

This intermediate product can react with the matrix in a cross-linking reaction which can improve the mechanical stability of the matrix:

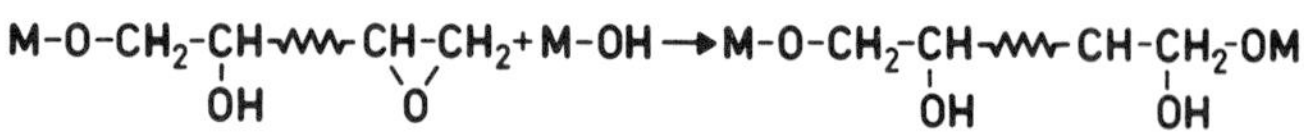

The reactivity toward nucleophilic groups in the ligand follows the usual order SH>NH>OH. Nucleophilic attack on aliphatic hydroxyl thus takes place at strongly alkaline pH, whereas SH will react even at lower pH (around neutrality).

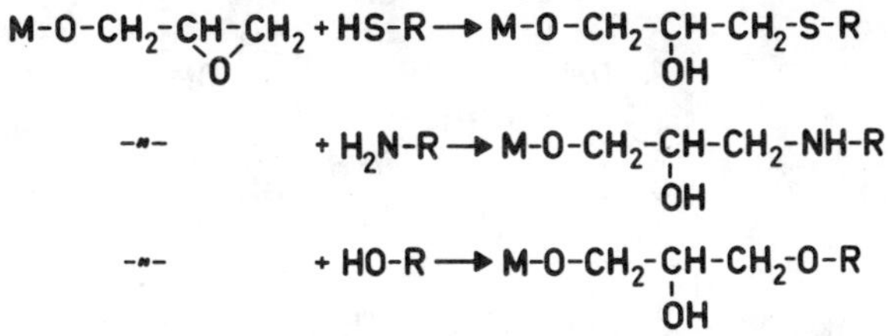

In cases where ligands are sufficiently reactive and stable the oxirane method offers some very important advantages. For example, specific affinity adsorbents can be produced by this method from non reducing saccharides without any alteration of the ligand but in the case of reducing sugars or amino sugars with free reducing groups, an isomerization is possible (32). Anyhow in the case of reducing polysaccharide, the epimerization of only the reducing terminal sugar may represent too minor an alteration in the ligand to affect seriously the specificity of the adsorbent. The O-C, N-C and S-C bonds formed are extremely stable due probably to the high pH used during the coupling. By selecting an appropriate long-chained bisoxirane one can introduce a long hydrophilic spacer and thereby favor the ligand-counterligand interaction in cases where sterical hindrance would otherwise be an obstacle.

2.2.3. Divinylsulfone

Divinylsulfone groups can be introduced into polysaccharides by a reaction mechanism analogous to that of the bisoxirane method. The vinyl derivative is much more reactive than the corresponding oxirane containing gels.

The activation takes place as follows:

$$M\text{-}OH + CH_2\text{=}CH\text{-}SO_2\text{-}CH\text{=}CH \longrightarrow M\text{-}O\text{-}CH_2\text{-}CH_2\text{-}SO_2\text{-}CH\text{=}CH_2$$

In the coupling step the rates of nucleophilic attack by the ligands follow in the same order as for oxiranes, but the rate of reaction is faster and coupling can be done 1-2 pH units lower. Upon storage a slow release of ligand occurs at pH above 8. Sairam and Porath (32) have used divinylsulfone-coupled immunosorbents and they effected desorption with 0.1 molar ammonia repeatedly without significant decrease in capacity.

2.2.4. Benzoquinone

Benzoquinone is very reactive and will attack hydroxylic polymers more readily than any of the other reagents mentioned above. This method is easy to apply and gives highly activated structures which can undergo coupling between pH 3 and 10 to yield very stable products (34).

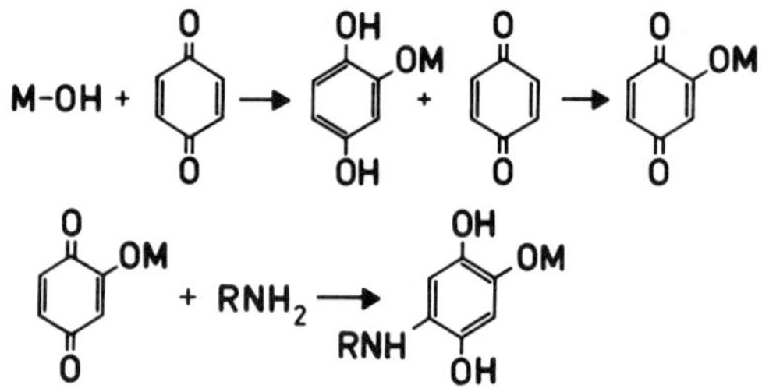

The method has not yet been optimized. The formation of strongly colored gels is presumably due to side reactions.

2.2.5. Thiol-difulfide interchange

This particularly attractive method introduced by Brocklehurst *et al.* (35) occurs via thiol-disulfide interchange with activated thiolated polysaccharide (36-37). In the first step, the thiol group in the matrix is activated

$$\text{M-SH} + \text{X-S-S-X} \longrightarrow \text{M-S-S-X} + \text{XSH}$$

(X can be the 2-pyridyl group). The thiol group-containing solute is then reacted with activated gel

$$\text{M-S-S-X} + \text{R-SH} \longrightarrow \text{M-S-S-R} + \text{XSH}$$

RSH can be desorbed by reduction.

This reaction can be used for reversible immobilization of enzymes. By this technique one can separate thiol-containing from nonthiol-containing proteins and peptides and isolate enzymes with active site thiol groups. Also, by fixation of thiolated polynucleotide one might use the method for isolating particular complementary polynucleotides by hybridization test (38).

2.2.6. Activated halogen

Highly reactive halogens can be introduced into hydroxylic polymers (39)

$$\text{MOH} + \text{Br-CO-CH}_2\text{-Br} \longrightarrow \text{M-O-CO-CH}_2\text{Br}$$

$$\text{M-O-CO-CH}_2\text{Br} + \text{RNH}_2 \longrightarrow \text{M-O-CO-CH}_2\text{-NHR}$$

2.2.7. Modification of polyacrylamide to attack enzymes

This modification occurs generally via azide formation (40). Reactions with appropriate compounds give amino ethyl derivatives as follows:

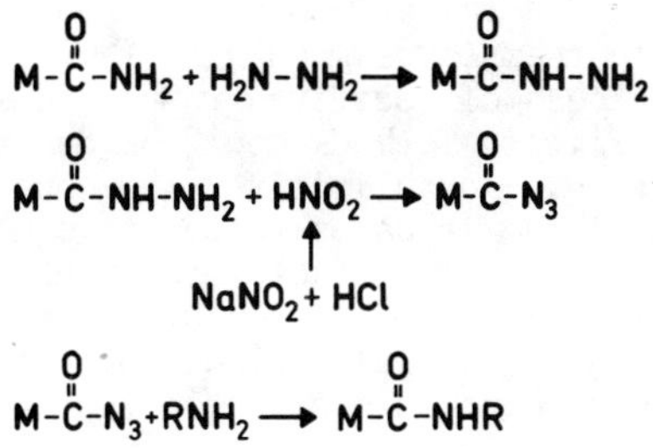

Proteins can also be coupled to polyacrylamide gel following treatment with glutaraldehyde (41), but neither the mechanism of the reaction of glutaraldehyde nor the nature of the products has been elucidated (42).

2.2.8. Azo-coupling to glass

The silanized carrier reacts with p-nitrobenzoylchloride followed by reduction with sodium dithionite (43).

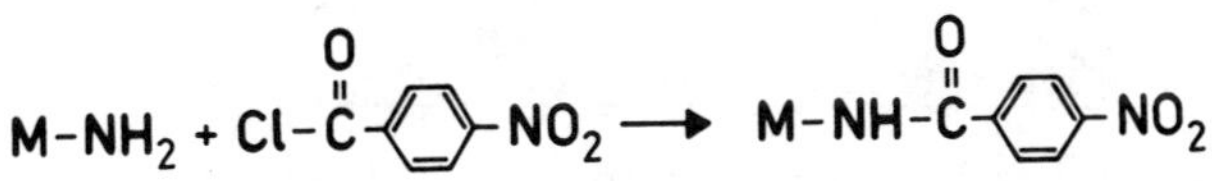

alkyl-amine glass   p-nitrobenzoyl-chloride

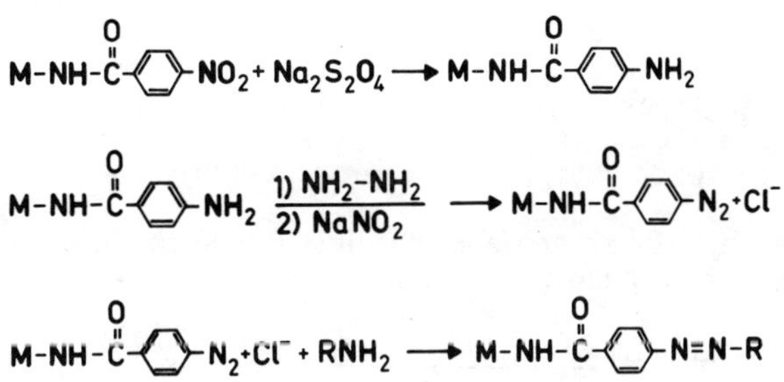

2.3. Concluding remarks to part 2.

It is often difficult *a priori* to determine which immobilization method should be best for the introduction of a particular ligand. The decision may be based on the fractionation conditions - whether chromatography or batch operations will be used, operational parameters such as pH, temperature, contact time, etc. - and on the required selectivity. In affinity chromatography it is particularly important to avoid side effects due to the presence of undesired adsorption centers in the adsorbent. The durability of the adsorbent and ease of regeneration must also be considered.

2.3.1. Importance of deactivation

Most researchers do not seem to trouble themselves much with the risk that the coupled product may still contain an excess of reactive groups. Some of the complications met in the application of bioaffinity methods could be due to the neglect of deactivation. The coupling yield is never quantitative. The concentration of excess reactive groups is particularly large in cases where large molecular substances are being fixed to the matrix. Upon the contact with solutes to be fractionated further coupling can occur resulting in fixation of the substance(s) to be recovered in purified form and, as an additional serious complication, of impurities that will act as non-specific adsorption centers in the subsequent use of the gel.

The problem of excess reactive groups is encountered in the use of all the coupling methods. It may be counteracted in two ways: 1) by solvolysis, 2) by coupling of a highly penetrable substance the corresponding ligand of which will not adversely affect the adsorption-desorption properties. The first method may be used when the ligands are stable in alkaline solution at elevated temperature, which conditions effect hydrolytic removal of the activated groups. The second method should be used when proteins and other labile substances have been coupled to the gel. Deactivating substances used include glycine and ethanolamine. Mercaptans can sometimes be used when introduction of nitrogen is to be avoided.

2.3.2. Introduction of undesired adsorption centers

In several reports, it has been pointed out that the ligand-polysaccharide conjugate contains unintentionally introduced hydrophobic (44) or ionic groups (45-46). The possibility cannot be excluded that the involuntarily introduced adsorption centers have been the deciding factor in some attempts at biospecific adsorption (11-47). The hydrophobic interaction, which is generally due to the introduced spacer arm, can be controlled by replacement of the hydrophobic spacer by polar or more hydrophilic ones, or by lowering the ionic strength (44). The cyanogen-bromide activation introduces cationic charges into the insoluble matrix (45-48). These charges may be sufficient to interfere with the use of these materials for affinity chromatography. The pH used during activation, and/or the pH and the concentration of the ligand in the coupling step are critical in determining the extent of the cationic charge of the product (49).

In our experience, bisoxirane and divinylsulfone methods minimize side reactions that create undesired adsorption centers.

2.3.3. Ligand stability

Release of ligand can occur under certain conditions in the case of CNBr (50-51) and divinylsulfone conjugates. The solvolysis of cyanogen-bromide coupled adsorbents could be facilitated at alkaline pH by neighbour groups in the carrier gel, such as hydroxyl or carboxyl groups (52).

The lability of DVS-coupled adsorbents in alkaline solutions has been pointed out by Sundberg and Porath (31). Ethers are much more stable than the amine derivatives. Divinylsulfone is therefore particularly suitable for coupling of phenols and alcohols including carbohydrates. Coupling of hydroxyl compound can also be effected easily by the oxirane method, in which case particularly stable products are formed with a minimum of side reactions.

3. CHARGE-TRANSFER CHROMATOGRAPHY (Electronic Donator-Acceptor Chromatography)

We shall present some recent results of charge transfer chromatography (we will use the expressions "charge transfer chromatography" and "EDA-chromatography" interchangeably) and thereby exemplify one particular type of affinity adsorption and the use of the previously described techniques for coupling ligand substances to hydrophilic matrices.

3.1. Theoretical aspects

The formation of neutral molecular compounds (*e.g.*, trinitrobenzene-antracene complex) under the influence of weak attracting forces has been known for quite a long time. If one of the complex-forming components is anchored to an insoluble matrix, the latter will be converted to an adsorbent for the other component of the molecular complex. The interaction may be more or less specific.

Donor and acceptor properties are functions of the conjugated system (type and number of aromatic rings) and of electron attracting or releasing substituents on the aromatic ring. It seems that the most important interaction should occur between two conjugated systems (π-π interaction), but other types of hypothetical charge transfer interaction might be involved, as for example, a n-π complex between the lone pair electron of oxygen or sulfur and the aromatic ring, or a π-σ complex between a polyhydroxylic gel and the aromatic ring (57,58).

There are different types of adsorbents for the formation of immobilized molecular compounds. Immunoadsorbents may be considered as one extreme as regards the specificity and molecular size of the interacting components. The charge transfer adsorbents we are now studying are located at the other extreme of the specificity scale, since they have rather low specificity. So far we have studied only small molecular size components by EDA-chromatography.

Only weak forces are involved in the molecular complex formation. While the covalent bonds of organic compounds usually have a length of less than 1.6 Å and hydrogen bonds about 1.8 Å, the contact distance in electronic charge transfer complexes is in the range 3.0-3.4 Å. The weak interaction is reflected in the low values of the enthalpy of dissociation (usually in the range 0.5-5 kcal/mol).

Charge transfer has been treated in a large number of monographs (53-56). An electron is in part or totally transferred from a donor D to an acceptor A:

$$D + A \leftrightarrow (D,A) \leftrightarrow (D^{+}:A^{-}) \qquad (1)$$

According to the Mulliken theory a stabilized ground state of the complex is produced having the wave function Ψ_N. This wave function is described in terms of interaction between a no bond state $\Psi_0(DA)$, and a polar state, $\Psi(D^{+}:A^{-})$:

$$\Psi_N = a\Psi(DA) + b\Psi(D^{+}:A^{-})\,(a>>b) \qquad (2)$$

A corresponding equation is valid for the exited complex.

Dispersion forces, dipole-dipole interaction and other forces may play an even more important role in the formation of molecular compounds.

The energy levels can be calculated by solving the Schroedinger equation. The energy of a delocalized π-electron in a molecular orbital is

$$E = \alpha + k\beta \qquad (3)$$

where α is the "coulomb integral" and β the "exchange integral". The energies of molecular orbitals are often expressed in terms of k in the diagrams for highest filled (usually positive k-values) and lowest empty orbitals (usually negative k-values). Such diagrams should be interpreted with care, since α and β are approximately constant only within the same chemical series. However, the diagrams may be used for crude predictions of charge transfer effects.

Important for the transfer of charge between two substances is the energy difference between the lowest unoccupied level of the acceptor and the highest occupied level of the donor. The smaller the difference the stronger is the interaction.

Potentially useful ligands can thus be chosen on the basis of published energy diagrams. It should be pointed out that a substance can act both as an acceptor and as a donor. This fact also explains the formation of "self-charge-transfer complexes" ("polymerization"). The importance of solvent effects cannot be neglected - a fact that makes it impossible to utilize charge transfer as the single adsorption factor. Since most of the satisfactory π-acceptor and π-donor molecules are aromatics, hydrophobic interaction can hardly be avoided. The charge transfer compounds are often coloured and moving coloured zones are a diagnostic of charge transfer interaction when the solutes are colorless.

Early attempts to develop charge tranfer adsorbents for biochemically interesting substances did not meet with much success (59). Also gas chromatography and thin layer chromatography have not as yet given very impressive results (60-62). Much more extensive systematic studies are needed to maximize the desired adsorption effects and minimize the disturbing factors. For this purpose we are now concentrating our efforts on the development of adsorbents that might ultimately be useful for chromatography of proteins and nucleic acids, although at present we are using them only for the fractionation of peptides, nucleotides and other types of low molecular weight organic compounds.

Before we discuss the use of charge transfer adsorbents it is appropriate to mention that the "aromatic adsorption" that is known to occur with phenols, tryptophan, uric acid and other π-electron-containing solutes on Sephadex can be regarded as a special case of charge transfer involving matrix-hydroxyl or matrix bound water (63). The introduction of acceptor (or donor?) molecules into the matrix often magnifies these chromatographic adsorption effects.

3.2. Preparation of charge transfer gels

Efficient electron acceptors may be found among nitro-aromatics, quinones, and certain heterocyclic compounds. In our studies we have tried quite a few compounds belonging to these classes. The solubility requirements limit the selection considerably.

The following two charge-transfer gels were synthesized from oxirane-Sephadex:

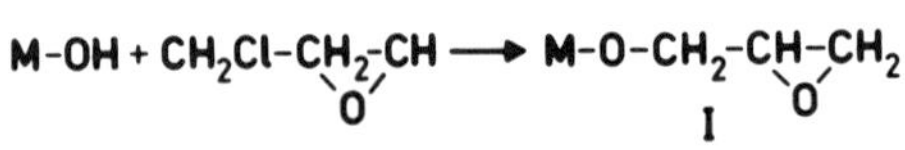

(Sephadex)

1) Dinitrophenyl-S-Sephadex, a yellow gel, was synthesized from I in two steps:

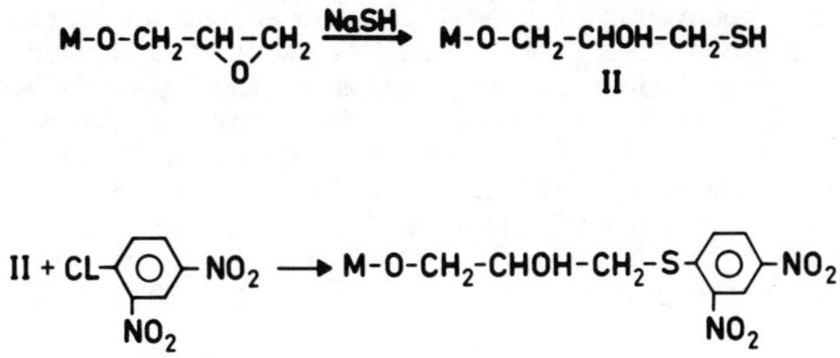

2) Acriflavin-Sephadex, a brown gel, was synthesized from I:

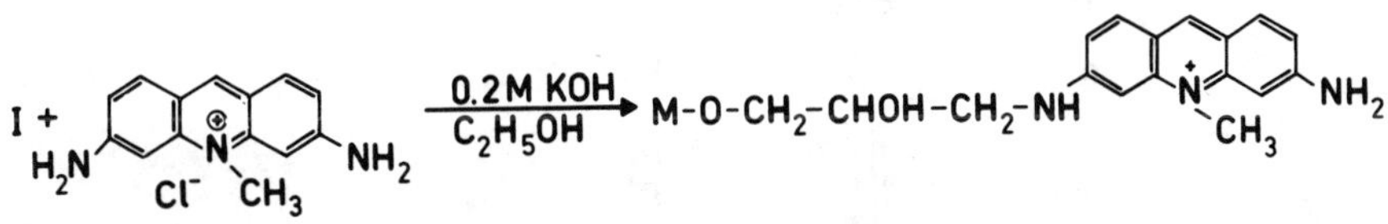

Dinitrophenyl-S-Sephadex is most probably acting only as an acceptor gel and this is likely to be the case also for pentachlorophenyl-S-Sephadex. The acriflavin gel is possibly an acceptor as well as a donor adsorbent.

The formation of yellow adsorption complexes is not easy to recognize on the coloured gel. However, at high concentrations tryptamine and pyrogallol move as red-orange zones on DNP-S-Sephadex (Fig. 1). They leave the column in colourless solutions. If adenosine monophosphate is mixed with acriflavin gel and dried a dark brown, almost black, product is formed - much darker than the dried acriflavin gel itself.

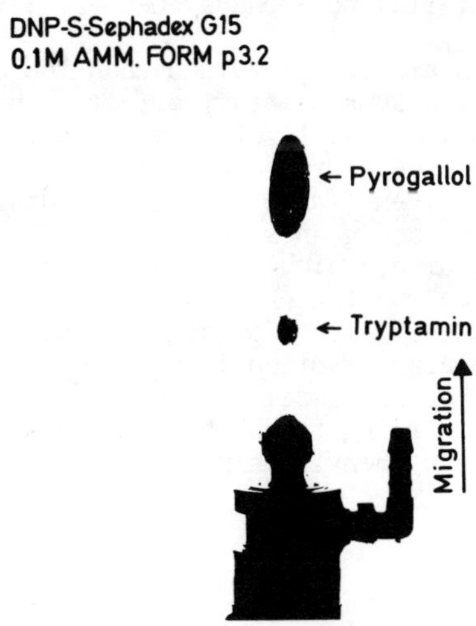

Fig. 1. Chromatographic column of 10 mg of pyrogallol and 10 mg of tryptamine on a DNP-S-Sephadex G-15. Elution buffer 0.1 M, ammonium acetate buffer pH 3.2 at 24°C.

3.3. Results of some chromatographic experiments

The chromatographic behaviour of aromatic amino acids, nucleotides and corresponding bases on these two different matrices has been studied under various conditions. We shall present reduced elution volumes (the V_E/V_T values) obtained with some model compounds. The results described here for the chromatography of some aromatic amino acids on Acriflavine Sephadex G-25 (Fig. 2) and some previous results (64) provide convincing evidence that the adsorption is due mainly to the π-electron acceptor capacity of the ligands. In the case of the aromatic amino acids we observe that variation of the pH of the elution buffer has little effect on the adsorption to the column (Fig. 2), which shows that ionic adsorption is not a major factor.

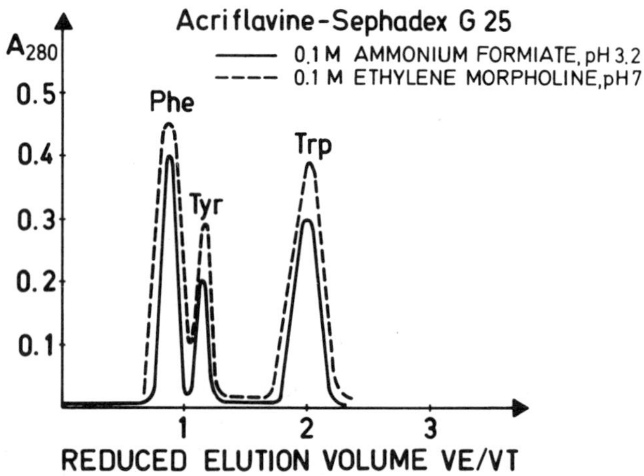

Fig. 2. Chromatographic resolution of a mixture of aromatic amino acids: phenylalanine (Phe), tyrosine (Tyr) and tryptophan (Trp) on Acriflavine Sephadex G-25 at 20°C.

The fact that the nucleosides are less retarded than the corresponding bases (Table 1) may be explained by a supplementary effect of hydrophobic adsorption or by steric hindrance. These values, together with published k-values (65) and the dissociation constants K_c (66) for the complexes of riboflavin and the respective compounds, are also given (Table 1). This finding is consistent with the k-values in as much as adenine and guanosine should be better electron donors than thymine, uracil and cytidine, which is also reflected in the order but not in the size of the association constant K_c. As expected, the reduced elution volumes fall into two categories, the purines being more retarded than the pyrimidines. The aromatic amino acids are also adsorbed in the order of decreasing k-values, as expected for charge transfer interaction. Adsorption thus seems to parallel the electron donor properties, according to which tyrosine should be less adsorbed than tryptophan but moreso than phenylalanine (Fig. 2).

Table 1. Comparison of different values characteristic for a charge transfer complex.

	k	K_c(M-1)	V_E/V_T
Adenine	0.486	7.86	3.29
Guanine	0.307	14.20	3.41
Thymine	-	2.44	1.25
Uracil	0.597	0.52	1.10
Cytidine	0.595	-	1.12
Adenosine	-	120	2.45
Cytosine	-	1.82	1.24
Tyrosine	0.792	-	1.10
Tryptophan	0.534	60	2.01
Phenylalanine	0.908	-	0.91

K_c = association constant for riboflavin complex

$$\text{Rib} + \text{X} \rightleftarrows (\text{Rib:X})$$

calculated in presence of riboflavin in 0.01 M sodium phosphate buffer pH 7.
V_E/V_T = reduced elution volume of samples chromatographed on an Acriflavin-ECD-Sephadex G-25 in 0.1 M ethyl morpholine buffer pH 7.0.
k = values of the highest filled molecular orbital (see equation 3).

When several adsorption centers are introduced into the solute molecule cooperative charge transfer effects will occur with an increase in the strength of adsorption. This is illustrated in Fig. 3 by the cooperative effect of charge transfer, electron charge transfer and hydrogen bonding interactions, which indicates the potential of the procedure for peptide separations. Mono-, di- and trinucleotides can be separated on the basis of the number of nucleotide units. Further separation occurs according to the relative content of purine and pyrimidine

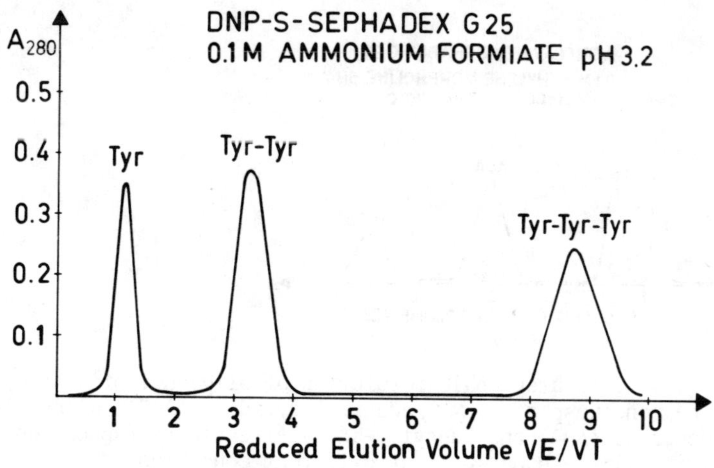

Fig. 3. Chromatographic resolution of a mixture of oligopeptides: tyrosine (Tyr), tyrosyl-L-tyrosine (Tyr-Tyr) and tyrosyl-L-tyrosyl-L - tyrosine on DNP-S-Sephadex G-15 at 20°C.

bases and is temperature dependent (Fig. 4a, 4b). Even when ionic adsorption is depressed at high ionic strength there is still considerable retention of the oligonucleotides (Fig. 4c). Additional results and pH independence studies to be published later show that ionic interaction makes no significant contribution.

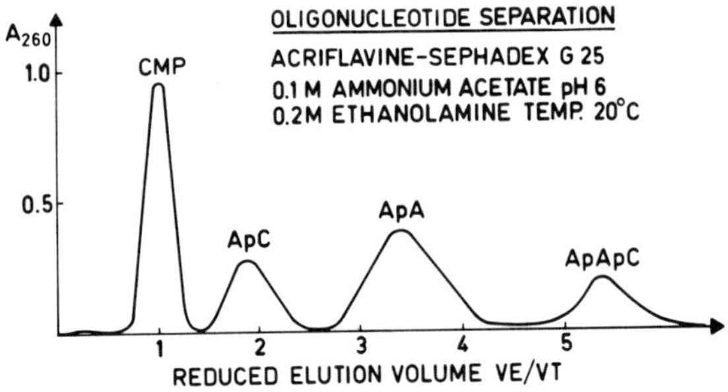

Fig. 4a

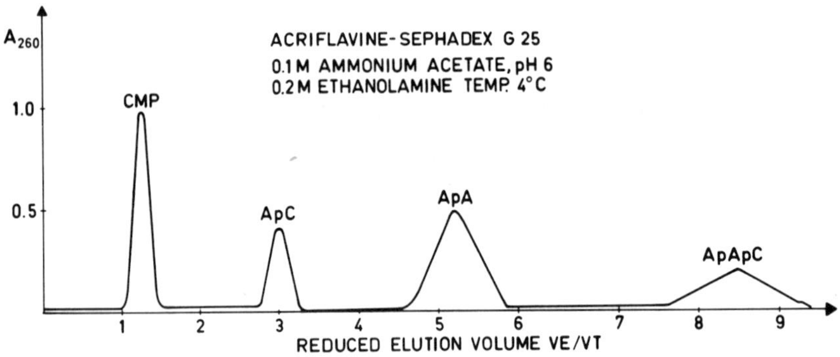

Fig. 4b

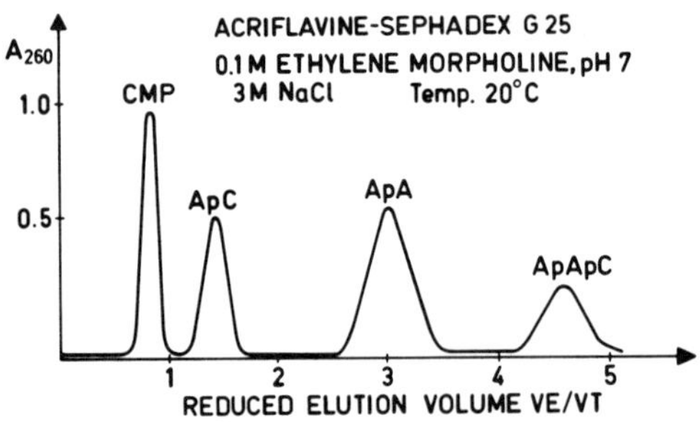

Fig. 4c

Fig. 4. Chromatographic resolution of oligonucleotides: cytosine monophosphate (CMP), adenylyl-cytidine (ApC), adenylyl-adenosine (ApA) and adenylyl-adenylyl-cytidine (ApApC) on Acriflavin-Sephadex G-25 at different conditions.

a) 0.1 M ammonium acetate buffer pH 6, 0.2 M ethanolamine at 20°C
b) 0.1 M ammonium acetate buffer pH 6, 0.2 M ethanolamine at 4°C
c) 0.1 M ethylene morpholine buffer pH 7, 3 M NaCl at 20°C.

In conclusion we will point out the following characteristic features of EDA-chromatography:

1) Low molecular compounds are linearly adsorbed
2) V_E/V_T-values for the adsorbents and monomeric solutes so far tested are in the range 1-15
3) Cooperative adsortion effects of many groups that may be involved in charge transfer have been discovered and are likely to be important in practical applications
4) Temperature, ionic strength and polarity of solvent have been found to have large effects on the charge-transfer interaction - a fact which most probably can be exploited to advantage
5) The adsorption capacity is surprisingly high.

Acknowledgements

The work on charge transfer adsorption has been covered by grants from the Swedish Board for Technical Development (dnr 75-3832) and the Natural Science Research Council (dnr K-0220-045) (to J.P.) and a schoolarship from EMBO and NATO (to J.M.E.) for which we express our gratitude.

We also like to thank Miss Birgitta Larsson for technical assistance.

4. REFERENCES

(1) J. Porath in Methods in Enzymology, Vol. 34, p. 13 (1974).
(2) P. Cuatrecasas and C.B. Anfinsen in Methods in Enzymology, Vol. 22, p. 31 (1972).
(3) J. Turkova, J. Chromatogr. 91, 267 (1974).
(4) H. Guilford, Chem. Soc. Rev. 2, 249 (1973).
(5) J. Porath and R. Axén in Methods in Enzymology, Vol. 44, p. 19 (1976).
(6) P.J. Robinson, Biochim. Biophys. Acta 242, 659 (1971).
(7) S.C. Venter, Proc. Natl. Acad. Sci. USA 69, 1141 (1972).
(8) B. Arkles and W.S. Brinigar, J. Biol Chem. 22, 8856 (1975).
(9) G.P. Royer, Chem. Technol. 411, 694 (1974).
(10) P. Cuatrecasas, J. Biol. Chem. 245, 3059 (1970).
(11) E. Steers, Jr., P. Cuatrecasas and H.B. Pollard, J. Biol. Chem. 246, 196 (1971).
(12) S.G. Doley, M.J. Harvey and P.D.G. Dean, FEBS Lett. 65, 87 (1976).
(13) D.H. Campbell, E.L. Luescher and L.S. Lerman, Proc. Natl. Acad. Sci. USA, 37, 575 (1951).
(14) J. Porath and P. Flodin, Nature 183, 1657 (1959).
(15) L. Wide and J. Porath, Biochim. Biophys. Acta 130, 257 (1966).
(16) A. Polson, Biochim. Biophys. Acta 50, 565 (1961).
(17) S. Hjertén, Biochim. Biophys. Acta 79, 393 (1964).
(18) J. Porath, J.C. Janson and T. Låås, J. Chromatogr. 60, 167 (1971).
(19) T. Låås, J. Chromatogr. 66, 347 (1972).
(20) J. Porath, T. Låås and J.C. Janson, J. Chromatogr. 103, 49 (1975).
(21) M. Zeichner and R. Stern, Biochemistry 7, 1378 (1977).
(22) W.R. Vieth and K. Venkatasubramanian in Methods in Enzymology, Vol. 44, p. 243 (1976).
(23) R. Axén and S. Ernback, Eur. J. Biochem. 18, 351 (1971).
(24) L. Sundberg and J. Porath, J. Chromatogr. 90, 87 (1974).
(25) R. Axén, J. Porath and S. Ernback, Nature 214, 1302 (1967).
(26) J. Porath, R. Axén and S. Ernback, Nature 215, 1491 (1967).
(27) P. Cuatrecasas, M. Wilchek and C.B. Anfinsen, Proc. Natl. Acad. Sci. USA 61, 636 (1968).
(28) H. Guilford, Chem. Soc. Rev. 2, 249 (1973).

(29) J. Kempf, N. Pfleger and J.M. Egly, J. Chromatogr. in press.
(30) M. Wilchek, T. Oka and Y.J. Topper, Proc. Natl. Acad. Sci. USA 72, 1055 (1975).
(31) L. Sundberg and J. Porath, J. Chromatogr. 90, 87 (1974).
(32) R. Uy and F. Wold, Anal. Biochem. 81, 98 (1977).
(33) M.R. Sairam and J. Porath, Biochem. Biophys. Res. Commun. 69, 190 (1976).
(34) J. Brandt, L.O. Andersson and J. Porath, Biochim. Biophys. Acta 386, 196 (1975).
(35) K. Brocklehurst, J. Carlsson, M.P.J. Kierstan and E.M. Crook, Biochem. J. 133, 573 (1973).
(36) J. Carlsson, R. Axén, K. Brocklehurst and E.M. Crook, Eur. J. Biochem. 44 183 (1974).
(37) R. Axén, H. Drevin and J. Carlsson, Acta Chem. Scand. B 29, 471 (1975).
(38) R.M.K. Dale and D.C. Ward, Biochemistry 14, 2458 (1975).
(39) A.T. Jagendorf, A. Patchornik and M. Sela, Biochim. Biophys. Acta 78, 516 (1963).
(40) J.K. Inman in Methods in Enzymology, Vol. 34, p. 30 (1974).
(41) J.L. Guesdon and S. Avrameas, J. Immunol. Meth. 11, 129 (1976).
(42) O.R. Zaborsky in Immobilized Enzymes (R.C. Weast, ed.) C.R.C. Press, 1974, Chapter 4.
(43) H.H. Weetall and A.M. Filbert in Methods in Enzymology, Vol. 34, p. 59 (1974).
(44) P. O'Carra, S. Barry and T. Griffin in Methods in Enzymology, Vol. 34, p. 108 (1974).
(45) R. Jost, T. Miron and M. Wilchek, Biochim. Biophys. Acta 362, 75 (1974).
(46) A.H. Nishikawa and P. Bailon, Arch. Biochem. Biophys. 168, 576 (1975).
(47) P. O'Carra, S. Barry and T. Griffin, Biochem. Soc. Trans. 1, 289 (1973).
(48) H.P. Jennissen and L.M.G. Heilmeyer, Biochemistry 14, 754 (1975).
(49) R.L. Schnaar, T.F. Sparks and S. Roseman, Anal. Biochem. 79, 513 (1977).
(50) J. Sica, E. Nola, I. Parikh, G.A. Pucca and P. Cuatrecasas, J. Biol. Chem. 248, 6543 (1973).
(51) M. Wilchek, FEBS Lett. 33, 70 (1973).
(52) G.I. Tesser, H.U. Fisch and R. Schwyzer, Helv. Chim. Acta 57, 1718 (1974).
(53) B. Pullmann and A. Polman, Rev. Modern Phys. 32, 428 (1960).
(54) G. Briegleb (1961) Elektronen Donator-Acceptor Complexes, Springer-Verlag, Berlin.
(55) R.S. Mulliken and W.B. Person (1969) Molecular Complexes. A lecture and reprint volume, Wiley Interscience, New York.
(56) M.A. Slifkin (1971) Charge Transfer Interactions of Biomolecules, Acad. Press, London.
(57) D. Eaker and J. Porath, Sep. Sci. 2, 207 (1967).
(58) J. Porath, 1st Tiselius Symposium, J. Chromatogr. in press.
(59) J. Porath, Proc. Nobel Symp. 3, 202 (1967)
(60) R.G. Harvey and M. Halonen, J. Chromatogr. 25, 294 (1968).
(61) G.H. Schenk, G.L. Sullivan and P.A. Fryer, J. Chromatogr. 89, 49 (1974).
(62) D.M. Jewell, Anal. Chem. 47, 2048 (1975)
(63) J.C. Janson, J. Chromatogr. 28, 12 (1967).
(64) J. Porath and K. Dahlgren Caldwell, J. Chromatogr. 133, 180 (1977).
(65) A. Szent-Gyorgyi (1960) Introduction to a Submolecular Biology, Acad. Press, New York.
(66) J.C.M. Tsibris, D.B. McCormick and L.D. Wright, Biochemistry 4, 504 (1965).

PART II

AFFINITY CHROMATOGRAPHY OF ENZYMES

AFFINITY CHROMATOGRAPHY OF ENZYMES

Peter D. G. Dean and David H. Watson

Department of Biochemistry, The University, Liverpool L69 3BX, England

ABSTRACT

The affinity chromatography of enzymes is reviewed in terms of the operational capacity of the ligand. Many group-specific ligands function in a rather inefficient manner. Reasons for this inefficiency are examined and the apparent K_As calculated from frontal analysis data. The effect of changing the support matrix is presented as well as the variation of the latter with ligand concentration. The direct relation between the capacity and apparent K_A is also shown. On the other hand, the capacities of two triazine-linked dyes, Procion Red and Cibacron Blue, are mugh higher than those of nucleotides. Cibacron Blue columns seem more effective in binding NAD^+-dependent dehydrogenases whilst Procion Red columns are better suited to the purification of $NADP^+$-dependent dehydrogenases. Yeast extract enzymes are separable using both blue and red columns with quite different elution profiles. The red column has been used to purify glutamate dehydrogenase from *N. crassa*. A novel method for the purification of *L. casei* dihydrofolate reductase on NADPH-Sepharose is described. A new method for the preparative electrophoretic desorption of proteins from affinity matrices and immunoadsorbents is illustrated.

INTRODUCTION

To date a large percentage of the known enzymes have been purified by biospecific adsorption using a very wide variety of affinity chromatography columns. In the course of this work almost every known aspect of enzymic activity has been exploited. Rather than attempt a large scale review, it is proposed to deal with selected problems in the hope that these will provoke a useful discussion.

SELECTION OF THE LIGAND

The problems of ligand selection are well illustrated using dehydrogenases as an example. Thus perhaps the most obvious ligand to prepare in a form suitable for immobilisation would be the dehydrogenase's substrate. However, it turns out that in many of the cases so far investigated that these enzymes are reasonably specific with respect to their substrate structures and that even minor alterations such as inclusion of a spacer arm can almost completely abolish the substrate binding to the enzyme. Perhaps the most well known example of a successful application of this approach is that of immobilised oxamate for the purification of lactate dehydrogenase (O'Carra *et al.* (1)). However, in the cases of almost all other dehydrogenases so far examined, alternative approaches have been made for their affinity purification.

Inhibitors of some dehydrogenases present rather the opposite effect, thus the folate analogue, methotrexate, binds to dihydrofolate reductase so tightly that high pH is required to elute the enzyme from columns of immobilised inhibitor and such conditions lead to enzyme inactivation (2).

NUCLEOTIDE GROUP-SPECIFIC ADSORBENTS

The use of immobilised cofactor as group-specific affinity adsorbents (particularly for the dehydrogenases), has been widely advocated (3,4,5). An example in the case of dihydrofolate reductase is shown in Fig. 1. This method was combined with the use of the oxidised coenzyme $NADP^+$ to purify the enzyme to homogeneity in a two-step procedure (Atkinson, Harvey & Dean, manuscript in preparation). The basis of the idea of coenzyme immobilisation lies in the very large percentage of enzymes which bind cofactors based on the adenine nucleotide 5'-AMP. Thus because immobilised adenine nucleotides such as AMP and NAD^+ will bind such a wide range of enzymes, devices have been sought either to narrow the specificity using specific elution techniques or to increase the resolution achieved by the adsorbent using gradients of eluant (pH, substrate, cofactor, temperature or salt) or by altering the adsorbent specificity by varying the point of attachment of the ligand (6).

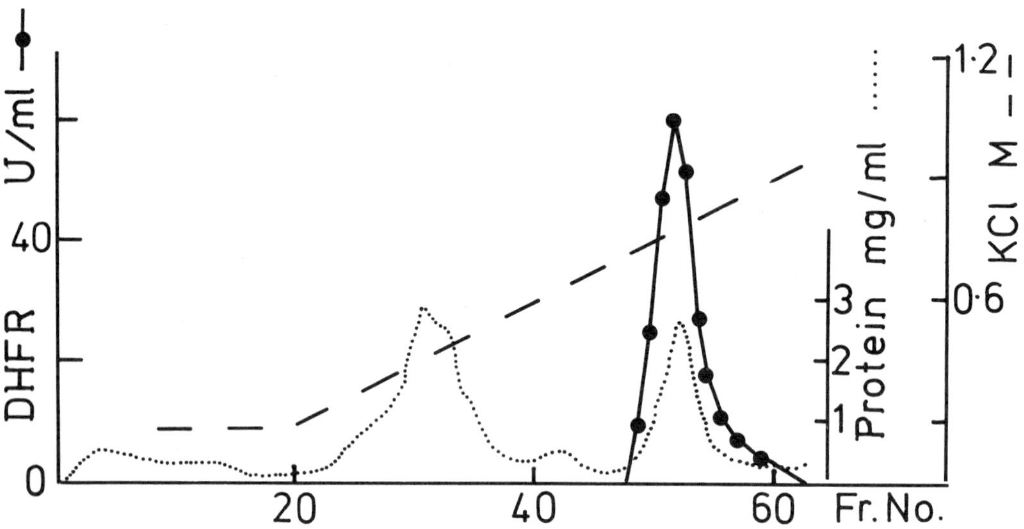

Fig. 1. The purification of dihydrofolate reductase from Lactobacillus casei using NADPH-Sepharose

The sonicated cells were extracted for enzyme, centrifuged and the supernatant applied to a column of $NADP^+$-Sepharose. The column was eluted with a gradient of KCl and the active fractions dialysed and applied to the same column after it had been treated with a solution of sodium dithionite to reduce the cofactor to NADPH. The enzyme was eluted with a gradient of KCl and had a specific activity of 20.2 I.U./mg.

Recently it has been shown that the success of some nucleotide adsorbents relies to a large extent on the synergistic combination of a hydrophobic component with an affinity component. The latter is a biospecific effect associated with coenzyme binding sites whilst the former probably involves the binding of the spacer arm to a fortuitously useful hydrophobic pocket adjacent to the nucleotide binding site.

Removal of this hydrophobic component in the case of N^6-hexyl-AMP adsorbents has been shown to abolish or diminish binding (7). Nevertheless, based on these methods a whole range of useful purifications are now appearing in the literature.

The assessment of group-specific adsorbents is best carried out by the separate consideration of the adsorption and desorption phases (8). A study of the desorption technique has obvious and immediate applications in any purification scheme for an enzyme. On the other hand, critical assessment of an adsorbent must also take into account the adsorption phase. A convenient approach is that of frontal analysis (9). It has been shown that apparent association constants may be calculated from data obtained from frontal analysis experiments (10). The following relationship demonstrates how this may be achieved.

$$K_A = \frac{V_e - V_e^*}{V_s[L] - [E](V_e - V_e^*)}$$

Fig. 2. The relationship between K_A (apparent), the elution volume for a protein, and the ligand concentration. V_e and V_e^* are determined as illustrated in Fig. 3. [L] is the ligand concentration, V_s is the volume of the gel available to the enzyme, [E] is the concentration of the enzyme.

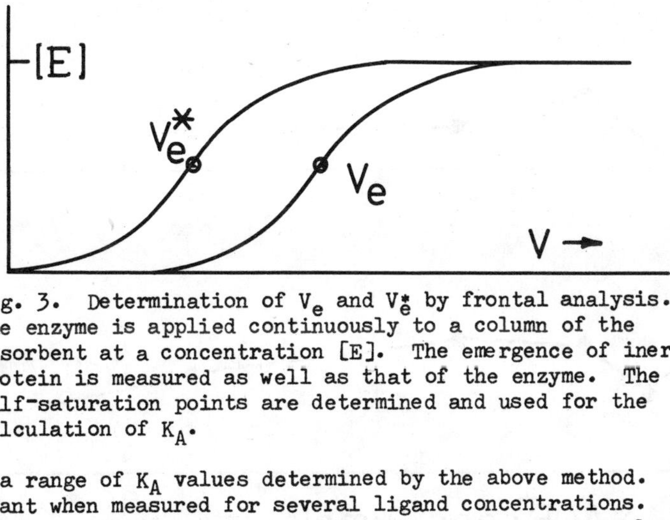

Fig. 3. Determination of V_e and V_e^* by frontal analysis. The enzyme is applied continuously to a column of the adsorbent at a concentration [E]. The emergence of inert protein is measured as well as that of the enzyme. The half-saturation points are determined and used for the calculation of K_A.

Table 1 shows a range of K_A values determined by the above method. These values were not constant when measured for several ligand concentrations. But they reached a maximum in all examples studied. This maximum was rarely in the range of concentration chosen by most workers (1-2 μmole/ml wet weight) (Fig. 4).

When the K_A values were measured at the same ligand concentration, it was found that K_A was proportional to the capacity (Fig. 5). The latter was determined by frontal analysis. A comparison of the capacities determined by frontal analysis and separately from desorption experiments showed good agreement.

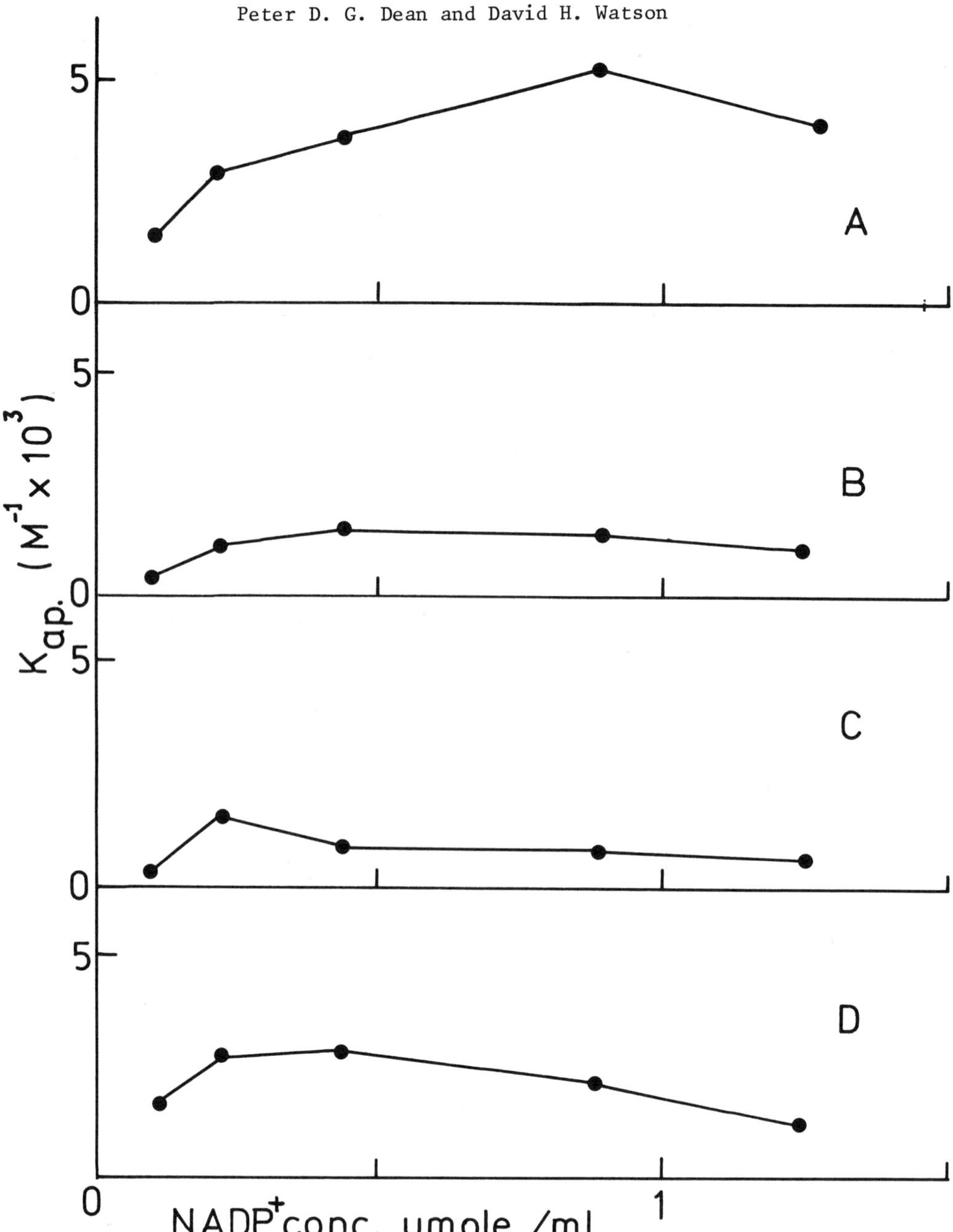

Fig. 4. The variation of K_A(apparent) with ligand concentration. A = *L. casei* Dihydrofolate reductase; B = yeast 6-phosphogluconate dehydrogenase; C = pig heart isocitrate dehydrogenase; D = *N. crassa* NADP$^+$-dependent L-glutamate dehydrogenase.

TABLE 1 K_A Values Determined for a Variety of Dehydrogenases and Different Nucleotides

		DHFR	6PGDH	ICDH	GDH	H_4 LDH
	$NADP^+$	210	710	360	480	-
Sepharose	2'AMP	160	320	190	nd	-
4B	5'AMP	30	20	40	50	270
	$NADP^+$	330	160	260	320	-
Ultrogel	2'AMP	440	120	320	260	-
22	5'AMP	160	10	60	<5	160
	$NADP^+$	3200	1400	1100	1600	-
Matrex	2'AMP	2600	1200	2600	800	-
201R	5'AMP	2800	<8	380	210	1500

The values were also determined for a range of different matrices. DHFR - L. casei dihydrofolate reductase; 6PGDH = yeast 6-phosphogluconate dehydrogenase; ICDH = pig heart isocitrate dehydrogenase; GDH = N. crassa L-glutamate dehydrogenase; H_4 LDH = pig heart lactate dehydrogenase.

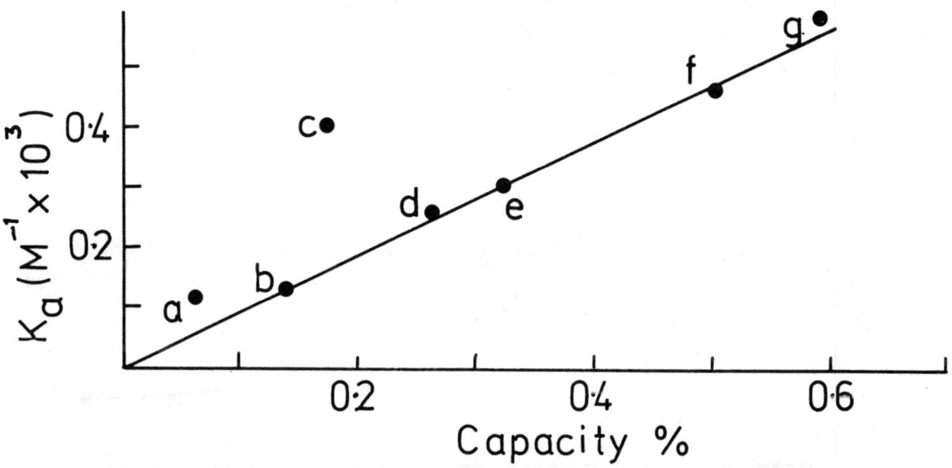

Fig. 5. The variation of capacity with the K_A(apparent) for various immobilised nucleotides. The enzyme studied was yeast 6-phosphogluconate dehydrogenase. The ligands used were: a) N^6-(3-aminopropylcarbamoylmethyl)-$NADP^+$; b) N^6-(6-aminohexyl)-5'-AMP; c) ribose-immobilised $NADP^+$ (modular solid state synthesis); d) N^6-(6-aminohexylcarbamoylmethyl)-2'-AMP; e) 8-(6-aminohexyl)-amino-2'-AMP; f) N^6-(-6-aminohexylcarbamoylmethyl)-$NADP^+$; g) ribose-immobilized-$NADP^+$. K_A(apparent) was measured by frontal analysis.

The relationship between the K_A for $NADP^+$-dependent dehydrogenases and the capacity is not apparent from free solution studies where the K_i's observed have been 1-2 orders of magnitude lower. This discrepancy could be explained on the basis of ligand crowding (for capacity values above the optimum ligand concentration). Diffusional limitations within the porous gel structure must also be considered to contribute to the above discrepancy. The latter may be partly removed by eliminating the porous nature of the gel. We have attached the same ligands to the pellicular support Matrex 201R (Table 1). The K_A is seen to be raised in all instances, by at least an order of magnitude. These measurements were made under conditions of extremely low flow rates and with closely similar ligand concentrations. When the flow rate of these columns is increased, and the behaviour of the gel (Sepharose) based columns compared with that of the pellicular (Matrex) equivalent then the diffusional effect can be seen to be increasingly important. Fig.6 shows that the capacity falls on increasing the flow rate much faster with the Sepharose ligands than with Matrex. It is expected that the K_A would show similar changes with respect to flow rate.

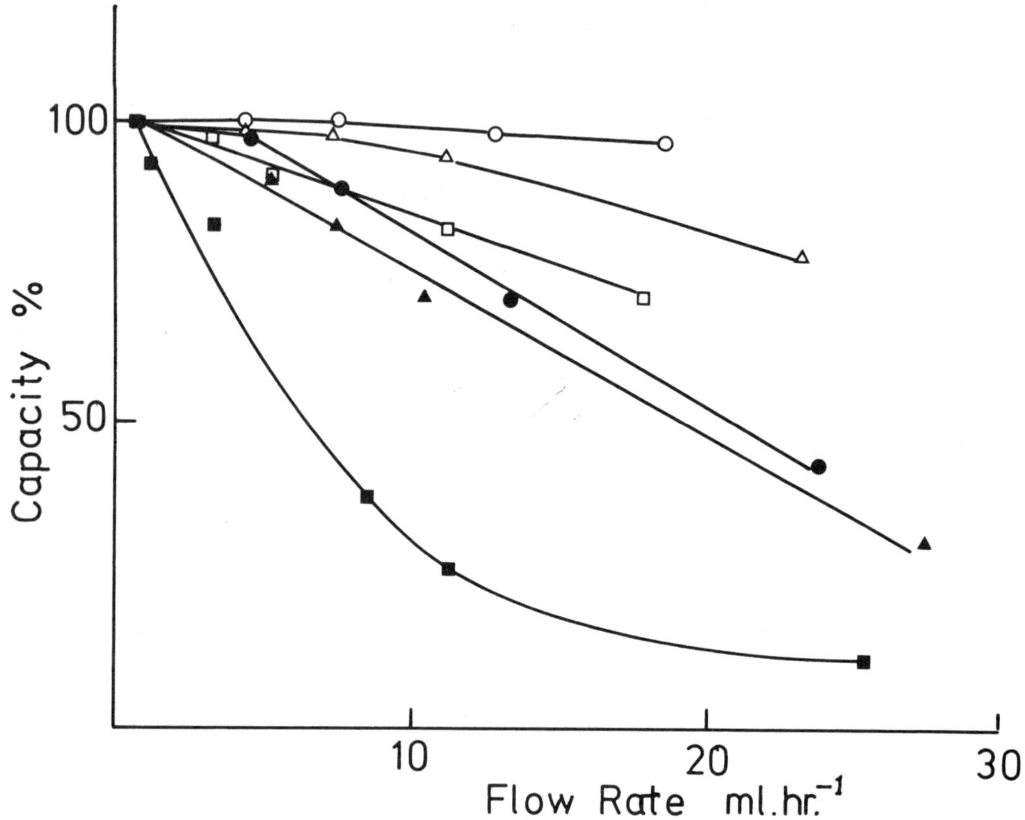

Fig. 6. The variation of capacity of $NADP^+$ matrices for various enzymes with increasing flow rate.

Capacities were measured by frontal analysis; closed symbols refer to Sepharose-based ligands, open symbols refer to Matrex 201R-based ligands; circles: dihydrofolate reductase (L. casei); triangles: isocitrate dehydrogenase (pig heart); squares: L-glutamate dehydrogenase (N. crassa).

In conclusion, the binding capacities of nucleotide affinity matrices can be measured by both frontal analysis and by traditional desorption methods. Both procedures show that these ligands rarely bind dehydrogenases in a manner which uses more than one percent (and never in our experience more than 5%) of the total ligand. This may place important restrictions on the large scale preparative applications of nucleotide-based affinity adsorbents. An examination of the behaviour of similar affinity columns tested in other laboratories shows that this low capacity is very commonly observed (Table 2).

TABLE 2 An Analysis of Some Papers using Nucleotide Affinity Columns.

	Enzyme	Immobilised Nucleotide	Capacity	Ref.
1.	Lactate dehydrogenase	NAD^+	0.1	(9)
2.	" "	5'-AMP	0.1	(6)
3.	NAD^+-dependent dehydrogenases	5'-AMP	0.05 - 0.2	(15)
4.	Thermophilic dehydrogenases	5'-AMP	0.05 - 0.1	(16)
5.	Lactate dehydrogenase, Glyceraldehyde 3-phosphate dehydrogenase	NAD^+ 5'-AMP	0.1 - 0.2	(11)
6.	Alcohol dehydrogenase , Xanthine dehydrogenase	NAD^+	0.1	(12)
7.	Alcohol dehydrogenase , Lactate dehydrogenase	NAD^+	0.004 - 0.02	(13)
8.	Glucose-6-phosphate dehydrogenase	$NADP^+$	0.05	(17)
9.	$NADP^+$-dependent DHs	2',5'-ABP	0.02 - 0.05	(14)
10.	Isocitrate dehydrogenase	$NADP^+$	0.05	(18)
11.	Glutamate dehydrogenase	2'-AMP	0.06	(19)

TRIAZINE-LINKED DYES AS GROUP-SPECIFIC LIGANDS

We have examined alternative adsorbents for the selective purification of dehydrogenases in an attempt to overcome the above problem of capacity associated with nucleotides as ligands. Cibacron Blue F3G-A has been widely used for the purification of a number of different proteins, including dehydrogenases (20). Another triazine-linked dye Procion Red HE-3B has recently been used by Baird *et al.* (21) to purify carboxypeptidase G. We have found that this dye is also capable of binding dehydrogenases.

The selectivity of Cibacron Blue and Procion Red for $NAD(P)^+$-dehydrogenases is illustrated in Fig. 7 where the capacities for these columns have been measured for both ligands and for a variety of enzymes.

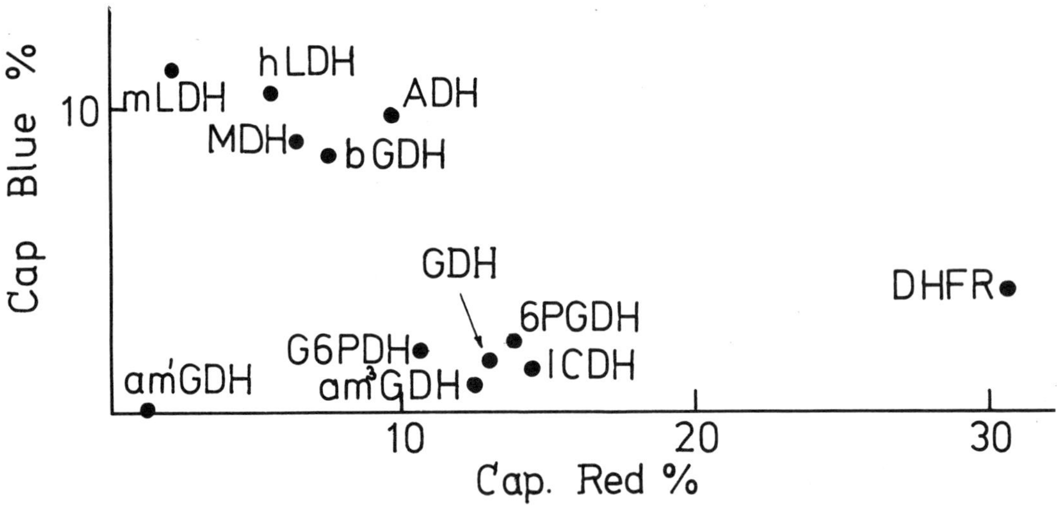

Fig. 7. The capacities of Cibacron Blue and Procion Red columns for various enzymes. The capacities were measured by frontal analysis.
GDH = N. crassa NADP$^+$-dependent L-glutamate dehydrogenase; bGDH = bovine liver NAD(P)$^+$-dependent L-glutamate dehydrogenase; am' and am^3-GDH = mutant N. crassa GDH(19).

We suggest that Cibacron Blue is more suited to the purification of NAD$^+$-dependent dehydrogenase whereas Procion Red is more effective for the purification of NADP$^+$-dependent dehydrogenases.

The capacities of both the red and blue adsorbents are markedly higher than the correspondingly specific nucleotide adsorbents (Table 3). This seems to be the case not only for the purified enzymes but also when these measurements are made in the presence of other proteins as in a yeast extract. In the latter case, however, the capacities are lower in both instances because of the lower concentration of enzyme (the capacity being directly related to the applied enzyme concentration).

The patterns of capacities of the red column and NADP$^+$-Sepharose columns are very similar when compared graphically (Fig. 8). The same may be observed when comparing the blue column with N^6-AMP-Sepharose.

The application of both the red and blue columns to the purification of NAD(P)$^+$-dehydrogenases from yeast is shown in Fig. 9. It can be seen that both columns provide methods for the separation of similar enzymes. Table 4 shows the incorporation of the red column into the purification of NADP$^+$-dependent L-glutamate dehydrogenase from N. crassa where it replaces two ion exchange columns and one G-200 column (22).

TABLE 3 Relative Capacities of Procion dyes and Nucleotide Affinity Columns for Enzymes in Yeast Extract

	Red	Blue	N^6-NADP	N^6-5'-AMP
ADH	2	1	<0.01	0.3
MDH	1.5	2	0	0.1
GR	11	0.3	0.1	<0.01
GDH	0.5	0.1	< 0.01	0
6PGDH	6	0.2	0.1	0.02
G6PDH	0.2	0.05	<0.01	0.01

ADH = alcohol dehydrogenase; MDH = malate dehydrogenase; GR = glutathione reductase; 6PGDH = 6-phosphogluconate dehydrogenase; G6PDH = glucose 6-phosphate dehydrogenase.

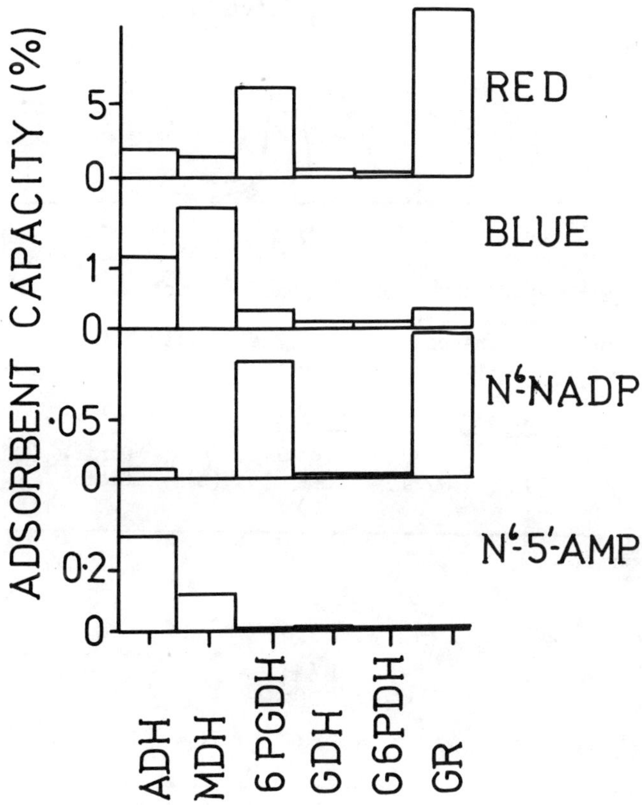

Fig. 8 Capacities of four adsorbents for various enzymes in terms of the percentage of ligand used at saturation. See Table 3.

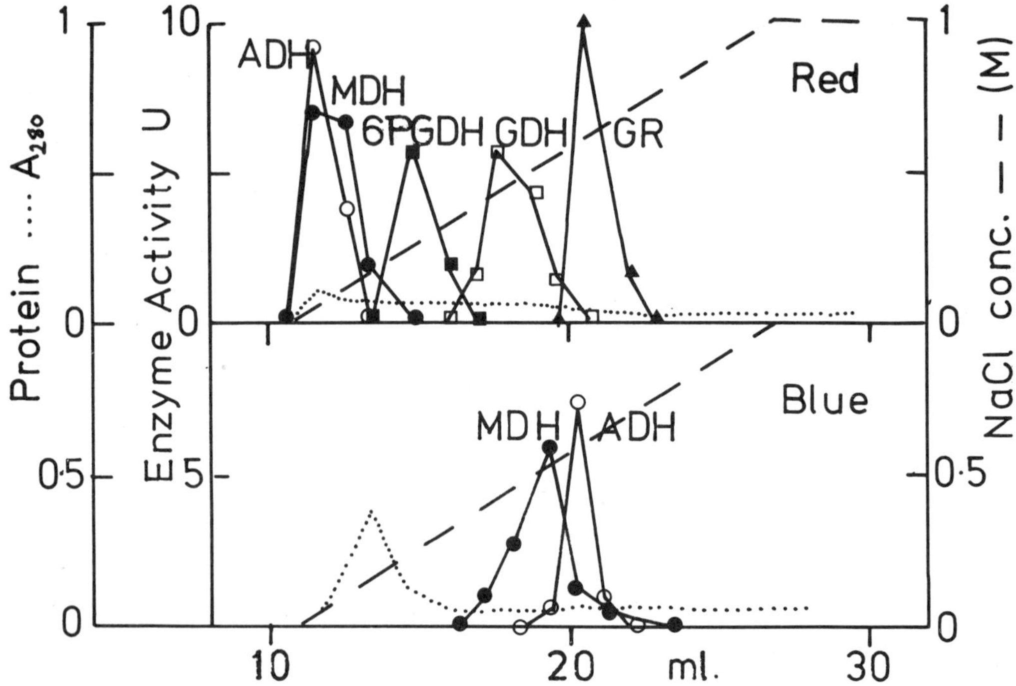

Fig. 9 Chromatography of crude yeast extract on Procion Red and Cibacron Blue-Sepharose.
After adsorption the columns were washed with buffer until all proteins were undetectable and an NaCl gradient then applied.

TABLE 4 N. crassa GDH Purification

	Vol. (ml)	Protein (mg/ml)	Total GDH (U)	Yield (%)	Purif. factor
Crude extract	250	-	872	100	-
46,000g SF	214	2049	656	75	-
heat step	208	1042	450	52	1.4
35%-60% $(NH_4)_2SO_4$	3.6	24	406	47	53
Red Seph. Column	3	0.9	73*	43	245**

* per 1ml column
** Specific activity of pooled active protein = 79 U/mg.

In conclusion, whilst it may be seen that the specificity of the dye columns is not as marked as the correspondingly specific nucleotide affinity support, nevertheless, two important advantages should be emphasised which are inherent in the dye columns. 1) the ease and economy in the preparation; 2) the improved operational capacities of the dye columns.

ELECTROPHORETIC DESORPTION OF AFFINITY MATRICES

In the course of our examination of the K_As of the dye columns we have consistently noted higher values for the latter when compared at similar concentrations with immobilised nucleotides. Indeed, it was observed in several instances that it was not possible to remove enzymes with pulses of nucleotide or salt from such columns. Thus in the case of both the red and blue dyes, when attached via polyethyleneimine spacers to Matrex 201R, the following enzymes could not be eluted either with nucleotides or with KCl: lactate dehydrogenase, malate dehydrogenase, glutamate dehydrogenase and dihydrofolate reductase. In these instances we have applied the technique of electrophoretic desorption to the column in order to recover the enzyme.

This technique consists of the direct electrophoresis of proteins from affinity adsorbents using modified gel electrophoresis equipment. The range of species that have been successfully desorbed includes antibodies and antigens as well as albumin from Cibacron Blue-Sepharose (Table 5). The technique appears to be generally applicable to a wide range of chromatography systems based on hydrophobic, hydrophilic, ionic and detergent interactions.

GENERAL CONCLUSIONS

In conclusion, nucleotide affinity matrices are likely to continue to be useful for highly specific protein purifications despite their low capacities. On the other hand, alternatives of high capacity are available in the form of red and blue triazine-based dyes; these appear to be relatively specific. Furthermore, the operational capacities of these gels is between one and two orders of magnitude greater than those of the equivalent nucleotide matrices.

A useful method for the assessment of such properties is that of frontal analysis. Using this method, apparent K_As may be determined and these values can assist in understanding some of the properties of affinity adsorbents.

Novel purification procedures are described for dihydrofolate reductase and glutamate dehydrogenase using nucleotide-Sepharose and Procion Red-Sepharose respectively.

ACKNOWLEDGEMENTS

The support of the Science Research Council, L'Industrie Biologique Française and The Amicon Corporation is gratefully acknowledged.

TABLE 5 Examples of Preparative Electrophoretic Desorption

Retarded molecule	Adsorbent	Conditions	Recovery
Ferritin	Anti-ferritin-Sepharose 4B	75 V 15 mA	95%
Oestriol-16α glucuronide	Anti-oestriol-16α-glucuronide Matrex-201R	120 V 10 mA	40%
Oestriol-16α* glucuronide antibody	Oestriol-16α-glucuronide-BSA-Sepharose	60 V 2 mA	90%
Lactate dehydrogenase	N^6-5'-AMP-Sepharose (or Matrex 201R)	40 V 10 mA	88%
" "	DEAE-cellulose (DE-52)	60 V 10 mA	92%
" "	Hexyl-Sepharose	200 V 10 mA	88%
" "	Cibacron Blue-(Sepharose or CB Matrex)	80 V 10 mA	83%
Human serum albumin	Cibacron Blue-Sepharose pH 9.5	90 V 40 mA	74%
Cytochromes	Cytochrome c-Sepharose (or Matrex)	300 V 5-10 mA	80%

*M.R.A. Morgan, personal communication (results obtained using desorption into a polyacrylamide gel).

REFERENCES

1. O'Carra, P. and Barry, S. (1974) Lactate dehydrogenase: Specific Ligand Approach, in Jakoby, W. B. and Wilchek, M. eds. Methods in Enzymology 34, 598.

2. Pastore, E. J., Plante, L. T. and Kisliuk, R. L. (1974) Pteroyllysine - Agarose in the Purification of Dihydrofolate reductase, ibid p.286; Kaufman, S. ibid, p.279.

3. Dean, P. D. G. and Harvey, M. J. Applications of Affinity Chromatography, Process Biochem. 10, 5 (1975)

4. Jakoby, W. B. and Wilchek, M. eds. (1974) Methods in Enzymology 34.

5. Lowe, C.R., Mosbach, K. and Dean, P. D. G. Affinity Chromatography of lactate Dehydrogenase on Immobilised Nucleotides, Biochem. Biophys. Res. Comm. 48, 1004, (1972).

6. Harvey, M. J., Lowe, C. R., Craven, D. B.and Dean, P. D. G. Affinity Chromatography on Immobilised Adenosine 5'-monophosphate II, Eur. J. Biochem. 41, 335 (1974).

7. Doley, S. G., Dean, P. D.G., Dietz, G., Harvey, M. J. and Neame, P.J. (1977) The Influence of Matrix and Spacer Arm on Group Specific Affinity Chromatography, in Advances in the Chromatographic Fractionation of Macromolecules. Ellis Harwood. In Press.

8. Dean, P. D. G. and Harvey, M. J.(1977) Affinity Chromatography, in Smith, H. (ed) Regulation of Enzyme Synthesis and Activity in Higher Plants. Acad. Press, London. In Press.

9. Lowe, C. R., Harvey, M. J., Craven, D. B. and Dean, P. D. G. Some parameters relevant to Affinity Chromatography on Immobilised Nucleotides, Biochem. J. 133, 499 (1973).

10. Nichol, L. W., Ogston, A. G., Winzor, D. J. and Sawyer, W. H. Evaluation of Equilibrium Constants by Affinity Chromatography, Biochem. J. 143, 435 (1974).

11. Mosbach, K., Guildford, H., Ohlsson, R. and Scott, M. General Ligands in Affinity Chromatography. Cofactor - Substrate Elution of Enzymes to the Immobilised Nucleotides AMP and NAD, Biochem. J. 127, 625 (1972).

12. Barry, S. and O'Carra, P. Affinity Chromatography of Nicotinamide-Adenine Dinucleotide - Linked Dehydrogenase on Immobilised Derivatives of the Dinucleotide, Biochem. J. 135, 595 (1973).

13. Schmidt, H. L. and Grenner, G. Coenzyme properties of NAD^+ bound to different matrices through the amino group in the 6-position, Eur. J. Biochem. 67, 295 (1976).

14. Brodelius, P., Larsson, P.-O. and Mosbach, K. The synthesis of Three AMP-Analogues and their Application as General Ligands in Biospecific Affinity Chromatography, Eur. J. Biochem. 47, 81 (1974).

15. Trayer, I. P. and Trayer, H. Affinity Chromatography of Nicotinamide Nucleotide-Dependent Dehydrogenases on Immobilised Nucleotide Derivatives, Biochem. J. 141, 775 (1974).

16. Comer, M., Craven, D. B., Harvey, M. J., Atkinson, A. and Dean, P.D.G. Affinity Chromatography on Immobilised Nucleotides. Some Applications to the Purification of Thermophilic Dehydrogenases and Kinases, Eur. J. Biochem. 55, 201, (1975).

17. Lamed, R., Levin, Y. and Wilchek, M. Covalent coupling of nucleotides to agaroses for affinity chromatography, Biochim. Biophys. Acta 304, 231 (1973).

18. Hy, M. and Reeves, H. C. $NADP^+$-Specific Isocitrate Dehydrogenase of *E. coli* III Two-Step purification Employing Affinity Chromatography, Biochim. Biophys Acta, 445, 280 (1976).

19. Watson, D. H. and Wootton, J. C. Affinity Chromatography of *N. crassa* $NADP^+$-Dependent L-Glutamate Dehydrogenase, Biochem. J. In Press (1977).

20. Easterday, R. L. and Easterday, I. M. Affinity chromatography of kinases and dehydrogenases on Sephadex and Sepharose dye derivatives in Immobilised Biochemicals and Affinity Chromatography, Dunlap, R. B. ed Plenum Publ. New York, 123.

21. Baird, J. K., Sherwood, R. F., Carr, R. J. G. and Atkinson, A. Enzyme Purification by Substrate Elution Chromatography from Procion Dye - Polysaccharide Matrices, FEBS Lett. 70, 61 (1976).

22. Ashby, B., Wootton, J. C. and Fincham, J. R. S. Slow conformation changes of a *Neurospora* glutamate dehydrogenase studied by Protein Fluorescence, Biochem. J. 143, 317 (1974).

AFFINITY CHROMATOGRAPHY ON IMMOBILISED COENZYMES

Christopher R. Lowe

School of Biochemical and Physiological Sciences, The University of Southampton, Bassett Crescent East, Southampton, Great Britain, SO9 3TU

Recent developments in enzyme purification technology have raised the status of the area from purely empirical to one displaying considerable rationale. These developments have been concerned with the preparation and application of chromatographic adsorbents based on biological specificity (Refs. 1 - 4). The technique of affinity chromatography is based on the unique ability of biological macromolecules to bind small complementary ligands specifically and reversibly and is realised by covalently attaching the ligand to an insoluble support. The adsorbent is packed into a column, the sample applied and only that protein that displays affinity for the immobilised ligand is retained; all others pass through unretarded. The biospecifically adsorbed species can subsequently be desorbed by altering the column irrigants to effect dissociation. In principle, biospecific adsorbents can be designed to purify or resolve enzymes, antibodies, receptor and transport proteins, isoenzymes, viruses and cells (Ref. 4).

The success of affinity chromatography depends largely on how closely the prevailing experimental conditions selected mimic the native or biological interaction between the ligand the macromolecule in free solution. Careful consideration must therefore be given to the hydrophilic matrix and the nature of the spacer molecule interposed between the ligand and the matrix. Furthermore, the ligand must be one that displays special and unique affinity for the macromolecule to be purified and bear functional groups suitable for modification without impairing its recognition. Many biological macromolecules, however, have several ligands which satisfy these requirements and the selection of the ligand for immobilisation thus becomes the personal choice of the investigator. Two approaches are discernible; the first employs a ligand which displays absolute specificity for the enzyme under study (Ref. 5) and the second utilises a "general ligand" with the capacity to adsorb a large number of enzymes or a group of enzymes (Refs. 6, 7). This so-called "group specific" or "general ligand" approach encompasses a number of adsorbents comprising immobilised coenzymes (Refs. 6, 7), nucleic acids and nucleotides (Refs. 8, 9), lectins (Ref. 10), dyestuffs (Ref. 11) and organomercurials (Ref. 12) and all of which display affinity for a group of complementary proteins. The advantages of "group specific" affinity matrices should be immediately apparent; the chemistry of ligand immobilisation is restricted to a single ligand and once the analogue is synthesised, the affinity gel is applicable to the purification of a number of complementary biomolecules.

The potential usefulness of immobilised nucleotides such as the adenine and pyridine nucleotide coenzymes (Refs. 4, 6-8) to affinity chromatography derives from their general applicability to the purification of a wide range of enzymes. For example, of the 2000 or so known enzymes, some 8% bind NAD^+, 8% bind $NADP^+$, 11% bind ATP and 4% coenzyme A. Not surprisingly, therefore, adenosine 5'-monophosphate (AMP) has been chosen as a general ligand, since in broad terms as a constituent of the above coenzymes, it binds to approximately one third of all known enzymes. However, the chemical complexity of the adenine nucleotide coenzymes has until recently hampered the development of methods suitable for their derivatisation and subsequent immobilisation. Nevertheless, despite acute synthetic problems, chemically characterised polymers have been described (Refs. 7, 8, 13-15). Several methods have concentrated on the derivatisation of the adenine moiety of the coenzyme (Refs. 7, 8, 13, 15). X-ray crystallographic studies of dogfish muscle lactate dehydrogenase have shown that the coenzyme NAD^+ is bound in a deep pocket with the adenine moiety outermost (Ref. 16). Data from this and other dehydrogenases has shown that the exocyclic N^6-amino group and the carbon atom at position 8 of the adenine nucleus are uncommitted to binding to the enzyme and protrude out of a shallow hydrophobic pocket in which the adenosine moiety is bound. Not surprisingly, therefore, early methods concentrated on substitution at the exocyclic N^6-amino group. For example, N^6-(6-aminohexyl)-5'-AMP has been synthesised by nucleophilic displacement of the thiol group of 6-mercapto-purine riboside 5'-phosphate by excess 1,6-hexanediamine (Ref. 13), chemically characterised, and subsequently coupled to CNBr-activated agarose. The 6-substituted analogue may also be synthesised by an equivalent procedure from 6-chloro-IMP (Ref. 17), whilst an 8-substituted derivative may be obtained by nucleophilic displacement of the bromine of 8-Bromo-AMP (Refs. 8, 15). Figure 1 illustrates the structures of two immobilised-AMP adsorbents substituted with a hexamethylene spacer molecule at the 6- and 8-positions of the adenine nucleus respectively.

(a)

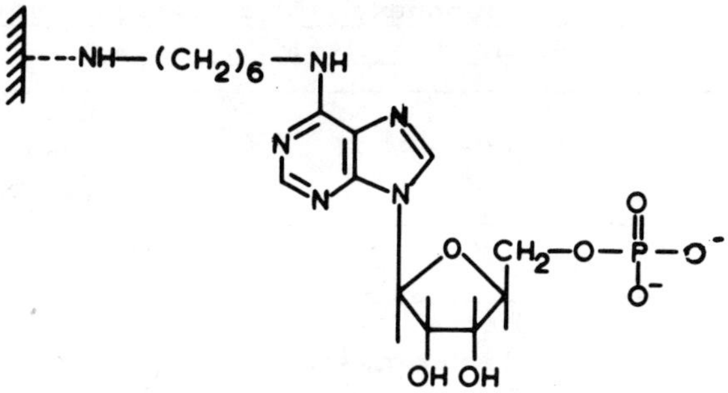

(b)

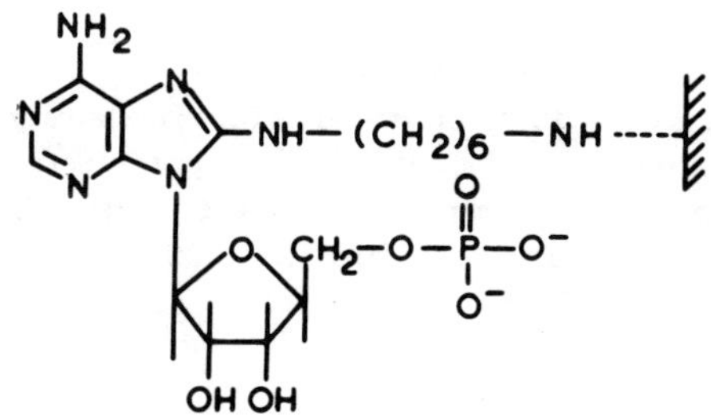

Fig. 1. The structures of two adenine-immobilised AMP adsorbents (a) N^6-(6-aminohexyl)-AMP-agarose and (b) 8-(6-aminohexyl)-AMP-agarose.

Both matrix-bound derivatives fulfill many of the requirements for a "general ligand" and bind a wide range of kinases and dehydrogenases. However, at present, very little comparative data is available to act as a guideline to ascertain which of the two derivatives would be more suitable for the isolation of a particular enzyme. Table 1 summarises the chromatographic behaviour of a number of AMP-dependent enzymes on agarose-bound N^6-(6-aminohexyl)-AMP and 8-(6-aminohexyl)-AMP under essentially identical experimental conditions. The affinity of the enzymes for the adsorbents is expressed in terms of the concentration of eluant required to elute the enzyme with peak activity on a linear gradient. Most of the enzymes studied preferred the 6-substituted to the 8-substituted adsorbent, although for a given adsorbent there was a considerable range of observed affinities. In contrast, both alanine dehydrogenase (EC 1.4.1.1) and malate dehydrogenase (EC 1.1.1.37) from B. subtilis displayed a marked preference for the 8-substituted AMP-derivative. Neither lipoamide dehydrogenase (EC 1.6.4.3) nor D-galactose dehydrogenase (EC 1.1.1.48) were adsorbed to either gel at the ligand concentration of 2.5 μmole AMP-analogue/g moist weight used.

TABLE 1 The Binding of Several AMP-dependent enzymes to N^6-(6-aminohexyl)-AMP-Agarose and 8-(6-aminohexyl)-AMP-Agarose

Enzyme	Source	Eluant Concentration (μM NADH)	
		C8	N6
Alanine Dehydrogenase	*B. subtilis*	80	8
Alcohol Dehydrogenase	Horse liver	30	75
Lactate Dehydrogenase	Rabbit muscle	30	190
Lactate Dehydrogenase	Pig heart	10	30
Malate Dehydrogenase	Pig heart	0	35
Malate Dehydrogenase	*B. subtilis*	10	0
Lipoamide Dehydrogenase	Pig heart	0	0
D-galactose Dehydrogenase	*Ps. fluorescens*	0	0
Phosphorylase b	Rabbit muscle	2.0*	3.0*

* Eluant was AMP (mM)

Samples of enzyme (1 unit) containing bovine serum albumin (0.8 mg)were applied to a column (0.5 g) of 6- or 8-substituted AMP-agarose in 50 mM potassium phosphate buffer pH 7.5 at 4°C, equilibrated for 15 min, eluted with a linear gradient of NADH (0 - 0.4 mM; 20 ml total volume) or AMP (0 - 10 mM; 20 ml total volume) and fractions 1.5 ml collected at a flow rate of 6 ml/h.

In an attempt to correlate the chromatographic behaviour of the enzymes on these adsorbents with a tangible free solution parameter, competitive inhibition constants (K_i) for each of the free 6- and 8-substituted AMP derivatives were determined from initial velocity measurements. Table 2 summarises the K_i values for several enzymes for free N^6-(6-aminohexyl)-AMP and 8-(6-amino-hexyl)-AMP. In most cases the K_i values for the 6-substituted analogue were

TABLE 2 Competitive Inhibition Constants for N^6-(6-amino-hexyl)-AMP and 8-(6-aminohexyl)-AMP

Enzyme	Source	K_i (mM)	
		C8	N6
Alanine Dehydrogenase	*B. subtilis*	0.08	2.0
Alcohol Dehydrogenase	Horse liver	1.5	0.4
Lactate Dehydrogenase	Rabbit muscle	2.0	0.1
Lactate Dehydrogenase	Pig heart	2.4	0.1
Malate Dehydrogenase	Pig heart	4.4	0.2
D-galactose Dehydrogenase	*Ps. fluorescens*	11.0	2.5

4-20 times lower than those for the 8-substituted derivative measured under the same conditions. In contrast, with alanine dehydrogenase, the K_i value for the 8-substituted derivative was 25-fold lower than that for the 6-substituted derivative. The data summarised in Table 2 confirm the general preference for 6-substituted derivatives found by other authors (Ref. 8). Furthermore, the data suggest that the K_i value for the free prospective ligand must be at least 2 mM in order to generate an effective adsorbent when the ligand concentration is 2.5 μmole/g moist weight gel. This conclusion appears to be independent of whether the gel comprises an immobilised 6- or 8-substituted derivative.

The chromatographic and kinetic data contained in Tables 1 and 2 indicate the preferred adsorbent for a particular enzyme purification. For example, 8-(6-aminohexyl)-AMP-agarose should prove an effective adsorbent for the purification of alanine and malate dehydrogenases from a crude extract of B. subtilis. Figure 2 shows that both enzymes are strongly bound to the immobilised-AMP adsorbent with all the detectable inert protein appearing in the void volume. The enzymes were quantitatively eluted as homogeneous proteins with a linear NADH gradient.

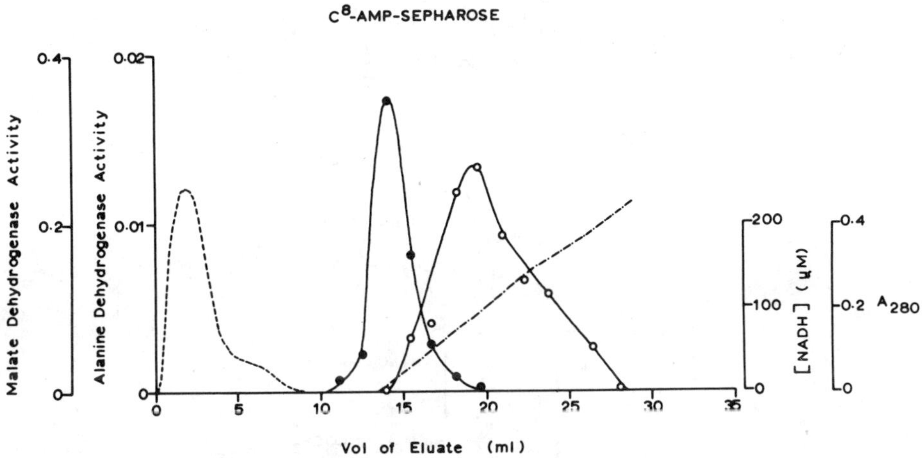

Fig. 2. Chromatography of a crude extract of B. subtilis on 8-(6-aminohexyl)-AMP-Sepharose. The bacterial extract (100 μl) was applied to a column (0.5 g) of immobilised -AMP and allowed to equilibrate for 15 min. The column was washed with 50 mM potassium phosphate buffer pH 7.5 (12 ml) and the enzymes eluted with a linear gradient of NADH (0 - 0.5 mM; 50 ml total volume) in the phosphate buffer. Alanine dehydrogenase activity (—O—), malate dehydrogenase activity (—●—), protein concentrations (----) and concentration of NADH in gradient (—·—·—).

Immobilised 8-substituted analogues were also found to be particularly useful for the purification of IMP dehydrogenase (EC 1.2.1.4) from E. coli. Inosine 5'-monophosphate (IMP) lies at the point of divergence of the purine biosynthetic pathway into the AMP branch via succinyl-AMP (SAMP) and the GMP branch via xanthosine 5'-monophosphate (XMP). IMP dehydrogenase catalyses the conversion of IMP into XMP with the concomitant reduction of NAD^+. In principle, therefore, the enzyme may be purified by affinity chromatography on immobilised NAD^+, IMP or AMP. Immobilised derivatives of all three nucleotides were prepared as follows. The 8-substituted AMP-analogue, 8-(8-amino-octyl)-AMP was synthesised by bromination of AMP for 24 h at pH4 followed by nucleophilic displacement of the bromine with 1,8-diamino-octane (Ref. 18). This AMP derivative was purified and characterised and converted into the 8-substituted IMP derivative by protection of terminal amino group of the spacer molecule by reaction with ethyltrifluoroacetylthiol acetate ($CF_3COSC_2H_5$) followed by diazotisation in the presence of nitrous acid at 0^oC for 3 hours. The terminal amino group of the 8-substituted IMP derivative was finally deprotected by alkaline hydrolysis at pH 12 for 3 h and subsequently coupled to CNBr-activated

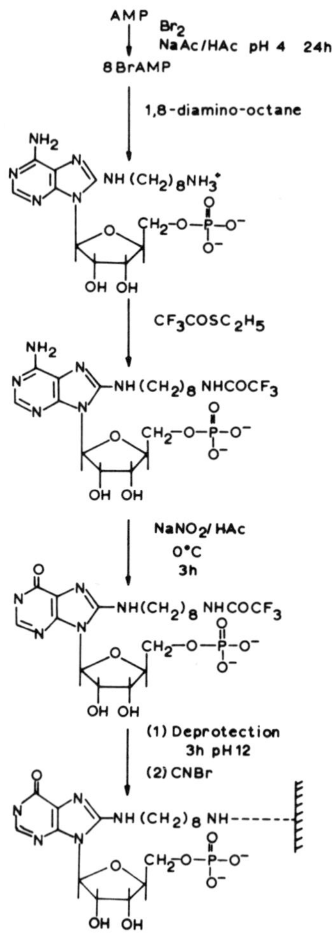

Fig. 3. The synthesis of the inosine 5'-monophosphate (IMP) derivative substituted in the 8-position of the purine ring

agarose. All intermediates in the synthesis outlined in Fig. 3 were purified to homogeneity and characterised prior to continuation with the synthesis. The final IMP derivative, 8-(8-amino-octyl)-IMP gave the expected reaction to a number of functional group specific reagents and had the anticipated spectral properties. Furthermore, attachment to CNBr-activated agarose yielded an adsorbent with a ligand concentration of approximately 2 μmole/g moist weight which had interesting properties. Most NAD^+-dependent dehydrogenases such as lactate dehydrogenase (EC 1.1.1.27) and B. subtilis alanine dehydrogenase displayed no affinity for the immobilised IMP adsorbent and were eluted along

(A)

(B)

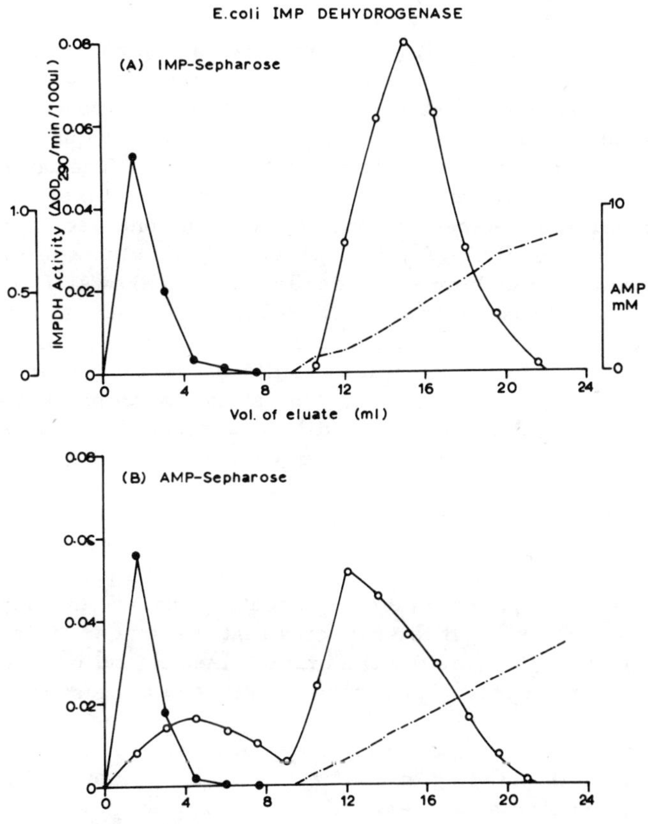

Fig. 4. Chromatography of a crude extract of E. coli on (A) 8-(8-amino-octyl)-IMP-Sepharose and (B) 8-(8-amino-octyl)-AMP-Sepharose. The bacterial extract (100 μl) was applied to a column (0.5 g) of the immobilised nucleotide and allowed to preincubate for 15 min. The column was washed with 50 mM potassium phosphate buffer pH 7.5 (12 ml) and developed with a linear gradient of AMP (0 - 20 mM; 20 ml total volume) in the phosphate buffer. IMP Dehydrogenase activity (—O—), protein concentration (mg/ml) (—●—) and concentration of AMP (—·—·—) were assayed in the eluant fractions (1.5 ml).

with inert protein in the void volume. In contrast, these enzymes displayed appreciable affinity for the parent immobilised-AMP adsorbent (Tables 1 and 2) and could be eluted quantitatively with an NADH gradient as expected. On the other hand, IMP dehydrogenase from E. coli binds tightly to IMP-Sepharose and requires either a pulse of 10 mM AMP or a linear gradient of AMP to effect elution (Fig. 4). The enzyme is recovered quantitatively in a state homogeneous by SDS electrophoresis whilst all the inert protein present in the original extract passes through unretarded and appears in the void volume of the column. E.coli IMP dehydrogenase also binds tightly to the immobilised-AMP adsorbent but is eluted at a lower concentration of AMP. Both of these procedures may be scaled up to yield 20 - 50 mg pure enzyme in a single day for a crude extract of E. coli. The enzyme displayed no affinity for an immobilised-NAD^+ derivative prepared by the method of Lindberg et al. (Ref. 19). Affinity chromatography on immobilised-AMP and IMP were therefore used regularly to purify IMP dehydrogenase for structural studies. Whilst AMP was routinely used as eluant, an NAD^+ gradient (0 - 5 mM) was equally effective in the quantitative release of enzyme, although NADH proved relatively ineffectual. The latter observations are in keeping with the known affinities of IMP dehydrogenase for the oxidised and reduced coenzymes respectively (Ref. 20). In contrast, despite its high affinity for the enzyme (Ref. 20), IMP proved a poor eluant of the enzyme both from the AMP and IMP-columns. Indeed as Fig. 5 shows, IMP dehydrogenase was eluted in a very broad band from IMP-Sepharose and required a pulse of 10 mM AMP to effect quantitative release. Furthermore, incorporation of 10 mM IMP in the equilibration buffer results in the appearance of the enzyme in a large number of fractions commencing at, and with highest activity in, the void volume. Quantitative elution could be expedited by pulsing with 10 mM AMP as found previously. These observations suggest the presence of non-specific interactions in addition to the biospecific interaction between the enzyme and the immobilised-IMP. It is becoming increasingly evident that non-specific adsorption phenomena of this type can augment or even grossly distort true biospecific behaviour (Ref. 21). It seems likely that such effects arise from the hydrophohic nature of spacer molecules used to separate the immobilised ligand from the matrix backbone. It is suspected that the spacer molecules interfere with the affinity of the enzyme for the immobilised ligand by interacting hydrophobically with accessible hydrophobic sites on the enzyme surface (Ref. 18).

Two approaches may be envisaged to assess the contribution of the hydrocarbon spacer molecule to the binding of an enzyme to an immobilised ligand. Firstly, the binding of a complementary enzyme may be investigated in the presence of low concentrations of organic solvents (Ref. 22). For example, the binding of pig heart lactate dehydrogenase to agarose-bound N^6-(6-aminohexyl)-AMP has been studied in the presence of several aqueous-organic cosolvent mixtures.

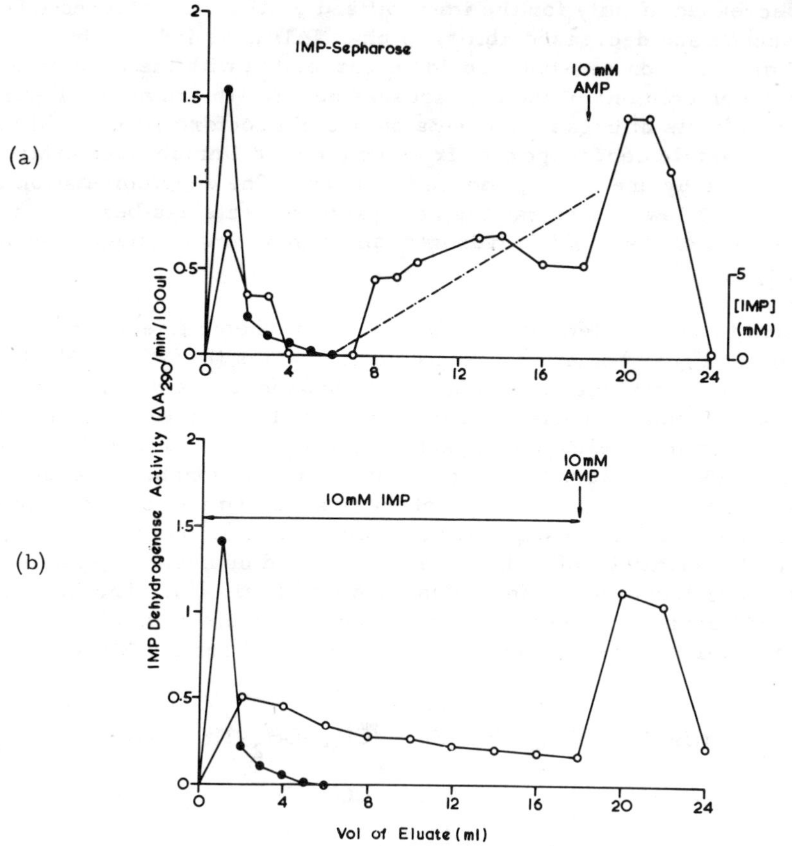

Fig. 5. Chromatography of E. coli IMP dehydrogenase on agarose-bound 8-(8-amino-octyl)-IMP. Conditions of adsorption are as described in the legend to Fig. 4 except that in (a) elution is effected with a linear gradient of IMP (0 - 10 mM; 20 ml total volume followed by a pulse (200 μl) of AMP (10 mM) at the arrow. In (b) 10 mM IMP was present throughout irrigation, the buffer was changed for 10 mM AMP at the arrow. IMP dehydrogenase activity (–O–), protein concentration (mg/ml) (–●–) and IMP (–·–·–) were assayed in the eluant fractions.

Supplementation of the irrigation buffers with low concentrations of ethylene glycol, dioxane or urea dramatically enhanced the recovery of lactate dehydrogenase from the immobilised-AMP. In the absence of organic solvents, recoveries of enzyme activity were as low as 60% under the conditions employed (Ref. 22) when the enzyme was eluted with a linear NADH gradient. The inclusion of 20-30% (v/v) ethylene glycol in column irrigants produced quantitative recovery of enzyme activity. Low concentrations of dioxane, N, N'-dimethylformamide and urea proved equally effective (Ref. 22). Concentrations of ethylene glycol above 35%, however, altered the native conformation of lactate dehydrogenase and lead to a decreased affinity for the immobilised-AMP. The altered intrinsic protein fluorescence and decreased ability to bind NADH, as indexed by the enzyme-NADH dissociation constant, could be correlated with the chromatographic behaviour on columns of the Sepharose-bound N^6-(6-aminohexyl)-AMP. These disruptive effects of organic solvents on protein conformation could be exploited to elute lactate dehydrogenase from columns of immobilised-AMP with a linear gradient of ethylene glycol, dioxane or urea. The supplementation of column irrigants with low concentrations of organic solvents has been utilised to effect a clean elution of a steroid dehydrogenase from an immobilised-steroid column (Ref. 23).

A second solution to the problem of non-specific hydrophobic effects relates to the synthesis of analogues bearing more hydrophilic arms (Ref. 24). Most of the spacer arms currently in use comprise aliphatic hydrocarbons bearing suitable terminal functional groups to permit anchorage of the ligand to the matrix. Such polymethylene spacer molecules are capable of interacting hydrophobically with exposed apolar regions on enzyme surfaces. It seems reasonable, therefore, to intersperse polar groups along the length of the spacer arms in order to break up hydrophobic enclaves and consequently discourage apolar interactions. Thus, four 8-substituted derivatives of AMP were synthesised comprising spacer molecules of similar length but differing composition (Ref. 18). The four analogues (I - IV) depicted in Fig. 6 were synthesised from suitable 8-(ω-aminoalkyl)-AMP derivatives prepared by reaction of 8-Bromo-AMP with the

$$\text{(AMP)}-8-NH-CH_2-CH_2-CH_2-CH_2-CH_2-CH_2-NH_2 \quad \text{-(I)}$$

$$\text{(AMP)}-8-NH-CH_2-CH_2-CH_2-NH-CO-CH_2-NH_2 \quad \text{-(II)}$$

$$\text{(AMP)}-8-NH-CH_2-CH_2-NH-CO-CH_2-CH_2-NH_2 \quad \text{-(III)}$$

$$\text{(AMP)}-8-NH-CH_2-\underset{\displaystyle OH}{\underset{|}{CH}}-CH_2-NH-CO-CH_2-NH_2 \quad \text{-(IV)}$$

Fig. 6. The structures of four 8-substituted analogues of AMP bearing spacer molecules of differing composition.

corresponding α,ω-diaminoalkanes. Condensation of the 8-(ω-aminoalkyl)-derivatives of AMP with N-trifluoroacetyl-glycine or β-alanine in the presence of a water-soluble carbodiimide yielded the four AMP analogues shown in Fig. 6. The intermediates and final derivatives were purified by Dowex ion exchange chromatography, characterised and coupled to CNBr-activated Sepharose 4B to yield gels containing 2.5 μmoles AMP analogue/g moist weight gel. The chromatographic behaviour of the immobilised-AMP adsorbents was tested by their ability to bind rabbit muscle lactate dehydrogenase (EC 1.1.1.27) and *B. subtilis* alanine dehydrogenase (EC 1.4.1.1) under essentially identical chromatographic conditions. Small columns (1.0 g adsorbent) were equilibrated with 10 mM KH_2PO_4-KOH buffer pH 7.5 and a 50 ml sample containing 5 units of the enzyme and 1.0 mg bovine serum albumin applied. Non-adsorbed protein was washed off with 7.5 ml buffer and elution effected with a linear gradient of NADH, 0 - 0.4 mM, 20 ml total volume at 4°C. Figure 7 shows that both

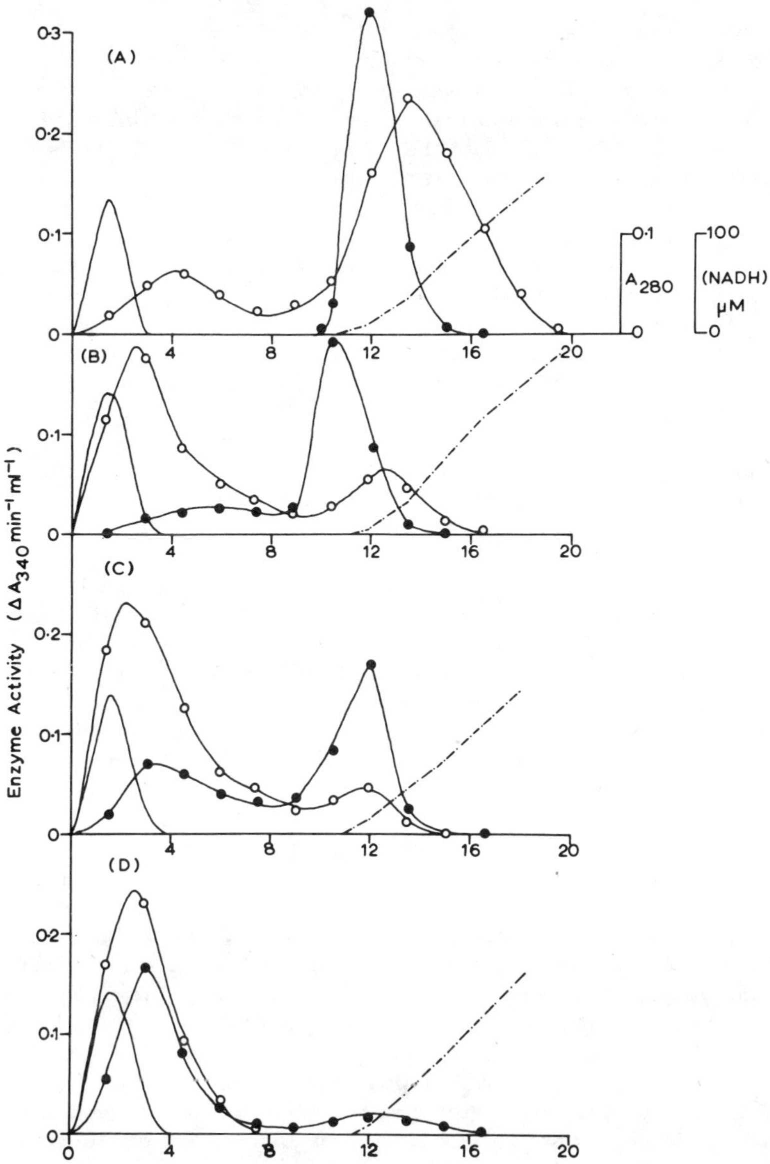

Fig. 7. Chromatography of L-lactate and L-alanine dehydrogenases on the Sepharose-bound AMP derivatives. Rabbit muscle lactate dehydrogenase (-●-), B. subtilis alanine dehydrogenase (-o-) and bovine serum albumin (—) were chromatographed on a column (1.0 g) of Sepharose-bound (A) 8-(6-aminohexyl)-AMP (I), (B) N-[3-(8-amino-AMP) propyl]-glycinamide (II), (C) N-[2-(8-amino-AMP) ethyl]-β-alaninamide (III) and (D) N-[2-hydroxy-3-(8-amino-AMP) propyl]-glycinamide (IV) in 10 mM potassium phosphate buffer pH 7.5. The abscissa represents the volume of eluate (ml) and (-·-·-) the NADH gradient.

dehydrogenases bound most tightly to the adsorbent whose ligand was separated from the matrix backbone by a hexamethylene spacer molecule (Fig 6, I) and least tightly to that adsorbent whose ligand was attached via a hydrophilic arm (Fig. 6, IV) of the same nominal length. Indeed, the ability of the adsorbents to bind the enzymes decreased in the same order (I > II > III > IV) as the hydrophobicity of the spacer molecules interposed between the ligand and the matrix. These results suggest that a hydrocarbon chain comprising three or more methylene groups proximal to the adenine nucleus is required to effect strong binding between the immobilised ligand and the complementary enzymes. These observations were confirmed (Ref. 18) by studying the effect of introducing a single hydroxyl group into the spacer molecule on the binding of several enzymes known to exhibit anomalous effects in affinity chromatography (Ref. 24). Thus, the chromatographic behaviour of phosphorylase b (EC 2.4.1.1), rabbit muscle D-glyceraldehyde 3-phosphate dehydrogenase (EC 1.2.1.12) and yeast alcohol dehydrogenase (EC 1.1.1.1) on adsorbents comprising immobilised II and IV (Fig. 6) were compared. These enzymes were bound more tightly to the ligand borne by the more hydrophobic arm (II) than to the adsorbent comprising AMP linked via the marginally more hydrophilic arm.

Two explanations for the differences observed between the behaviour of hydrophobic and hydrophilic adsorbents seem plausible. Firstly, the biospecific interaction between the enzyme and the 8-immobilised AMP is compounded with a significant non-specific hydrophobic interaction with at least part, if not all, of the spacer moiety. It is well known for X-ray diffraction studies of a number of enzymes containing the so-called 'dinucleotide' fold that the adenosine moiety of NAD^+ is bound in a pocket lined by apolar residues and that hydrophobic forces play a role in binding this part of the coenzyme to the enzyme (Ref. 16). It is not inconceivable therefore that that part of the spacer molecule immediately adjacent to the adenine ring system is also bound to the hydrophobic pocket in a similar fashion. If this explanation is correct one would anticipate that the inhibitor constants (K_i) for the free derivatives would mirror the chromatographic behaviour observed when the same derivatives are covalently attached to Sepharose 4B. Thus, whilst there is a significant difference between the chromatographic behaviour of rabbit muscle lactate dehydrogenase on adsorbents comprising immobilised I and IV (Figs. 6, 7), the K_i values for the strongly hydrophobic analogue 8-(6-aminohexyl)-AMP (I) and the hydrophilic derivative N-[2-hydroxy-3-(8-amino-AMP)-propyl] glycinamide (IV) were similar at 2.0 and 1.8 mM respectively. Similar observations have been made with the binding of rat hepatic glucokinase to immobilised derivatives of glucosamine bearing both hydrophobic and hydrophilic arms (Ref. 25). These observations suggest a second more plausible explanation for the decreased binding of enzymes to ligands attached to Sepharose via hydrophilic spacer molecules. The more hydrophilic derivatives of AMP may hydrogen bond to the polysaccharide backbone and therefore may not be physically available for interaction with their complementary enzymes. This would apparently decrease the effective ligand concentration (Ref. 25) and may provide an explanation for the relatively low capacities observed for both affinity (Ref. 26) and hydrophobic adsorbents (Ref. 27).

The capacities of this type of immobilised nucleotide affinity adsorbent have rarely been found to exceed 0.1 nmole lactate dehydrogenase/g moist Sepharose

(Refs. 26, 28), suggesting that only about 0.1% of the total theoretical amount of ligand is utilised in the binding of the enzyme. Recent findings suggest that the capacity of an affinity adsorbent increases with time (Ref. 29) over a period of several hours. There appear to be two clear stages in the adsorption of lactate dehydrogenase and several other dehydrogenases to N^6-(6-aminohexyl)-AMP-agarose under batchwise conditions. First, a very fast uptake of some of the enzyme onto the gel within 1 - 2 min, followed by a much slower uptake of enzyme at a rate dependent on the concentration of the AMP-Sepharose. Furthermore, the two distinct phases in the adsorption to the gel must be attributed to interactions between the immobilised-AMP and specific sites on the enzymes since both adsorption processes are inhibited by the presence of free nucleotides in the incubation medium. These two processes may reflect the different environment or reactivities of at least two species of immobilised-AMP.

Thus, whilst there is general appreciation of the utility of immobilised nucleotides as affinity adsorbents, little is known about the effect of immobilisation of the ligand on subsequent interactions with the complementary enzymes. It has almost invariably been assumed that there is a close parallel between the interaction of the enzyme and free ligand in homogeneous aqueous solution with that between the enzyme and immobilised ligand. However, it is clear that steric restrictions imposed by proximity of the matrix backbone to the ligand and the possibility of non-specific hydrophobic interaction with the spacer molecule may adumbrate this simple assumption. Therefore, it is imperative to re-evaluate the nature of the interaction between the immobilised ligand and complementary enzymes in order to rationalise affinity chromatography.

REFERENCES

(1) P. Cuatrecasas, Protein purification by affinity chromatography. Derivatisations of agarose and polyacylamide beads, J. Biol. Chem. 245, 3059 (1970).

(2) P. Cuatrecasas & C.B. Anfinsen, Affinity Chromatography, Ann. Rev. Biochem. 40, 259 (1971).

(3) J. Turkova, Affinity Chromatography, J. Chromatog. 91, 267 (1974).

(4) Lowe, C.R. & Dean, P.D.G. (1974) Affinity Chromatography, Wiley, London.

(5) P. O'Carra & S. Barry, Affinity Chromatography of lactate dehydrogenase. Model studies demonstrating the potential of the technique in the mechanistic investigation as well as in the purification of multi-substrate enzymes, FEBS Letters 21, 281 (1972).

(6) C.R. Lowe & P.D.G. Dean, Affinity chromatography of enzymes on insolubilised cofactors, FEBS Letters 14, 313 (1971).

(7) K. Mosbach, H. Guilford, R. Ohlsson & M. Scott, General ligands in affinity chromatography. Cofactor-substrate elution of enzymes bound to the immobilised nucleotides adenosine 5'-monophosphate and nicotina-adenine dinucleotide, Biochem. J. 127, 625 (1972).

(8) I. P. Trayer, H.R. Trayer, D.A.P. Small & R.C. Bottomley, Preparation of adenosine nucleotide derivatives suitable for affinity chromatography, Biochem. J. 139, 609 (1974).

(9) H. Schaller, X. Niiuaalwin, F. J. Bonhoeffer, C. Kurz & I. Nietzchmann, Affinity Chromatography of DNA-binding enzymes on single-stranded DNA-agarose columns, Eur. J. Biochem. 26, 474 (1972).

(10) T. Kristiansen, Group-specific separation of glycoproteins, Methods of Enzymology, 34, 331 (1974).

(11) S.T. Thompson, K.H. Cass & E. Stellwagen, Blue dextra-sepharose: an affinity column for the dinucleotide fold in proteins, Proc. Nat. Acad. Sci. U.S.A. 72, 669 (1975).

(12) A. Ruiz-Carrillo & V.G. Aufrey, A method for the purification of histone fraction F3 by affinity chromatography, Archs. Biochem. Biophys. 154, 185 (1973).

(13) D.B. Craven, M.J. Harvey, C.R. Lowe & P.D.G. Dean, Affinity chromatography on immobilised adenosine 5'-monophosphate 1. A new synthesis and some properties of an N^6-immobilised 5'-AMP, Eur. J. Biochem. 41, 329 (1974).

(14) R. Lamed, Y. Levin & M. Wilchek, Covalent coupling of nucleotides to agarose for affinity chromatography, Biochim. Biophys. Acta 304, 231 (1973).

(15) C-Y. Lee, D.A. Lappi, B. Wermuth, J. Everse & N.O. Kaplan, 8-(6-Aminohexyl)-Amino-Adenine Nucleotide Derivatives for Affinity Chromatography, Archs. Biochem. Biophys. 163, 561, (1974).

(16) M.J. Adams, M. Buehner, K. Chandrasekhar, G.C. Ford, M.L. Hackert, A. Liljas, M.G. Rossmann, I.E. Smiley, W.S. Allison, J. Everse, N.A. Kaplan & S.S. Taylor, Structure-function relationships in lactate dehydrogenase, Proc. Nat. Acad. Sci. U.S.A. 70, 1968 (1973).

(17) H. Guilford, P.O. Larsson & K. Mosbach, On adenine nucleotides for affinity chromatography, Chemica Scripta 2, 165 (1972).

(18) C.R. Lowe, The synthesis of several 8-substituted derivatives of adenosine 5'-monophosphate to study the effect of the nature of the spacer arm in affinity chromatography, Eur. J. Biochem. 73, 265 (1977)

(19) M. Lindberg, P.O. Larsson & K. Mosbach, A new immobilised NAD^+ analogue, its application in affinity chromatography and as a functioning coenzyme, Eur. J. Biochem. 40, 187 (1973).

(20) E. Heyde, A. Nagabhushanan, M. Vonarx & J.F. Morrison, Studies on inosine monophosphate dehydrogenase. Steady state kinetics, Biochim. Biophys. Acta 429, 645 (1976).

(21) S. Barry & P. O'Carra, Affinity chromatography of nicotinamide-adenine dinucleotide-linked dehydrogenases on immobilised derivatives of the dinucleotide, Biochem. J. 135, 595 (1973).

(22) C.R. Lowe & K. Mosbach, Biospecific affinity chromatography in aqueous-organic cosolvent mixtures. The effect of ethylene glycol on the binding of lactate dehydrogenase to an immobilised-AMP analogue, Eur. J. Biochem 52, 99 (1975).

(23) L.E. Aukrust, K.R. Norum & B.A. Skalhegg, Affinity chromatography of 3α-hydroxysteroid dehydrogenase from Pseudomonas testosteroni. Use of N,N-dimethylformamide to prevent hydrophobic interactions between the enzyme and the ligand, Biochim. Biophys. Acta. 438, 13 (1976).

(24) P. O'Carra, S. Barry & T. Griffin, Spacer arms in affinity chromatography Use of hydrophilic arms to control or eliminate nonbiospecific adsorption effects, FEBS Letters 43, 169 (1974).

(25) M.J. Holroyde, J.M.E. Chesher, I.P. Trayer & D.G. Walker, Studies on the use of Sepharose-N-(6-Aminohexanoyl)-2-Amino-2-Deoxy-D-Glucopyranose for the large scale purification of hepatic glucokinase, Biochem. J. 153, 351 (1976).

(26) C.R. Lowe, M.J. Harvey, D.B. Craven & P.D.G. Dean, Some parameters relevant to affinity chromatography on immobilised nucleotides, Biochem. J. 133, 499 (1973).

(27) H.P. Jennissen & L.M.G. Heilmeyer, 'General aspects of hydrophobic chromatography. Adsorption and elution characteristics of some skeletal muscle enzymes, Biochemistry 14, 754 (1975).

(28) M.J. Harvey, C.R. Lowe, D.B. Craven & P.D.G. Dean, Affinity chromatography on Immobilised Adenosine 5'-monophosphate. 2. Some parameters relating to the selection and concentration of the immobilised ligand, Eur. J. Biochem. 41, 335 (1974).

(29) C.R. Lowe & M.G. Gore, The interaction of some dehydrogenases with N^6-(6-aminohexyl)-adenosine 5'-monophosphate Sepharose, FEBS Letters 77, 247 (1977).

IMMOBILIZED ADENINE COENZYMES IN GENERAL LIGAND AFFINITY CHROMATOGRAPHY AND THEIR USE AS ACTIVE COENZYMES

Klaus Mosbach

Biochemical Division, Chemical Center, University of Lund, S-220 07 Lund 7, Sweden

There is considerable advantage in affinity chromatography in a versatile ligand or "general ligand" with a broad biological affinity-binding spectrum in that a new synthetic route leading to suitable affinity matrices does not have to be devised for every putative biospecific purification. Adenine coenzymes meet this requirement and the first examples that appeared in the literature on their application involved model separations of dehydrogenases using immobilized NAD (refs. 1, 2), at which time the term "general ligand" was first coined (ref. 3). (It should be added that the examples to be discussed here are taken primarily from our own work; for more information on the subject see other contributions in this volume and a review in Advances in Enzymology, K. Mosbach (ed. A. Meister) 1977, John Wiley & Sons, New York). By virtue of its generality, however, this approach appears at first sight to sacrifice something of the very specificity which is the keystone of affinity chromatography. However, it has become apparent that in particular bioelution provides adequate compensation in retaining specificity, as will be discussed in some depth later. In scheme 1 the relationship between adsorption and elution is given indicating that to obtain a pure enzyme, the demands for specificity of the eluent usually decrease the more specific the affinity ligand is and vice versa.

Scheme 1

	(Bio)adsorption	(Bio)elution	
	Nonspecific ligand (e.g., ion exchanger)	(e.g., inhibitor, substrate or effector)	
	Semi-specific ligand (e.g., cibachron blue)		
		Ternary complex coenzyme-inhibitor dead-end complex adduct	
	General ligand (e.g., AMP)		
		Binary complex coenzyme	
increasing specificity of ligand	"Monospecific" ligand	(e.g., salt elution)	decreasing specificity of eluent

Of the about 2000 different enzymes which have been assigned a specific number (Enzyme nomenclature 1973 + supplement), about one third require that one of the following four adenine coenzymes, NAD (~ 165), NAD(P) (~ 50), NADP (~ 115), CoA (non-substituted) (~ 80), and ATP (~ 225), participate in the enzymic reactions. In addition, there are enzymes with masked nucleotide binding sites and further there is a number of enzymes with binding sites for nucleotide effector molecules, which further increases the number of enzymes that may be affinity-bound by the above four general ligands.

In many general ligand affinity chromatography studies reported, it has been shown that the entire coenzyme molecule is not essential for binding. Fragments of the coenzyme, usually the "half-molecule" thereof, were found to retain the affinity and specificity of the intact parent moiety. For instance AMP has been found to be an adequate substitute in many situations for the more labile NAD.

Major emphasis will be given here to so called "preassembled" adenine coenzymes, i.e. ligands modified by substitution with a spacer carrying a functional group suitable for binding to a matrix, to which it is then coupled as opposed to ligands coupled to a matrix to which spacers have already been attached.

A. GENERAL LIGAND AFFINITY CHROMATOGRAPHY

I. Some Adenine Nucleotides Used as General Ligands

5'-AMP, 2',5'-ADP and 3',5'-ADP-analogs. A considerable amount of effort has been expended on synthesizing preassembled analogs of 5'-AMP, 2',5'-ADP, and 3',5'-ADP, which constitute the corresponding "half-molecules" of NAD, NADP, and CoA, respectively. The first general ligand thus tailored for immobilization and of unambiguous, defined structure was N^6-(6-aminohexyl)-AMP (ref. 3) (Fig. 1). In the subsequent studies the corresponding N^6-(6-aminohexyl)-2',5'-ADP and N^6-(6-aminohexyl)-3',5'-ADP were prepared (ref. 4). All these analogs are modified at the same site, i.e. position N^6, the exocyclic nitrogen on the adenine moiety, and by the same spacer molecule.

A number of affinity chromatographic studies have also been reported with other AMP analogs, including 8-(6-aminohexyl)-amino-AMP, substituted at position 8 of the adenine moiety. The corresponding 2',5'-ADP substituted in position 8 has also been synthesized as well as ribosyl-linked AMP prepared by reaction of periodate-treated AMP with a dihydrazide, followed by coupling to CNBr-activated agarose (alternatively by reaction of periodate-treated AMP with a spacer-matrix).

ATP (ADP)-analogs. Four different points of attachment have been chosen for ATP or ADP derivatization, including the N^6- and C-8-positions of the adenine moiety, the ribose and in addition the terminal phosphate group. These preparations have found application in the affinity chromatography of kinases.

NAD and NADP-analogs. Considerable efforts have concentrated on the synthesis of NAD-analogs with a spacer molecule and carrying a functional group suitable for covalent binding. Both N^6-, C-8- and ribose-substituted NAD-analogs have been prepared. To obtain a N^6-substituted NAD-analog, the N^1 position of the adenine ring has been quaternized with iodoacetic acid followed by a Dimroth rearrangement leading to a carboxymethyl group at position N^6 (ref. 5). The spacer can be extended by carbodiimide condensation with hexamethylenediamine (Fig. 2).

Analogous to NAD, the same points of attachment for derivatization of NADP have been chosen. The synthetic procedures applied for the synthesis of N^6-, C-8- and ribose-substituted NADP were also equivalent. For attachment to the ribose moieties however, only the NMN-ribose is available.

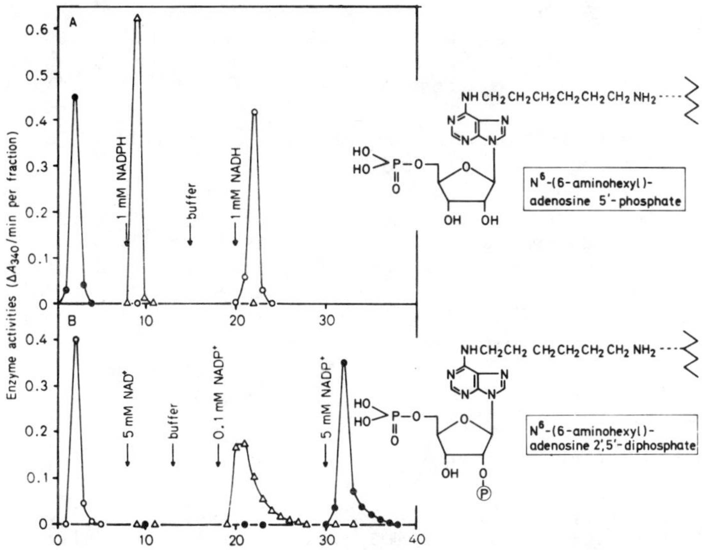

Fig. 1. Affinity chromatography of mixtures of lactate dehydrogenase, glucose-6-phosphate dehydrogenase and phosphogluconate dehydrogenase on Sepharose-bound N^6-(6-aminohexyl)-adenosine 5'-monophosphate (A) and on Sepharose-bound N^6-(6-aminohexyl)-adenosine 2',5'-diphosphate (B). Applications of NAD(H) and NADP(H) dissolved in irrigant buffer are indicated by arrows. (o-o) Lactate dehydrogenase; (Δ-Δ) Glucose-6-phosphate dehydrogenase; (●-●) 6-phosphogluconate dehydrogenase.

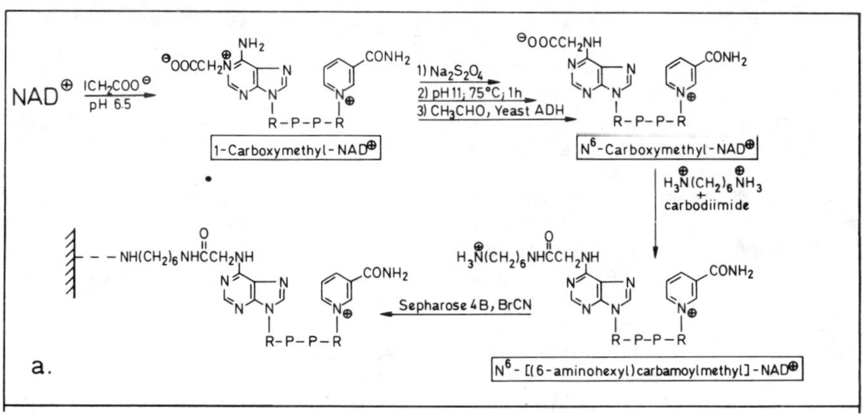

Fig. 2. Synthesis of a Sepharose-bound NAD-analog N^6-[(6-aminohexyl)-carbamoylmethyl]-NAD.

Recently, a "preassembled" CoA-analog has been synthesized carrying the same spacer molecule at position N^6 as the above mentioned NAD- and NADP-analogs (E. Rieke and K. Mosbach, unpublished results). In addition, recently the synthesis of a bifunctional dinucleotide AMP-ATP has been reported on, of potential interest in general ligand affinity chromatography (ref. 6).

In general it can be said that the preassembled NAD and NADP analogs and their corresponding half molecules 5'-AMP and 2',5'-ADP, derivatized in the adenine moiety either at position N^6 or C-8 have proven to be excellent general affinity chromatography ligands for NAD- and NADP-dependent dehydrogenases. By contrast substitution at the nicotinamide portion, gave, as expected, poor affinity material for this class of enzymes. Ribose substitution, however, yields useful bioaffinity material although with weaker affinity. For the other major class of enzymes utilizing ATP, i.e. kinases and ATPases, substitution at all the positions of ATP (ADP) mentioned gave in general good affinity material.

II. (Bio)adsorption

Despite non-specific factors often involved in the binding of biological macromolecules to ligands, e.g. through hydrophobic interaction with a polymethylene spacer arm, biospecific adsorption is probably involved and of importance in most of the reported purifications of enzymes using general ligand affinity chromatography. Two examples from our own work displaying "true" biospecificity serve to illustrate how enzymes are able to recognize subtle variations in structures of immobilized ligands. The ligands N^6-AMP and N^6-2',5'-ADP, comprising the half molecules of NAD and NADP, respectively, were immobilized on Sepharose (4) (Fig. 1). The only difference between the two preparations was the additional phosphate residue on the 2'-position of the ribose. When a mixture of NAD and NADP dependent dehydrogenases was applied to the latter, the NAD-dependent enzyme lactate dehydrogenase was washed through unretarded as expected. The two NADP-dependent enzymes present were bound and could subsequently be eluted off. Conversely lactate dehydrogenase would bind to the 5'-AMP Sepharose gel, but the NADP-dependent 6-phosphogluconate dehydrogenase would not. The NADP-dependent glucose-6-phosphate dehydrogenase did, however, have a very weak affinity for the AMP gel, consistent with the observation that this enzyme is inhibited in free solution to some degree by AMP. Nevertheless, these gels are useful not only for purification of individual enzymes but also for quick group separation of NAD- from NADP-dependent enzymes.

In the other example to be mentioned here, the enzyme succinate thiokinase, a CoA-dependent enzyme, was shown to have affinity for the Sepharose-bound 3',5'-ADP-analog corresponding to the half molecule of coenzyme A, but not for the closely related 2',5'-ADP analog (ref. 7).

The examples given hitherto deal primarily with the "half-molecules" of their corresponding coenzymes. A number of general ligand affinity chromatography studies are also known in which the entire coenzyme molecule has been utilized. For example, in a recent study, we were able to demonstrate the affinity binding of lactate dehydrogenase to the aforementioned immobilized NAD-analog, which could be further increased in the presence of pyruvate, known to form a strong dead-end complex. On addition of a dimer of the same NAD-analog, which is interconnected via its terminal aminogroups, additional lactate dehydrogenase did bind when pyruvate was present, strongly indicating the occurrence of affinity cross-linking of the tetrameric enzyme while (bio)adsorbed to the Sepharose beads (P.O. Larsson and K. Mosbach, to be published).

Alternative to the binding to catalytic sites, biospecific binding has also been utilized in the purification of enzymes affinity-bound to the effector site. Thus, glycogen phosphorylase b would bind to N^6-AMP immobilized to Sepharose, allowing its efficient purification (AMP is known to be an allosteric effector of the enzyme).

Related is the reported purification of transhydrogenases obtained by binding the enzyme to the nucleotide effector site using an immobilized 2',5'-ADP-analog (ref. 8) to be discussed later.

III. (Bio)elution

In the two-step process involved in affinity chromatography, i.e. adsorption followed by elution, the importance of the latter's inherent potential for introducing even greater specificity has been increasingly recognized. Using a general ligand with affinity for a large number of enzymes, makes greater demands on the specificity of the elution process. Elution from general ligands has been affected by both specific and non-specific means. To the latter category belong changes of temperature and pH, elution through changes in ionic strength, and elution with hydrophobic agents such as ethylene glycol. Many successful purifications of enzymes from general ligands applying such non-specific elution conditions have been reported. In the author's opinion however, bioelution seems to be the procedure of choice. This implies the use of either a soluble counterligand or other ligands or combinations of ligands for which the enzyme has higher affinity. The eluting nucleotide, which is usually employed, is simply removed by dialysis or gel chromatography and the cost is rather low since the concentrations of the eluting agent is usually in the millimolar or less range. Normally short pulses suffice, although by applying gradients when necessary, additional resolving power can be gained. The method is more subtle, more additional information of theoretical interest as to the nature of enzyme binding sites can be obtained, and most important the specificity and therefore resolving power is far higher.

To illustrate the potential and power of bioelution, a few examples will be given from work in the author's laboratory on the elution of dehydrogenases affinity-bound to the general ligand N^6-(6-aminohexyl)-AMP.

An example on bioelution utilizing binary complex formation is the separation of the five isozymes of lactate dehydrogenase with a gradient of NADH (Fig. 3) (ref. 9). All five isozymes could be separated, illustrating the high resolving power and sensitivity of the system considering that the various K_{diss} constants of the LDH-NADH binary complex are within the narrow range of $0.4\text{-}2 \cdot 10^{-6}$ M. To obtain completely pure isozymes however, rechromatography of the fractions from each peak area had to be performed. The utilizing of ternary complex formation to elute enzymes has also been applied. In the first example, a coenzyme-inhibitor combination was used for the specific elution of the SS isozyme of alcohol dehydrogenase (a steroid-active form) by applying a pulse of NAD and cholic acid, a steroid inhibitor (ref. 10). Simultaneous addition of NAD and pyruvate allows formation of a strong ternary dead-end or abortive complex, specifically eluting lactate dehydrogenase (ref. 11). Finally, application of specific NAD-adducts led to specific elution of the three enzymes malate dehydrogenase, alcohol dehydrogenase and lactate dehydrogenase (ref. 12).

IV. Application in Enzyme Purification and Separation

In this paragraph a few examples will be given dealing with the application of general ligand affinity chromatography including (a) purification and enrichment of enzymes from crude samples, (b) purification of enzymes on a preparative scale (to illustrate the resolving power of the procedure), (c) resolution of mixtures of isozymes on a preparative scale, (d) removal of contaminants, and (e) separation

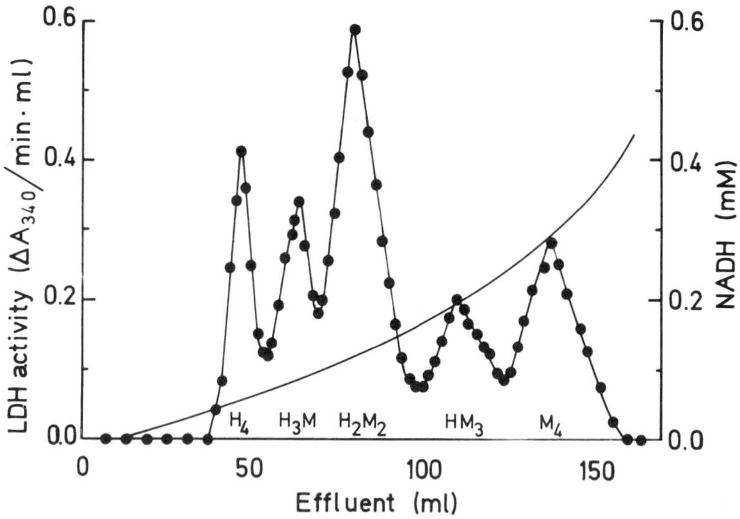

Fig. 3. (Bio)elution of LDH isozymes affinity-bound to N^6-(6-aminohexyl)-adenosine 5'-monophosphate using a concave gradient of NADH.

of active from non-active species. (a) As an example on purification of enzymes from crude extracts the case of transhydrogenase from Pseudomonas aeruginosa can be given, which has been purified 1.300-fold in one step using N^6-substituted 2', 5'-ADP-Sepharose and NADPH as eluant (ref. 8). A 17.000-fold purification of human serum lactate dehydrogenase has been obtained in a single step using immobilized N^6-AMP, with an NAD-pyruvate adduct as eluant (ref. 12). These examples suffice to illustrate the potential of general ligands not only in "conventional" affinity chromatography but also to their power to allow enrichment of enzymes present at very low levels in fluids or tissues. (b) As an example on enzyme purification on a preparative scale will be mentioned the purification of dogfish lactate dehydrogenase using immobilized N^6-AMP followed by elution with a pulse of 0.12 mM reduced NAD-pyruvate adduct (ref. 12). After removal of the pyridin nucleotide on Sephadex G-200, the enzyme crystallized readily. About 30 mg of pure enzyme were obtained in this single-step procedure from 100 grams of dogfish muscle using 30 ml of the gel. It is rather seldom though that the affinity chromatography technique alone yields a homogeneous protein in a single step direct from tissue extracts as in this case, especially if this involves purifications of greater than 1.000-fold. For such cases it is more advisable to apply a gross fractionation procedure prior to the biospecific step. (c) Isozyme purification on preparative scale has been applied in the next example. Following ion-exchange chromatography (ref. 10), a liver extract was applied to a N^6-Sepharose column. Elution with 0.2 mM NAD and 1.5 mM of the steroid inhibitor cholic acid, yielded homogeneous, steroid-active alcohol dehydrogenase SS isozyme with a yield of 30 mg from 260 g of liver. (d) An example on the usefulness of affinity chromatography in separating contaminants is the removal of various dehydrogenases associated with cytochrome oxidase. As a result of their adsorption onto a NAD-matrix, the activity of the oxidase was enhanced substantially when calculated as a function of heme aa_3 content (ref. 13).

V. Techniques

In the predominant number of cases, affinity chromatography has been carried out applying column procedures. Recently, a procedure has been reported on involving the use of "magnetic" affinity gels prepared by treatment of affinity gel preparations with magnetic ferrofluids. With this technique both 5'-AMP- and 2',5'-ADP-Sepharose were "magnetized". These preparations retained their biospecific binding properties. Due to the magnetic properties of the gel the separation procedure is simplified obviating the usual centrifugation and column chromatography steps (K. Mosbach and L. Andersson, Nature, 1977, in press).

VI. Applications to Fundamental Studies

Apart from their use in the separation and purification of biological macromolecules, in particular enzymes, adenyl general ligands have recently been applied to a few more fundamental investigations. As will be seen from the examples, it has been of the utmost importance that the ligands have a chemically well-defined "natural" structure.

1. Binding site. Bioaffinity ligands have recently been used to distinguish between the catalytic and effector sites of nicotinamide nucleotide transhydrogenase from Pseudomonas aeruginosa (ref. 8). Since NADP acts both as a substrate and an effector, conventional kinetic experiments were difficult to interpret. However, differences in behaviour of this enzyme in presence of N^6-AMP-Sepharose or N^6-2',5'-ADP-Sepharose cast some light on the problem. The enzyme could be eluted from a column of the former with either 1 mM NADPH or NADH but from the 2',5'-ADP column it could be displaced only with NADPH. This was taken as a strong indication that there are at least two different binding sites, one for both NADH and NADPH, (as substrates), the other binding NADPH only (as an allosteric effector).

Rapid screening of unknown enzymes for their binding properties and possible coenzyme participation is also feasible with the various coenzyme (or coenzyme fragment) columns now to hand.

2. Enzyme mechanism. An excellent example of the use of affinity chromatography in exploring enzyme mechanisms, although not involving a general immobilized ligand, is the following: In the absence of an appropriate cofactor, lactate dehydrogenase was shown to have no affinity for a Sepharose linked oxamate derivative. However, this enzyme is known to form a strong ternary complex with NADH and pyruvate with which oxamate is closely analogous. In the presence of NADH the enzyme was indeed strongly adsorbed to the column. These data strongly imply that NADH induced a binding site of the enzyme in line with a proposed compulsory-ordered mechanism (ref. 14).

Bioelution has also been involved in the study of enzyme mechanism. Following from an early preliminary study on elution of lactate dehydrogenase from AMP-columns using a tight dead-end complex, NAD and pyruvate, a thorough study on rat heart citrate synthase bound to a N^6-ATP-column was carried out using a number of combinations of substrates and substrate analogs (ref. 15). Although neither oxaloacetate nor CoA alone would effect elution, in combination they sharply eluted the enzyme. A closely analogous putative dead-end complex mixture, however, citrate and acetyl-CoA, was ineffective in eluting the enzyme. (R)-malate together with CoA eluted the enzyme rapidly whereas the (S)-malate form and CoA led to slow elution.

Affinity chromatography of mammalian phosphofructokinase on N^6-AMP, ribose-linked ATP, and the above N^6-ATP-Sepharose, has been compared (ref. 16). It was found that the capacity of the N^6-ATP ligand for this enzyme was 90 times greater than that of the N^6-AMP or the ribose-attached ATP; blue dextran-Sepharose was also less effective. The kinase was biospecifically eluted with ADP plus fructose-6-phosphate, apparently forming a strong dead-end complex. These studies demonstrate the usefulness of affinity chromatography in confirming mechanistic evidence adduced from steady state kinetics, on for instance the order in enzyme mechanisms and in allowing convenient screening for dead-end complexes.

3. Kinetic constants. The range of application of general ligand affinity chromatography has recently been extended to quantitative investigations of the interactions between enzymes and specific ligands. Various isozymes of lactate dehydrogenase were adsorbed on a column of immobilized N^6-AMP. On eluting with a gradient of NADH, a linear relation was observed between the concentration of nucleotide in the eluant required to displace a particular isozyme and the reported dissociation constants for the corresponding binary enzyme-nucleotide complexes (ref. 17). This relation was then utilized to determine unknown K_{diss}-constants of other isozymes. Using this gradient affinity chromatography technique, K_{diss}-constants of various coenzyme fragments used as eluants for N^6-AMP-affinity bound alcohol dehydrogenase could also be determined.

4. Enzyme "freezing". Attempts have been made to achieve "permanent activation" of the enzyme phosphorylase *b* by its immobilization to a matrix in the presence of an excess of the effector molecule AMP in the hope that it would be "frozen" in its active conformation (ref. 18). After dialysis, however, the enzyme still required addition of AMP for activity. The enzyme was therefore covalently bound to a matrix concomitantly with the aforementioned N^6-AMP-analog; It was shown that such preparations no longer required externally added effector for activity.

B. IMMOBILIZED ACTIVE COENZYME PREPARATIONS

1. Retention and Regeneration Methods

Perhaps the most intriguing facet of the subject of immobilized coenzymes is their capacity to retain activity and biospecificity when bound either to a water-soluble or a particulate support. The development of efficient enzyme-catalyzed processes has been recognized as being a prerequisite to the use of enzyme technology on an industrial scale. Many of these enzyme-catalyzed chemical processes require the participation of readily dissociable cofactors. As these are often very expensive, their economical utilization requires techniques both for their retention in the reaction mixture and for their regeneration. In addition, in cases where it is essential not to perturb the medium, for instance in certain medical and analytical applications, coenzyme immobilization is essential. Various retention and regeneration methods have hitherto been applied to coenzymes (see Scheme 2). The choice of regenerating procedure is of course intimately connected with the retention method of choice. Amongst the alternatives, coenzyme analogs immobilized on water soluble supports and used entrapped within semi-permeable membranes together with a recycling enzyme system constitute the most closely studied procedure to date. Alternative procedures include electrochemical regeneration, a method still in its infancy, like the chemical approach utilizing artificial electron acceptors and donors. In a recent example of the latter method, an electron acceptor, acriflavin, was immobilized on a gel to allow subsequent facile removal from the assay mixture (ref. 19). In this form it was used for the continuous nonenzymic oxidation of NADH. It was found necessary to couple acriflavin to epoxy-activated gels, which yielded stable bonds, since for such recycling systems prevention of leakage is more critical than in normal affinity chromatographic procedures.

Scheme 2

Retention	Regeneration
"tight" membrane	enzymic recycling
covalent binding to soluble support	chemical electrochemical

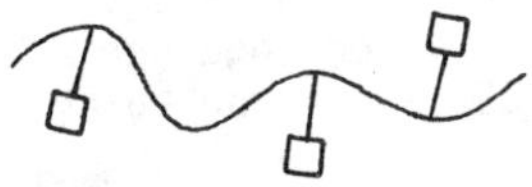

covalent binding to particulate support

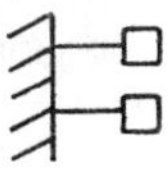

enzyme-coenzyme complex — coupled substrate

(a) 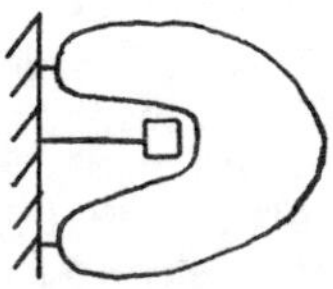 (b) 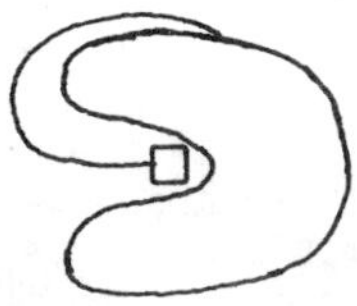

2. Application in Enzyme Electrode and Enzyme Reactor

Two examples from our own studies on the application of immobilized coenzymes in analysis and enzyme technology serve as illustrations of the work in this field (ref. 20). The aforementioned N^6-substituted NAD analog, N^6-[(6-aminohexyl)-carbamoylmethyl]-NAD, was immobilized on dextran and coentrapped with soluble lactate dehydrogenase and glutamate dehydrogenase within a dialysis bag encasing a cation-sensitive enzyme-electrode (see Scheme 3).

Scheme 3

An enzyme electrode system using recycling NAD.

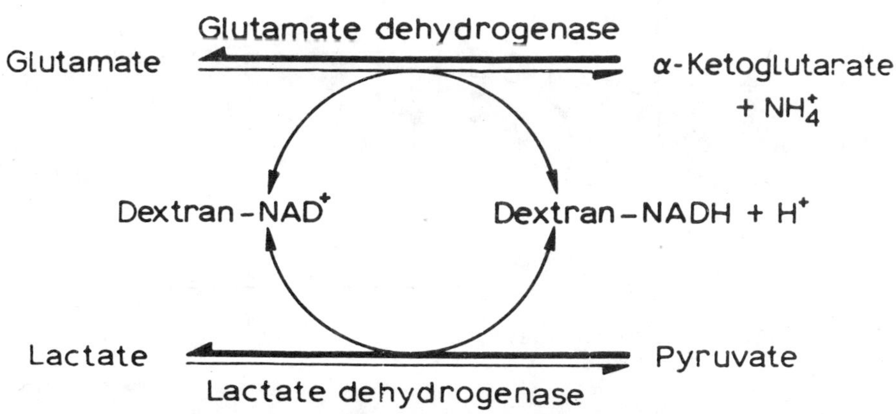

In the presence of glutamate in the assay medium, which also contained pyruvate as substrate for the recycling enzyme lactate dehydrogenase, ammonium ions were generated which were recorded by a NH_4-sensitive electrode. A linear relationship was found between the glutamate present and the current response obtained.

Using the same dextran-NAD preparation, a model enzyme reactor reaction was carried out. Inside an ultrafiltration apparatus the NAD-preparation was kept in proximity with soluble lactate dehydrogenase and alanine dehydrogenase. On continuously pumping substrate through the unit, alanine was produced from pyruvate (ref. 20). A similar two-enzyme system was also applied to alanine production on a small scale, but using a polyethyleneimine-coupled NAD-analog, entrapped together with soluble enzymes in cellulose triacetate fibres with the NAD-preparation being recycled at a rate of 222 cycles/h (ref. 21).

A rather different enzyme-coenzyme arrangement has been reported for regeneration purposes, which simultaneously solves the retention problem (see Scheme 2 a). Here an immobilized enzyme-coenzyme complex is built up on a matrix so that the coenzyme analog of NAD, N^6-[(6-aminohexyl)carbamoylmethyl]-NAD, is sitting permanently in the active site of the enzyme, liver alcohol dehydrogenase (ref. 22). On adding a second substrate, the coenzyme is regenerated in a coupled substrate recycling manner, as outlined in Fig. 4. The recycling rate in this type of immobilized NAD-molecule is high and has been estimated to be about 3.400 cycles/h. However, the relatively small number of such coupled systems known, which may be of practical interest, seems to restrict a wider use of this type of preparation. Attempts to couple the coenzyme NAD directly onto the enzyme lactate dehydrogenase have also been reported on (see Scheme 2 b) (ref. 23). The coenzyme could be recycled to some extent with artificial electron donors/acceptors while bound to the enzyme. In the obtained preparation the normally dissociable coenzyme NAD has thus been artificially modified to become a bound "prosthetic" group.

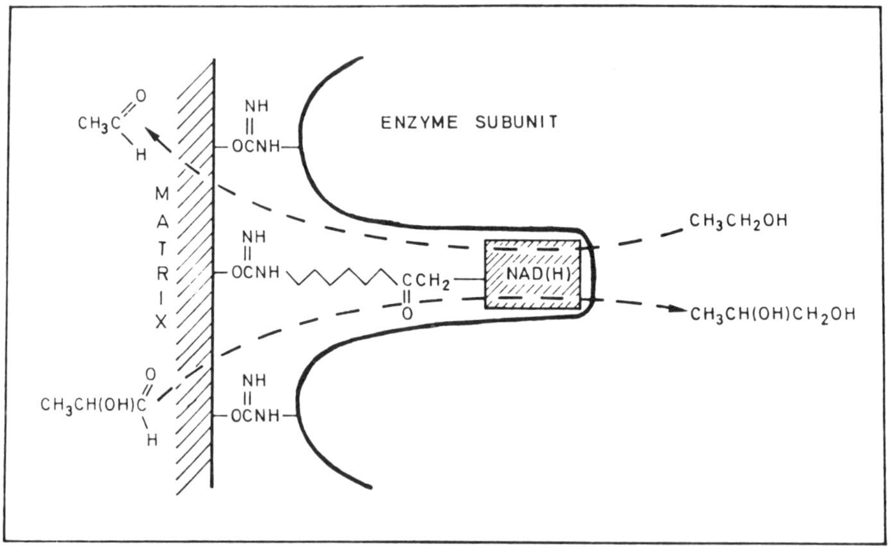

Fig. 4. Schematic drawing of an active alcohol dehydrogenase/NAD/Sepharose.

3. Discussion on Coenzymic Activity of Immobilized Coenzyme Preparations

In most of the publications on the use of immobilized coenzymes attention has been focused either on the aspect of affinity chromatography or on the use of coenzymically active preparations, but very seldom have both been considered together. Nevertheless both aspects are intimately related to one another and much useful information can be gained from such joint appraisal. Reports on coenzymically active preparations are scarce and of rather recent date. In most cases preassembled coenzymes, carrying a spacer with a functional group, have been used for immobilization. The coenzymes tested include in particular NAD but also to some extent NADP and ATP. Generally speaking the "coenzymic activities" of such analogs are somewhat lower than those of the parent coenzyme under comparable conditions and decrease further upon binding due in part to diffusional hindrance exerted by the support, showing the following general relationship in rate: coenzyme>coenzyme analog>coenzyme bound to soluble support>coenzyme bound to particulate support. As an example, yeast alcohol dehydrogenase can be showing the following relative "coenzymic activities": NAD = 1, N^6-NAD = 0.61, water soluble dextran-bound = 0.16, and Sepharose 4B-bound = 0.07 (measured at constant nucleotide, substrate, and enzyme concentrations; judged from the few data available on K_m-values of these coenzyme analogs, they appear to be rather close to those of the parent nucleotide).

In closing I wish to point again to the fact that the type of general ligands discussed in the last section, i.e. the various adenine coenzymes, are unique in allowing an answer to the often raised question of whether true biospecific adsorption is involved in an affinity chromatographic system. For instance in a recent study the properties of the above mentioned Sepharose-bound NADH analog were tested using front-face fluorimetry. It is known that horse liver alcohol dehydrogenase forms a strong fluorescent complex with NADH in the presence of excess of isobutyramide. A similar strong fluorescence emission spectrum was obtained also with the immobilized NADH preparation on addition of the enzyme and isobutyramide, strongly indicating that the entire structure of NADH needed for formation of this complex is retained and available (L. Andersson and K. Mosbach, to be published). Another proof for biospecific adsorption is the simple consideration that if a coenzyme analog as NAD is "coenzymatically active" when immobilized to soluble or particulate supports, thus allowing the enzyme to reach to the nicotinamide position, where the oxidation/reduction takes place, then it is extremely likely that the same biospecific interaction is prevailing when the enzyme is affinity-bound to its immobilized coenzyme. No such definite proof for biospecific binding can be obtained from an enzyme's elution behaviour alone. Even if an enzyme is bioelutable with soluble biospecific ligands, this may simply be the result of conformational changes introduced, or changes in charge obtained near the active site on ligand binding, both of which would lead to an enzyme's elution also from a non-biospecific matrix. Nevertheless in the predominant number of cases, an observed effective bioelution is a very strong indication that the enzyme also has been biospecifically bound either through complete or at least partial biospecificity.

REFERENCES

1. K. Mosbach, H. Guilford, P.O. Larsson, R. Ohlsson, and M. Scott, Biochem. J. 125, 20 (1971).
2. C. R. Lowe, and P. D. G. Dean, FEBS Lett. 14, 313 (1971).
3. K. Mosbach, H. Guilford, R. Ohlsson, and M. Scott, Biochem. J. 127, 625 (1972).
4. P. Brodelius, P.O. Larsson, and K. Mosbach, Eur. J. Biochem. 47, 81 (1974).

5. M. Lindberg, P. O. Larsson, and K. Mosbach, Eur. J. Biochem. 40, 187 (1973).

6. C. Y. Lee, P. O. Larsson, and K. Mosbach, J. Solid-Phase Biochem. 2, no. 1, 31 (1977).

7. S. Barry, P. Brodelius, and K. Mosbach, FEBS Lett. 70, 261 (1976).

8. B. Höjeberg, P. Brodelius, J. Rydström, and K. Mosbach, Eur. J. Biochem. 66, 467 (1976).

9. P. Brodelius, and K. Mosbach, FEBS Lett. 35, 223 (1973).

10. L. Andersson, H. Jörnvall, and K. Mosbach, Analyt. Biochem. 69, 401 (1975).

11. R. Ohlsson, P. Brodelius, and K. Mosbach, FEBS Lett. 25, 234 (1972).

12. N. O. Kaplan, J. Everse, J. E. Dixon, F. E. Stolzenbach, C. Y. Lee, C. L. T. Lee, S. S. Taylor, and K. Mosbach, Proc. Nat. Acad. Sci. 71, 3450 (1974).

13. J. Holbrook, J. Bucher, and R. Penniall, Hoppe-Seyler's Z. Physiol. Chem. 357, 623 (1976).

14. P. O'Carra, and S. Barry, FEBS Lett. 21, 281 (1972).

15. A. Mukherjee, and P. A. Srere, J. Biol. Chem. 251, 1476 (1976).

16. C. S. Ramadoss, L. J. Luby, and K. Uyeda, Arch. Biochem. Biophys. 175, 487 (1976).

17. P. Brodelius, and K. Mosbach, Analyt. Biochem. 72, 629 (1976).

18. K. Mosbach, and S. Gestrelius, FEBS Lett. 42, 200 (1974).

19. M. O. Månsson, B. Mattiasson, S. Gestrelius, and K. Mosbach, Biotechnol. Bioeng. 18, 1145 (1976).

20. P. Davies, and K. Mosbach, Biochim. Biophys. Acta 370, 329 (1974).

21. W. Marconi, S. Prosperi, S. Giovenco, and F. Morisi, J. Mol. Catalysis 1, 111 (1976).

22. S. Gestrelius, M. O. Månsson, and K. Mosbach, Eur. J. Biochem. 57, 529 (1975).

23. D. Venn, P. O. Larsson, and K. Mosbach, Acta Chem. Scand. 133, 141 (1977).

DISSOCIATION CONSTANTS OF PHOSPHORYLASES TO OLIGO- AND POLYGLUCANS STUDIED BY THE AFFINITY ELECTROPHORESIS

Kazusuke Takeo and Shojiro Nakamura

Department of Biochemistry, Yamaguchi University, School of Medicine, 755 UBE, Yamaguchi-ken, Japan

Using polyacrylamide gel disc electrophoresis, we reported previously the interaction of glycogen with various rabbit tissue phosphorylases (Ref. 1). With electrophoresis gel containing glycogen, the mobilities of the phosphorylases were observed to be retarded. This retardation was substrate specific. On the basis of variations in mobility as a function of glycogen concentration, the dissociation constants of the reaction of phosphorylase with glycogen were calculated. This procedure was called "affinity electrophoresis" (Ref. 2). According to this procedure, we reported the dissociation constants of the reactions of various phosphorylases and α-amylases with glycogen and starch (Ref. 2).

When the affinity electrophoresis was carried out in the presence of glycogen together with an oligoglucan such as maltotriose or β-cyclodextrin, the phosphorylase mobility was recovered from the retardation. Assuming that the oligoglucan combines competitively with the glycogen site of phosphorylase and the mobility of the phosphorylase-oligoglucan complex is the same as that of the free phosphorylase, the dissociation constant of the phosphorylase to the oligoglucan can be calculated. In this report, we have presented the procedure to calculate the dissociation constant of rabbit muscle phosphorylase and potato glycogen phosphorylase to various oligoglucans.

METHODS

Electrophoresis. Affinity electrophoresis was carried out by the slightly modified procedure of Takeo and Nakamura (Ref. 1). Glass tubes, 7.5 cm long and 0.5 cm in diameter, were used. Separating gels of 5.0% acrylamide with or without glycogen together with varying concentrations of oligoglucan were prepared to a height of 5.0 cm. In order to secure a uniform concentration of glycogen and oligoglucan throughout the gel, the separating gel was prepared by layering a solution which contained the same concentration of glycogen and/or oligoglucan as that of the separating gel. To minimize dilution of the oligoglucan due to diffusion in the separating gel during the electrophoresis, a spacer gel was prepared containing the same concentration of the oligoglucan as that in the separating gel. A sample solution, 10% sucrose solution containing 20 mM thioglycolate at pH 6.7, was then applied. Electrophoresis was carried out at 150 V and 1.5 mA per tube at room temperature. After electrophoresis, the gel was removed from the glass tube and a fine wire was inserted at the position of the tracking BPB band and incubated at 37°C in a glass tube, 10 cm long and 0.8 cm in diameter, containing a substrate solution, composed of 2.0 M acetate buffer, pH 5.9, containing 0.5 mg of 5'-AMP,

2.5 mg of G-1-P, and 0.2 mg of shellfish glycogen per ml. The gel was then immersed in 10 ml of 7% acetic acid solution containing 30 μg of I_2 and 20 μg of KI per ml. A strong violet band appeared corresponding to the phosphorylase fraction. The relative migration distance of the phosphorylase was the value of the ratio of the migrating distance of the phosphorylase fraction to that of the tracking BPB band.

Theory of the affinity electrophoresis. In a previous paper (Ref. 1), we reported on the dissociation constant (K) in the reaction of phosphorylase with glycogen. The theory is based upon the fact that the mobility of the phosphorylase-glycogen complex is zero, and the concentration of glycogen is very large in comparison to that of the enzyme, so that the total concentration of glycogen, c, was almost equal to that of the free glycogen, [G]. In accordance to this fact, it will be

$$Rm_i/Rm_o = [P]/([P] + [PG]), \quad (1)$$

where Rm_o and Rm_i are the relative migration distances of phosphorylase in the absence or the presence of glycogen in the gels. [P] and [PG] are the concentration of the free phosphorylase and the phosphorylase-glycogen complex. Since $K = ([P]\cdot[G])/[PG]$, and $[G] = c - [PG] \simeq c$,

$$K = c\cdot Rm_i/(Rm_o - Rm_i). \quad (2)$$

Equation 2 can be transformed into

$$1/Rm_i = (1/Rm_o)[1 + (c/K)]. \quad (3)$$

Therefore, if reciprocal values of the Rm_i are plotted against the concentration of glycogen, c, a straight line will be obtained. The intercept of the line on the c-axis gives the negative K value.

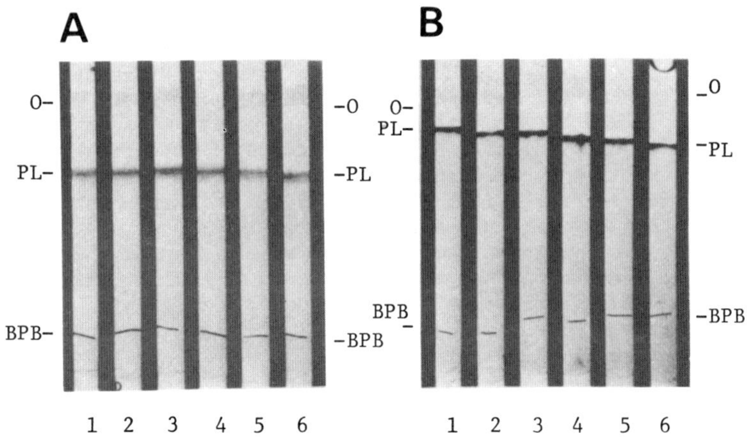

Fig. 1 Activity staining patterns of rabbit muscle phosphorylase in the presence of glycogen and/or maltotriose. A: Glycogen free separating gels, B: separating gels containing 0.2% glycogen. The gels of 1, 2, 3, 4, 5, and 6 contained 0, 10, 20, 30, 40, and 50 mM maltotriose, respectively. O: The origin of the separating gel, PL: fraction of phosphorylase, and BPB: position of the tracking BPB band.

When an oligoglucan was added to the separating gel containing glycogen, phosphorylase mobility was recovered from the retardation by the glycogen. Assuming that oligoglucan and glycogen combine competitively to the glycogen site of phosphorylase, and the mobility of the phosphorylase-glycogen complex is zero, while that of the phosphorylase-oligoglucan complex and the free phosphorylase are the same as will be seen in the section of the results, it will be

$$Rm_i/Rm_o = ([P] + [PI])/([P] + [PG] + [PI]), \quad (4)$$

where [PI] is the concentration of the phosphorylase-oligoglucan complex. Since $K_i = ([P]\cdot[I])/[PI]$, and $[I] = i - [PI] \simeq i$,

$$1/Rm_i = (1/Rm_o)[1 + (c/K)]/[1 + (i/K_i)], \quad (5)$$

where i and [I] are concentrations of the total and the free oligoglucan, and K_i is the dissociation constant of phosphorylase to oligoglucan. Equation 5 can be transformed into

$$Rm_i/(Rm_o - Rm_i) = (K/c)[1 + (i/K_i)]. \quad (6)$$

Therefore, if $Rm_i/(Rm_o - Rm_i)$ is plotted against the concentration of oligoglucan, i, in a constant concentration of glycogen, a straight line will be obtained. The intercept of the line on the i-axis gives the negative K_i value.

RESULTS AND DISCUSSION

In Fig. 1, we demonstrated the activity staining patterns of rabbit muscle phosphorylase by the affinity electrophoresis. In separating gels containing varying concentrations of maltotriose (Fig. 1-A, 2-6), the mobility of the phosphorylase was not retarded. It was almost identical with that of the free phosphorylase (Fig. 1-A, 1), as assumed in the theory. In the separating gel contained glycogen (Fig. 1-B, 1), mobility of the phosphorylase was markedly retarded. But in the separating gels contained glycogen together with maltotriose (Fig. 1-B, 2-6), the mobility of the phosphorylase was recovered in relation to the increase of the concentration of maltotriose. In Fig. 2, we presented the plots calculated by equation 6 from the data of Fig. 1 (Fig. 2-A), and that of β-cyclodextrin (Fig. 2-B). The results were in agreement with the theory, and straight lines were obtained. From the plots, the dissociation constant of phosphorylase to maltotriose in the presence of 0.2% glycogen was estimated to be 11.5 mM, and that of β-cyclodextrin was estimated to be 6.7 mM using the gels containing 0.1% or 0.175%

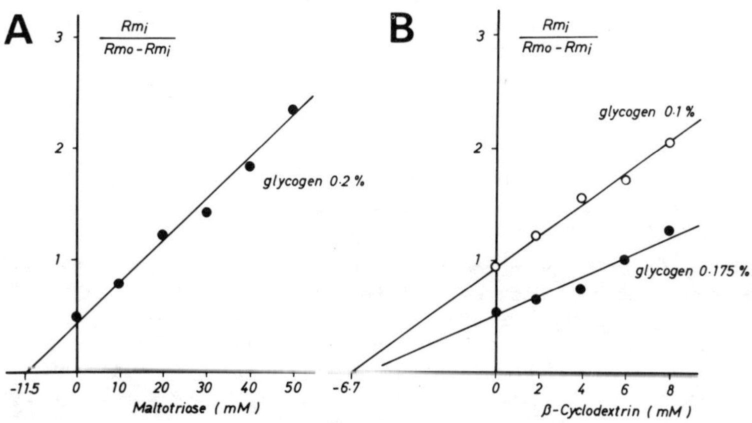

Fig. 2 Calculation plots of phosphorylase-inhibitor interactions. A: Plot for rabbit phosphorylase-maltotriose interaction. Glycogen 0.2% and maltotriose 0 - 50 mM. B: Plots for the phosphorylase-β-cyclodextrin interaction. Glycogen 0.1% or 0.175% and β-cyclodextrin 0 - 8 mM.

glycogen, respectively. In the same way, we calculated the dissociation constants of rabbit muscle phosphorylase and potato glycogen phosphorylase to various oligoglucans. In Table I, the obtained dissociation constants were summerized. From the results, it can be concluded that:

1) Glucose showed a very weak affinity to rabbit and potato phosphorylase, while maltose, maltotriose, and β-cyclodextrin showed strong affinity. The affinity of

these oligoglucans are progressively increased according to the number of glucosyl residues. The increasing rate is estimated to be about 5 for the rabbit skeletal muscle phosphorylase and about 15 for the potato glycogen phosphorylase.
2) In contrast with maltose, other diglucans, such as cellobiose (β-1:4) and trehalose (α,α-1:1'), showed very low affinity to the phosphorylases. As reported previously (Ref. 1), phosphorylase showed a strong affinity to glycogen and starch, but not to dextran and inulin. The facts indicate that both the retardation of phosphorylase mobility by polyglucan and the recovery from the retardation by oligoglucan are due to a specific affinity of the α-1:4-D-glucans to phosphorylase.
3) 5-Gluconolactone showed no affinity to rabbit phosphorylase in the affinity electrophoresis, as seen in Table I. It was reported (Ref. 3) that 5-gluconolactone is a competitive inhibitor of the rabbit muscle phosphorylase against G-1-P. It may be assumed that the G-1-P site of the phosphorylase is located at a different position in the enzyme molecule from that of the glycogen site.

Table I Calculated Dissociation Constants of Phosphorylases to Various Poly- and Oligoglucans by the Affinity Electrophoresis

Phosphorylase		Rabbit muscle phosphorylase	Potato glycogen phosphorylase
Polyglucan	Rabbit liver glycogen	0.11%	-
	Shellfish glycogen	0.13%	0.0017%
	Potato soluble starch	-	0.033%
	Pullulan	0.54%	1.1%
Oligoglucan	D-Glucose	350 mM	210 mM
	D-Maltose	82 mM	14 mM
	D-Cellobiose	196 mM	180 mM
	D-Trehalose	no affinity	200 mM
	Maltotriose	11.5 mM	0.61 mM
	β-Cyclodextrin	6.7 mM	0.075 mM
	5-Gluconolactone	no affinity	-

In conclusion, it is to be noted that the dissociation constants of an enzyme-substrate complex and an enzyme-inhibitor complex can be calculated not only kinetically, but also electrophoretically. This procedure can be applied to calculate the dissociation constant of the concanavalin A-carbohydrate interaction (Ref. 4). Since the calculation was based on the theory derived from the similar phenomena of chromatography, it may also be possible to calculate the dissociation constant of interactions in the affinity chromatography.

REFERENCES

(1) K. Takeo and S. Nakamura, Dissociation constants of glucan phosphorylases of rabbit tissues studied by polyacrylamide gel disc electrophoresis, Arch. Biochem. Biophys. 153, 1-7 (1972).
(2) K. Takeo, A. Kuwahara, H. Nakayama, and S. Nakamura, Affinity electrophoresis of phosphorylases and α-amylases, Protides of Biol. Fluids, 23, 645-649 (1976).
(3) A. M. Gold, E. Legrand, and G. R. Sánchez, Inhibition of muscle phosphorylase a by 5-gluconolactone, J. Biol. Chem. 246, 5700-5706 (1971).
(4) K. Takeo, M. Fujimoto, R. Suzuno, A. Kuwahara, H. Ogata, and K. Hironaga, Calculation of dissociation constants of concanavalin A-carbohydrate interaction by the affinity electrophoresis. In preparation.

AFFINITY CHROMATOGRAPHY AND CONFORMATIONAL ISOMERS OF dCMP-AMINOHYDROLASE

Carlo Antonio Raia, Roberto Nucci, Carlo Vaccaro, Santo Sepe, Eduardo Scarano and Mosé Rossi

Intern. Inst. of Genetics & Biophysics, 80125 Naples, Italy

ABSTRACT

Matrices containing ligands which interact with catalytic or allosteric sites have been successfuly used for the purification of the allosteric enzyme dCMP-aminohydrolase. This enzyme is eluted specifically by a competitive inhibitor from a matrix coated with ligands which interact with the catalytic sites of the enzyme. The allosteric activator elutes the enzyme from a matrix coated with hydrazide-UTP which behaves as an allosteric inhibitor exploiting the allosteric conformational changes of dCMP-aminohydrolase.

INTRODUCTION

Important control mechanisms of DNA-biosynthesis operate at the level of the synthesis of deoxynucleoside phosphates (1, 2). Understanding the allosteric properties of the related enzymes is mandatory to clarify the integrated biosynthesis of the deoxynucleoside triphosphates.

Spleen dCMP-aminohydrolase catalyzes the hydrolytic deamination of deoxycytidine 5'-monophosphate to deoxyuridine 5'-monophosphate (3). The reaction is at a branching point of the pathway leading to deoxycytidine triphosphate (dCTP) and deoxythymidine triphosphate (dTTP), which are the immediate pyrimidine deoxynucleotide precursors of DNA biosynthesis. Mg-dCTP acts as allosteric activator and Mg-dTTP as allosteric inhibitor. The activation or the inhibition of the enzyme depends on the relative concentrations of the two effectors (4).

By inhibiting dCMP-aminohydrolase, dTTP favours the formation of dCTP, whereas dCTP favours the formation of dTTP by reversion of the dTTP inhibition and activation of the enzyme. Thus, the concentration of dCTP and dTTP are controlled. dTMP, dUMP, dGMP and dAMP are competitive inhibitors of the enzyme (5).

The allosteric effectors induce conformational changes in the enzyme molecule, thus modifying the affinity of the catalytic sites for the substrate (4,5), without changing its molecular weight (3). The occurrence of at least three conformational isomers of dCMP-aminohydrolase has been proposed (6) on the basis of kinetic data, U.V. differential spectra, fluorescence quenching, digestion with proteolytic enzymes, titration of the exposed sulfhydryl groups, etc. The proposed conformers are: the enzyme without ligands, the enzyme-Mg-dCTP complex (activated conformation) and the enzyme-Mg-dTTP complex (inhibited conformation).

MATERIALS AND METHODS

dCMP-aminohydrolase at different levels of purification was prepared as previously described (3). The enzyme activity was measured at 37°C using a Zeiss DMR-10 and is expressed as ΔOD_{280}/minute (4). Sepharose 4B and Blue Dextran were obtained from Pharmacia. Sepharose-adipic hydrazide-CMP and Sepharose-adipic hydrazide-UTP were prepared according to Wilchek and Lamed (7) and contained 4-5 μmoles of nucleotide/ml of the packed gel. The synthesis of hydrazide-UTP was performed by reaction of equimolecular solutions of oxidized UTP and adipic dihydrazide in 0.1 M acetate buffer pH 5, at 4° C for 2 hours. The nucleotide derivative was purified by DEAE chromatography.

Blue Dextran-Sepharose was prepared according to Ryan and Vestling (8).
Hg-dCTP was prepared as described by Dale et al. (9) and was purified by DEAE chromatography. Affi-gel 401 was obtained by Biorad and was saturated with Hg-dCTP.

RESULTS AND DISCUSSION

The purification of dCMP-aminohydrolase by affinity chromatography has been approached in two ways: a) synthesis of matrices containing ligands which interact with the catalytic sites of the enzyme; b) synthesis of matrices containing ligands which interact with the regulatory sites of the enzyme.

CMP, an analogue of the substrate dCMP, is appreciably deaminated by dCMP-aminohydrolase only in the presence of the allosteric activator dCTP (7), namely by the activated conformer of the enzyme. In fact, the enzyme binds to the Sepharose-hydrazide-CMP only in the presence of the allosteric activator. The elution is achieved with a competitive inhibitor (dAMP or dTMP) at a concentration 0.1 mM. However, the capacity of the gel synthesized by us is rather low.

Better results have been obtained with Blue Dextran-Sepharose, prepared according to Ryan and Vestling (8). Blue Dextran complexes with some proteins because of its affinity for a super-secondary structure called "dinucleotide folding" present in some enzymes (11). This type of folding forms specific nucleotide binding sites. Since dCMP-aminohydrolase should have this super-secondary structure, we have used Blue Dextran-Sepharose to purify the enzyme. dCMP-aminohydrolase binds strongly to a column of Blue Dextran-Sepharose and is eluted by the competitive inhibitor dTMP at a concentration of 0.1MM (Fig. 1).

The allosteric effectors do not release the enzyme from the column, but dTTP favours the binding while dCTP does not affect it. This is an additional evidence, though indirect, of different binding sites on the enzyme molecule for isosteric and allosteric ligands.

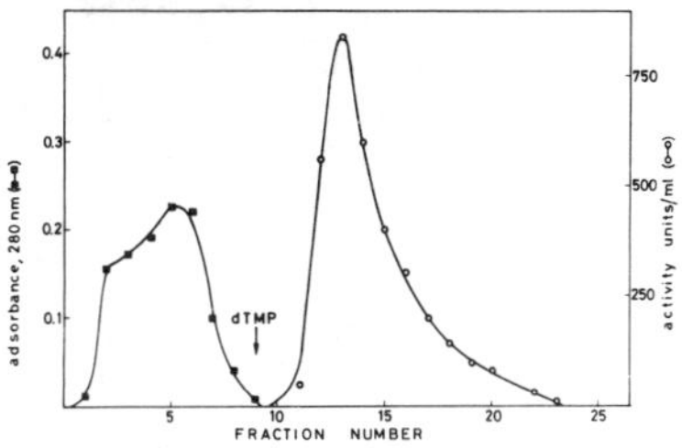

Fig. 1. Binding of dCMP-aminohydrolase by Blue Dextran-Sepharose and elution by dTMP (0.1 mM).
(-■-■-) absorbance at 280 nm;
(-o-o-) dCMP-aminohydrolase activity

The purification achieved with the Blue Dextran-Sepharose chromatography depends on the starting specific activity of the enzyme. Purification up to 1000-fold can be achieved. In every instance, the recovery is about 60 %. An essential requirement for this step is that the enzyme has to be free of nucleic acids.

These data indicate that Sepharose-CMP and Blue Dextran-Sepharose bind dCMP aminohydrolase through the catalytic sites.

Taking into account the specificity of the allosteric ligands (4-6), the ideal resin designed to bind the enzyme through the allosteric sites should have the spacer arm bound to the 5 position of the pyrimidine ring of the ligand. In fact, only modifications in the 5 position give allosteric ligands with an activity similar to that of dCTP or dTTP. Indeed, CH_3-dCTP, I-dCTP, Br-dCTP are allosteric activators and dUTP, I-dUTP, Br-dUTP are allosteric inhibitors. However, CTP is not an allosteric activator and UTP is not an allosteric inhibitor. We tried to prepare a thioether ($NH_2CH_2CH_2$-S-dCTP) which could be coupled to activated Sepharose, by reaction of 5-bromo-dCTP with cysteamine. However, the recovery of the thioether is very low because under the alkaline conditions of the reaction hydrolysis of dCTP to dCMP is occuring. In addition, dCTP was mercurated in the 5 position (9) and subsequently reacted with Affi-gel 401 (Biorad) containing an SH terminal group. Affi-gel 401-S-Hg-dCTP proved not to be useful because it inactivates the enzyme probably by an exchange reaction of some of the 24 sulfhydryl groups of the enzyme with the Hg-dCTP bound to the resin.

Sepharose-adipic hydrazide-UTP is at present the best support for the purification of the enzyme. Adipic hydrazide-UTP is an analogue of the allosteric inhibitor dTTP and inhibits the enzyme.
Fig. 2 shows a Dixon plot of the effect of adipic hydrazide-UTP, UTP and dCTP on the dCMP-aminohydrolase activity. UTP is a poor inhibitor, whereas adipic hydrazide-UTP inhibits the enzyme in a cooperative manner. The inhibition is released by the addition of the activator dCTP at a concentration of 1 μM.

As shown in Fig. 3, dCMP-aminohydrolase binds strongly to the Sepharose-UTP resin and is released by the addition of dCTP but not

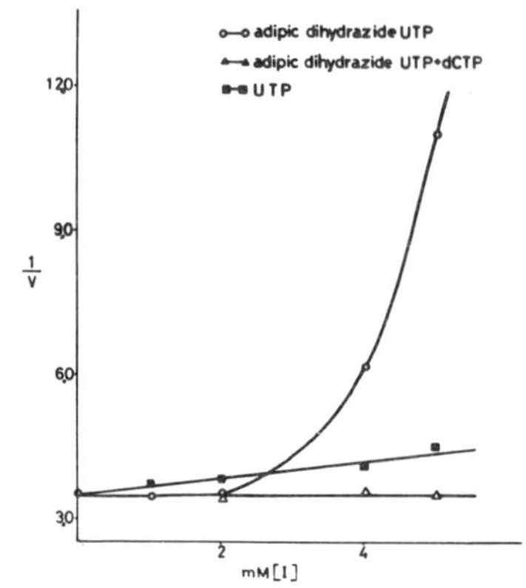

Fig. 2. Dixon Plot of the inhibition of hydrazide-UTP and UTP and reversal of the inhibition by dCTP. Incubation mixture contained in 0.3 ml 0.05 M Tris, 2 mM $MgCl_2$, 1 mM mercaptoethanol, 1 mM dCMP and 1 enzyme unit. V = OD_{280}/min. Cuvettes 1 mm light path. T = 37°C

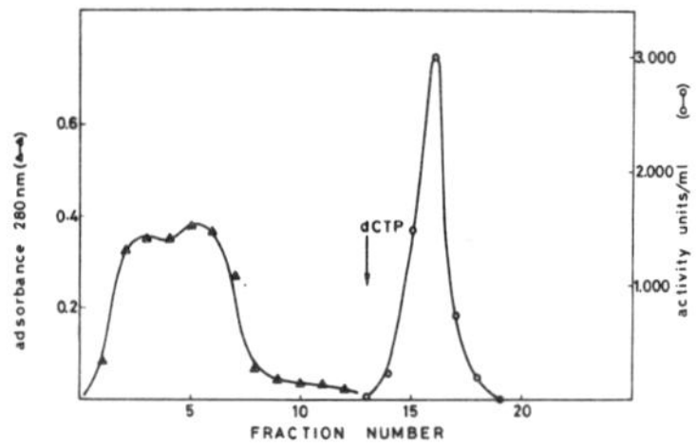

Fig. 3. Binding of dCMP-aminohydrolase by Sepharose-adipic hydrazide-UTP and elution with dCTP (10 μM). (-▲-▲-) adsorbance at 280 nm; (-o-o-) dCMP-aminohydrolase activity

by the addition of competitive inhibitors. Depending on the starting specific activity the purification obtained is 100- to 1000-fold with a 90 % recovery. These data indicate that Sepharose-UTP binds dCMP-aminohydrolase through the allosteric site. The elution of the enzyme by dCTP can be explained by the induction of the activated conformation upon binding of the regulatory nucleotide. Thus a highly specific and efficient elution of the enzyme is realized.

We can conclude that the chromatography on Sepharose-hydrazide-UTP exploites the allosteric conformational changes of the enzyme.

REFERENCES

(1) P. Reichard, The biosynthesis of deoxyribose, Ciba Lectures in Microbial Biochemistry, Wiley, New York (1967).

(2) E. Scarano, M. Rossi, G. Geraci, Regulation of the activity of deoxycytidylate aminohydralase, IV FEBS-Meeting Oslo, 1967, Symp. "The regulation of enzyme activity and allosteric interactions", p. 145-165, Universitetsverlaget, Oslo (1968).

(3) G. Geraci, M. Rossi, E. Scarano, Deoxycytidylate aminohydralase. I: Preparation and properties of the homogeneous enzyme, Biochemistry 6, 184 (1967).

(4) E. Scarano, G. Geraci, M. Rossi, Deoxycytidylate aminohydrolase. II: Kinetic properties, The activatory effect of dCTP and the inhibitory effect of dTTP. Biochemistry 6, 192 (1967).

(5) M. Rossi, G. Geraci, E. Scarano, Deoxycytidylate aminohydrolase. III: Modifications of the substrate sites caused by allosteric effectors, Biochemistry 6, 3640 (1967).

(6) M. Rossi, I. Dosseva, M. Pierro, M.G. Cacace, E. Scarano, Studies on the conformational isomers of deoxycyticylate aminohydrolase, Biochemistry 10, 3060 (1971).

(7) M. Wilchek, R. Lamed, Immobilized nucleotides for affinity chromatography, Methods in Enzymology 34, part B, p. 475, Acad. Press, New York (1974).

(8) L.D. Ryan, C.S. Vestling, Rapid purification of lactate dehydrogenase from rat liver hepatoma: a new approach, Arch. Biochem. and Biophys. 160, 279 (1974).

(9) R.M.K. Dale, E. Martin, D.C. Linvingston, D.C. Ward, Direct covalent mercuration of nucleotides and polynucleotides, Biochemistry 14, 2447 (1975).

(10) M. Rossi, R.L. Momparler, R. Nucci, E. Scarano, Studies on analogs of isosteric and allosteric ligands of deoxycytidylate aminohydrolase, Biochemistry 9, 2539 (1970).

(11) S.T. Thompson, K.H. Cass, E. Stellwagen, Blue Dextran-Sepharose: An affinity column for the dinucleotide fold in proteins, Proc. Nat. Acad. Sci. 72, 669 (1975).

OPTIMISATION OF CONDITIONS FOR THE AFFINITY CHROMATOGRAPHY OF HUMAN ENTEROKINASE

David A. W. Grant, Anthony I. Magee and John Hermon-Taylor

Department of Surgery, St. George's Hospital Medical School, London, SW17 ORE

INTRODUCTION

The purification of enterokinase by affinity chromatography on immobilised p-aminobenzamidine has been reported with mixed success from different laboratories (1,2,3,4). This study investigates the conditions necessary for the successful isolation of the human enzyme and compares its behaviour with that of trypsin towards the same immobilised ligand.

MATERIALS AND METHODS

Aliquots of Sepharose 4B were mixed with different concentrations of CNBr and glycylglycine spacer-arm in the initial activation and coupling stages. The concentration of coupled spacer-arm was measured by including a 1-^{14}C glycylglycine marker. This was also the immobilised ligand concentration as its attachment with carbodiimide was shown to be complete by the abolition of cytochrome C adsorption. The enterokinase and trypsin binding capacities were estimated by passing 1ml samples of partly-purified enterokinase (1) or commercial bovine trypsin through 0.5ml beds of each gel in 50mM-Tris/HCl pH 8.3 (buffer A). Loading was continued to saturation which was taken as the point when less than 95% of the enzyme was retained when assayed by frontal analysis. Adsorbed enzyme was eluted with 70mM-sodium succinate pH 5.6 containing 0.5M-KCl; each eluate was assayed for enterokinase, aminopeptidase and maltase (1).

RESULTS AND DISCUSSION

There was no apparent relationship between the immobilised ligand concentration and enterokinase binding capacity (Table 1), whereas there was good correlation with the trypsin capacity (p=0.001). The critical factor was the concentration of spacer-arm in the coupling slurry; only at 2mmol per g Sepharose and above was there significant enterokinase adsorption. One interpretation of this would be that at high CNBr concentrations the gel polymer would be

TABLE I
Enterokinase and Trypsin
binding capacities of various gels

Spacer-arm concentration mmol/g Sepharose	CNBr concentration mg/g Sepharose	Degree of substitution µmol/glycyl-glycine per ml Sepharose	Enterokinase capacity pmol/ml Sepharose	Trypsin capacity nmol/ml Sepharose
3.0	300	22.8 ± 1.4	121	252
2.5	200	22.3 ± 1.8	210	280
2.5	100	21.5 ± 1.5	136	246
2.5	50	17.8 ± 1.5	101	210
2.0	300	20.5 ± 1.4	75.5	269
1.75	250	36.2 ± 4.0	0	295
1.5	250	25.8 ± 1.6	0	260
1.0	300	19.6 ± 1.0	<18	170
0.5	150	10.2 ± 1.5	0	147

extensively cross-linked through cyclic and acyclic imidocarbonates effectively sealing the bead interior. In addition the high density of coupled spacer-arm on the bead surface would further discourage penetration of glycylglycine to the interior. Presentation of the ligand into the interstitial spaces between the beads would be sterically favourable for the binding of a macromolecule like enterokinase (296,000 daltons). At lower concentrations glycylglycine may have been more evenly distributed and the interior sites were inaccessible to enterokinase but not to the smaller globular trypsin.

When aliquots of the best adsorbents were serially diluted 1/4, 2/4 and 3/4 with unlabelled Sepharose 4B the trypsin capacity fell in a linear fashion consistent with 1 to 1 binding between the enzyme and the ligand (Fig.1a). Enterokinase capacity fell exponentially which was linearised in a double logarithmic plot (Fig.1b). At very dilute ligand concentrations the slope of this plot approximated to the slope of a Hill plot (5) and the Hill coefficient was estimated to be ⩾3. This suggested that there was positive co-operativity between enterokinase and an undefined number of binding sites. The highest affinity was clearly for the active site and this was augmented by non-specific secondary interactions with the gel matrix, ionic groups or spacer-arm not necessarily located on the same bead.

On a preparative scale the best adsorbent separated 99% of both intestinal maltase and aminopeptidase from enterokinase following

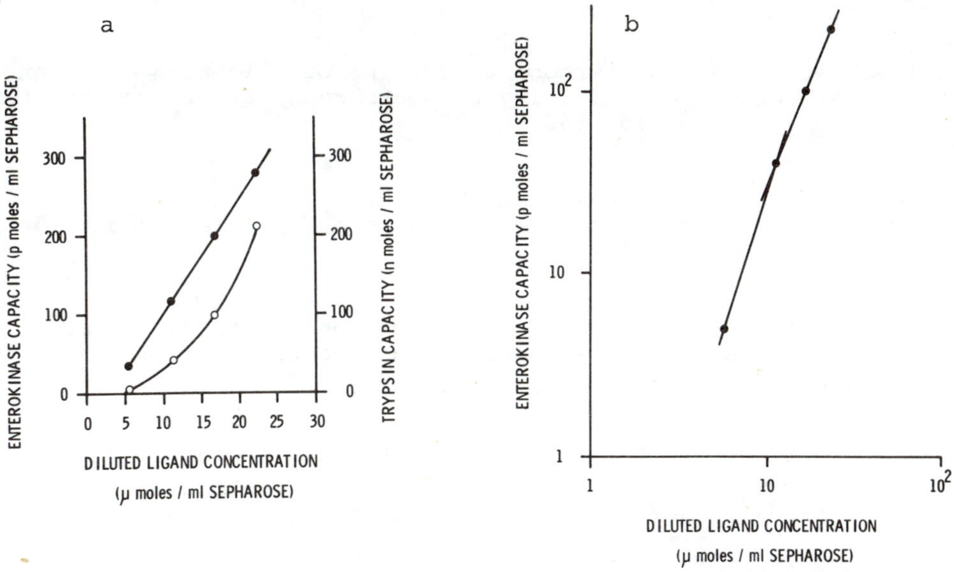

Fig. 1 (a) Reduction in binding capacity with dilution
(b) Double logarithmic plot of the adsorption of enterokinase

prolonged washing (96 hours) with buffer A after loading; but the desorbed enzyme was less than 4% pure. The persistence of so much uncharacterised contamination suggested that the interactions responsible for their retention were strong. The interpretation of the mode of enterokinase adsorption had suggested that its secondary interactions although co-operatively important were by themselves collectively weak. Further purification might be achieved by exploiting this and using a gel of reversed polarity in which the biological specificity of the ligand for the active site had been eliminated whilst retaining the number and configurations of the secondary sites. The simplest manoeuvre was to remove the essential guanidinium group by substituting aniline for p-aminobenzamidine. On glycylglycyl-aniline Sepharose 4B enterokinase peak activity was retarded by about 6 column volumes and 90% of residual contamination was retained. High affinity binding had been lost since the Ki of aniline for enterokinase was $>10^{-2}$M in free solution. The enzyme was now 45% pure and could be completely purified by immuno adsorption (1).

REFERENCES

(1) D.A.W. Grant and J. Hermon-Taylor, The purification of human enterokinase by Affinity Chromatography & Immunoadsorption, Biochem.J. 155, 243 (1976)

(2) L.E. Anderson, K.A. Walsh and H. Neurath. Bovine enterokinase, Purification, Specificity, and some Molecular Properties. Biochem. 16, 3354 (1977)

(3) R.J. Barns and R.G. Elmslie, Preparation of Porcine Enteropeptidase free of Intestinal Aminopeptidase activity with Con-A Sepharose, Biochim.Biophys.Acta. 445, 815 (1976)

(4) J. Baratti and S. Maroux, On the catalytic and binding sites of porcine enteropeptidase Biochim.Biophys.Acta. 452, 488 (1976)

(5) A.V. Hill, The possible effects of the aggregation of the molecules of haemoglobin on its dissociation curves Proc.Physiol.Soc. 40 iv (1910)

AFFINITY CHROMATOGRAPHY OF BRAIN GLUTAMATE DECARBOXYLASE (EC 4.1.1.15) USING IMMOBILIZED PYRIDOXAL PHOSPHATE

Alfred Fleissner

Neurochem. Abteilung der Psychiatrischen Universitätsklinik Martinistrasse 52, 2000 Hamburg 20, BRD

Glutamate decarboxylase (GAD) is known to have a high affinity to its coenzyme pyridoxal phosphate (PLP). The Michaelis constant is about 100 nM. This has encouraged us to attempt the purification of the enzyme using immobilized PLP as affinity adsorbent.
For the porcine enzyme we have found K_s = 200 nM in respect of pyridoxal phosphate (Fig. 1).

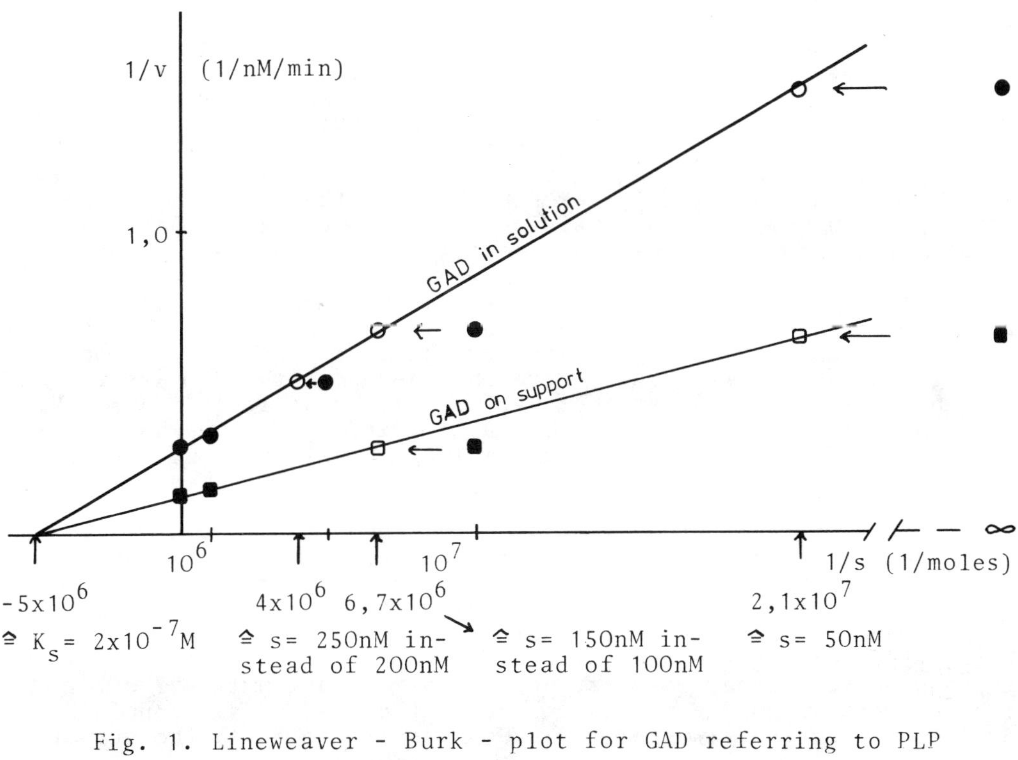

Fig. 1. Lineweaver - Burk - plot for GAD referring to PLP

The coenzyme is bound so tightly that it is not possible to remove it quantitatively with usual methods like dialysis, gel filtration or even treatment with 0.1 M hydroxylamine. There seems to be a PLP-concentration of 50 nM, when the activity determination of dialysed enzyme is carried out PLP-free, and also the other points roughly come to lie on the straight line when we add 50 nM to the concentrations used in the test. It cannot be ruled out that there exists an isozyme with partly fixed coenzyme responsible for the apparent PLP-free activity. The bound enzymatic activity can be detected on all tested affinity adsorbents showing the same kinetics as in solution. The material with bound enzyme may be washed several times with buffer at neutral pH without significant loss of activity under PLP-free conditions demonstrating the strong fixation of some coenzyme.

The affinity adsorbents used were prepared in the following well-known way (Ref. 1): Epoxy activated Sepharose 6B was transformed to the amino derivative and then coupled with p-nitrobenzoylazide to give after reduction the starting material I (Fig. 2). With pyridoxal phosphate the Schiff base is formed, yielding the pyridoxamine phosphate derivative III after reduction with sodium borohydride. This material was tested as affinity adsorbent without success, and therefore derivative I was diazotized and coupled to PLP in position 6 to give material IV, which was more effective.

The binding studies were carried out in batch experiments: 2 g of material were agitated for an hour with 20 ml of enzyme solution and washed with the same volume of buffer. One half was eluted with 2 M sodium chloride, and washed with buffer again. 0.5 g of support after washing (a), 0.5 g of support after elution (b), the starting activity, and the activity after exposure to the material was determined in duplicate. The activities both bound and passed through amounted to 80 - 100 % of the starting activity in all experiments. The material IV was so effective in binding the glutamate decarboxylase that up to 70 % of the starting activity was still bound to it after washing at high ionic strength, and even 4 M urea did not detach the activity. The elution with buffer, containing 10 mM PLP yielded 25 % of the bound enzymatic activity with a purification factor of about 3. But there is a loss of another 25 % of the activity, because only 50 % could be found again on the support. This residual activity was not eluted even with 100 mM PLP. Washing the gel with 100 mM hydroxylamine removes the remaining traces of coenzyme, and there is no activity under PLP-free test conditions, but in the presence of 0.1 mM PLP, 60 % of the activity is restored. After 200 mM hydroxylamine there is no longer any activity on the support, but only about 20 % of the original activity is found in the eluate after desalting. In our opinion the affinity to the immobilized coenzyme was too strong, but no better results were obtained when the PLP was replaced by a pyridoxamine phosphate derivative (V), by salicylaldehyde (VI), or even by phenol (VII). As expected the binding is somewhat lower, but no considerable purification resulted. In all binding studies there was little difference when using dialysed "PLP-free" enzyme or a preparation at 0.5 mM PLP. We assume that the azo group is responsible for some unspecific binding, which may also explain the relatively high remaining impurities on the enzyme after elution with its coenzyme. Moreover a strong association of proteins to glutamate decarboxylase is well-known and has prevented its purification with other methods for a long time (Ref. 2).

The unexpected tight binding of the enzyme to the Schiff base II led us to hope for better results, but the purification factor after elution with 10 mM PLP was as low as in the case of azo coupled PLP. 50 % of the activity was eluted, 20 % was recovered on the support, so

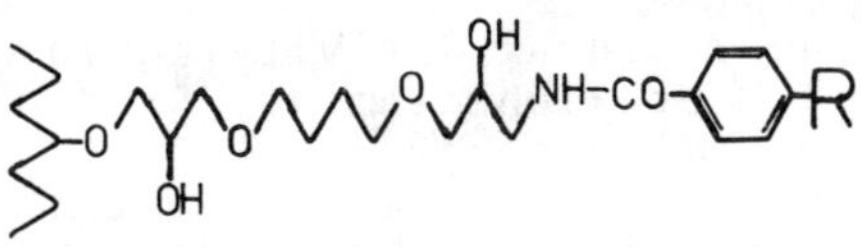

2 g support + 20 ml GAD-solution with an activity of 4 nM/min/ml:

Bound activity (nM/min/g support)

R:		a)	b)
$-NH_2$	I	7 (18 %)	4 (10 %)
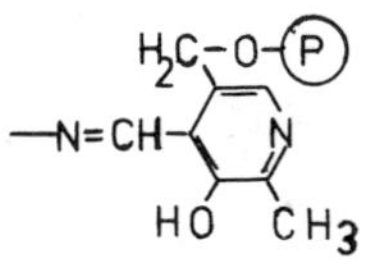	II	18 (45 %)	16 (40 %)
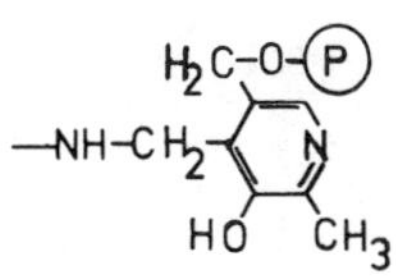	III	1 (3 %)	-- --
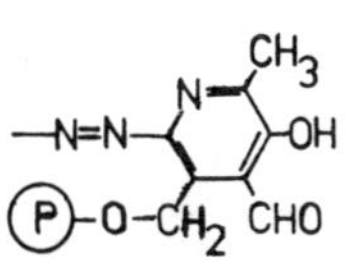	IV	31 (78 %)	28 (70 %)
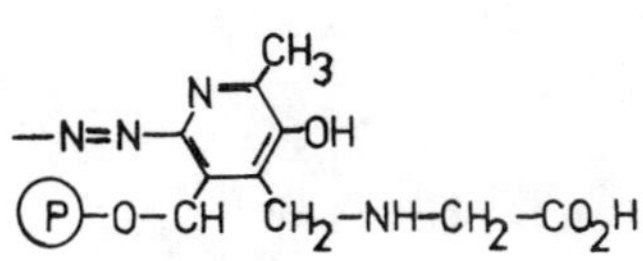	V	20 (50 %)	12 (30 %)
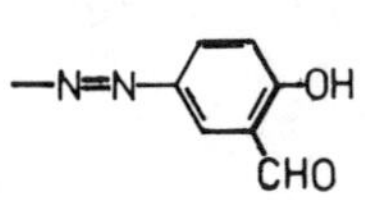	VI	22 (55 %)	20 (50 %)
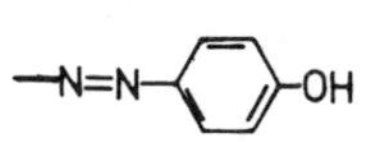	VII	18 (45 %)	16 (40 %)

Fig. 2. Affinity adsorbents used in binding studies of GAD

a) after washing with buffer, pH 7.0
b) after elution with 2 M NaCl, pH 7.0

that a loss of 30 % had to be accepted. The residual activity could be eluted to some extent with a solution of 1 M ethylenediamine, 5 mM PLP. 30 % of the activity was obtained with a purification factor of 14. Our conclusion is that purification of glutamate decarboxylase with affinity adsorbents using immobilized coenzyme is possible, but not very efficient because of the strong association of the accompanying proteins to the matrix and to the enzyme itself. The affinity adsorbents studied are all weak inhibitors of glutamate decarboxylase, but under PLP-free test conditions an enhancement of the activity in the presence of immobilized amino groups is found (Fig. 3). Perhaps the affinity adsorbents are valuable tools for investigating some properties of the enzyme in a bound state, as it is assumed to be when it is active in synaptic membranes, transforming an excitatory transmitter into an inhibitory one.

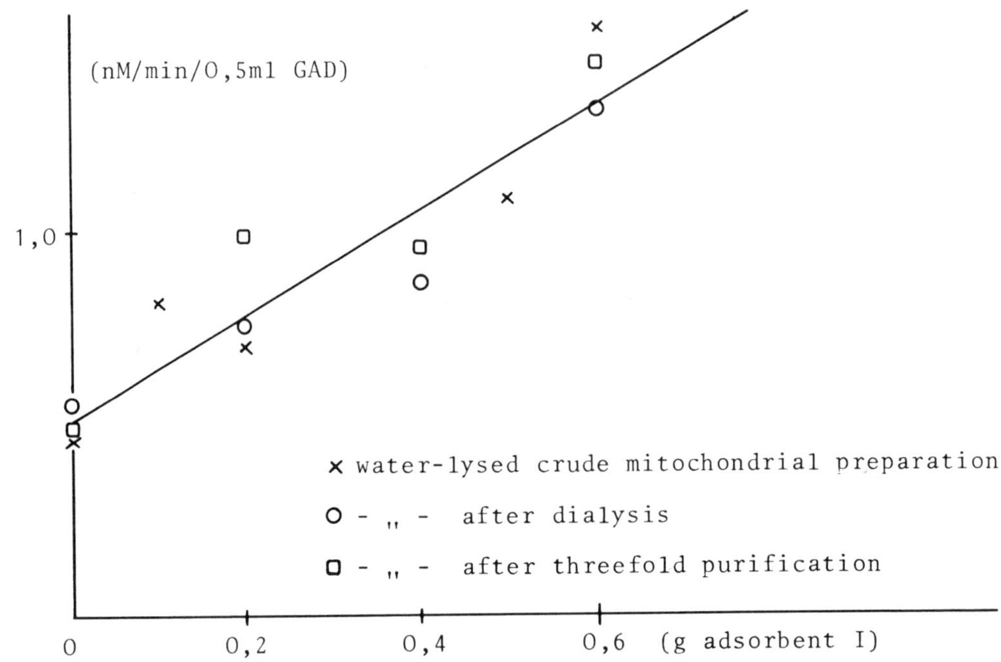

Fig. 3. Activity of "PLP-free" GAD-solutions in the presence of affinity adsorbent I

REFERENCES

(1) L. Sundberg, J. Porath, Preparation of adsorbents for biospecific affinity chromatography, J. Chromatogr. 90, 87 (1974).

(2) J.-Y. Wu, T. Matsuda, E. Roberts, Purification and characterization of glutamate decarboxylase from mouse brain, J. Biol. Chem. 248, 3029 (1973).

AFFINITY CHROMATOGRAPHY OF TRYPTOPHAN SYNTHASE FROM *E. COLI*

H. P. Gschwind and K. Kirschner

Abteilung Biophysikalische Chemie Biozentrum der Universitaet, Basel, Switzerland

ABSTRACT

Nonreactive substrate analogues of indoleglycerol phosphate have been synthesized, starting from tryptophanol phosphate. The parent compound does not inhibit tryptophan synthase from E.coli. However, a number of N-acyl derivatives are excellent competitive inhibitors. These compounds are the basis of Sepharose derivatives used for affinity chromatography of tryptophan synthase.

INTRODUCTION

Tryptophan synthase from E.coli catalyzes the last two steps of the biosynthesis of L-tryptophan (Trp) from indoleglycerol phosphate (IGP) and L-serine (Ser) (1).

$$\text{Indoleglycerol-P} \rightleftharpoons \text{Indole + D-Glyceraldehyde-P} \quad [1]$$

$$\text{Indole + L-Ser} \longrightarrow \text{L-Trp} + H_2O \quad [2]$$

TSase from E.coli is a multienzyme complex composed of two α-subunits and one β_2-subunit. It can be purified from overproducing mutants following standard procedures (3) but problems arise when isolation is attempted from organisms with low levels of enzyme. Several materials have been tested for affinity chromatography, but none was particularly efficient (4,5). This work is concerned with the systematic development of a more efficient material. The enzyme from E.coli was used as a test system.

MATERIALS AND METHODS

ENZYMES Assays and analytical methods have been published (2).

SUBSTRATES AND LIGANDS Indoleglycerol phosphate and indolepropanol phosphate (IPP) were obtained as described (2), tryptophanol phosphate (TRPP), N-acetyl-TRPP, N-(ε-CBZ-aminocaproyl)-TRPP, N-(hydrazido-adipoyl)-TRPP and N-[(N'-CBZ-aspartyl)-ε-aminocaproyl]-TRPP were synthesized and well characterized (6).

SUBSTITUTED GELS Epoxy-activated Sepharose-CL-4B was prepared by the method of Sundberg & Porath (7) and coupled with adipic acid dihydrazide (ADH) and then substituted with N-[(N'-CBZ-aspartyl)-ε-aminocaproyl]-TRPP by the aqueous carbodiimide method (8).

AFFINITY CHROMATOGRAPHY A 2.5 x 10 cm column of the affinity material was equilibrated with 300 mM potassium phosphate buffer, pH 7, containing 1 mM EDTA, 0.2 mM dithioerythritol and 0.04 mM pyridoxal-5'-phosphate. 99'000 units of TSase, partially purified by DEAE-cellulose chromatography of a crude extract from E.coli ($trpR^-$ trpΔED102/F'trpΔED102) (3) and dialyzed against the equilibration buffer, were applied to the column. After washing with 150 ml of buffer, the TSase complex was desorbed by a linear gradient of L-serine in buffer.

RESULTS

Indolepropanol phosphate is a competitive inhibitor of the indoleglycerol phosphate cleavage reaction (cf. reaction 1). We therefore decided first to study the optimal site for attachment of the spacer arm on indolepropanol phosphate. The constants of competitive inhibition (as obtained from a kinetic analysis of reaction 1 in the steady state) served as a quantitative measure of affinity ($1/K_i$) of the various derivatives for the complex. Derivatives of tryptophanol phosphate proved to have smaller inhibition constants than those in which various side chains were attached either to N(1) or to C(5) of the indole moiety or to the phosphate group of indole propanol phosphate (6). The inhibition constants obtained for derivatives with different side chains coupled to the amino group of tryptophanol phosphate are listed in Table 1.

TABLE I Derivatives of tryptophanol phosphate

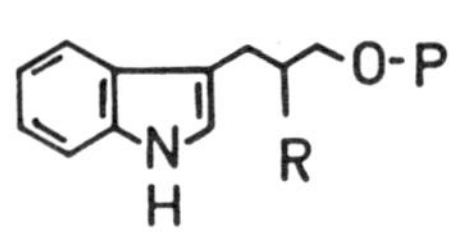

No.	R	K_I (μM)	REMARKS
I	-H	4.0	Kirschner et al (2)
II	$-NH_3^+$	-	No inhibition
III	$-NHCOCH_3$	150	
IV	$-NHCOCH_2(CH_2)_3CH_2NH_3^+$	-	No inhibition
V	$-NHCOCH_2(CH_2)_3CH_2NHCOCH_3$	260	
VI	$-NHCOCH_2(CH_2)_2CH_2CONHNH_2$	240	
VII	$-NHCOCH_2(CH_2)_3CH_2NH\text{-}CBZ$	4.5	

Introduction of an uncharged side chain at C(2)' of indolepropanol phosphate increases the inhibition constant 40- to 60-fold (cf. I with III,V,VI). A positive charge on the molecule abolishes the binding affinity (cf. I with II,IV). To our surprise the introduction of an hydrophobic residue on an uncharged side chain increases the affinity (VII) to that of indolepropanol phosphate (I).

The change in the free energy of binding is -2.5 kcal/Mol, a value in accord with the free energy of transfer of a phenyl group from water to an hydrophobic environment (9).

OPTIMUM AFFINITY GEL The deleterious effect of a positive charge on the side chain of the ligand eliminates the cyanogen bromide method for immobilization (10). After several trials with divinyl sulfone as coupling agent we chose 1,4-butanedioldiglycidylether instead because it leads to a very stable chemical bond (11). The affinity gel with which the following experiments were conducted, has the structure given below:

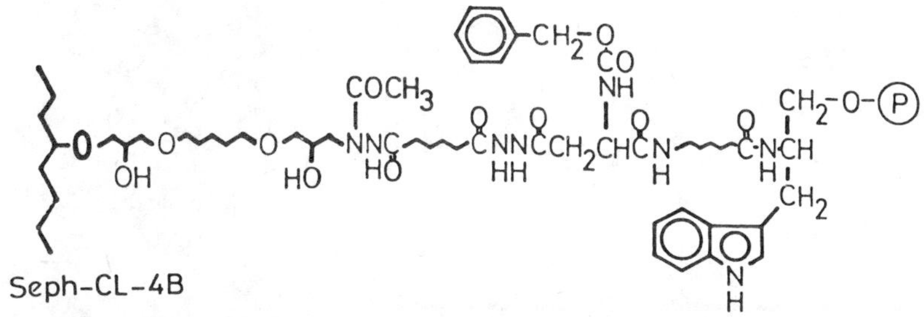

PRELIMINARY STUDIES Figure 1 shows that both the α-subunit and the complex of tryptophan synthase are merely retarded when chromatographed in 50 mM Tris buffer. It is surprising that the complex is retarded even less than the α-subunit because it has a lower dissociation constant for indolepropanol phosphate (4 μM, cf.Table 1) than the α-subunit (K_i = 46 μM, (2)).

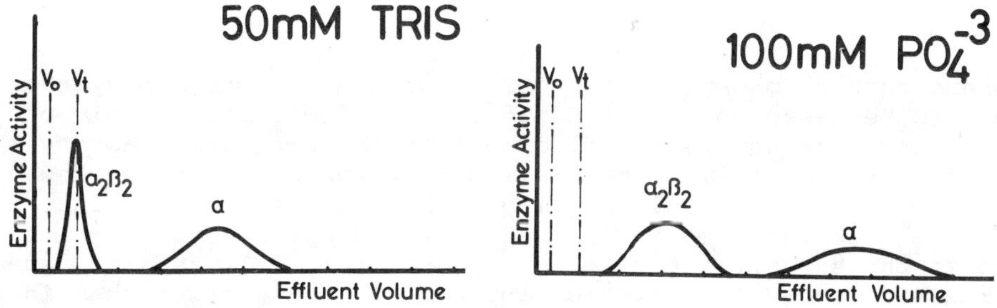

Fig. 1. Schematic elution profile of the α-subunit and the complex in Tris- and phosphate buffers at pH 7 and 4°C
V_o: Void volume, V_t: Total volume of the gel

Replacement the Tris buffer by 0.1 M phosphate leads to stronger retardation of both enzyme species (cf. Fig. 2, phosphate). Complete binding is observed in 300 mM phosphate, but the enzyme is specifically released when IGP or IPP are present in the eluting buffer. In contrast, addition of 0.5 M potassium chloride or sodium sulfate to Tris buffer does not significantly alter the elution profile of Fig. 1, Tris.

AFFINITY CHROMATOGRAPHY OF THE TRYPTOPHAN SYNTHASE COMPLEX The enzyme can be purified to homogeneity in a single step, starting from a partialy purified extract obtained from an overproducing strain of E.coli (Fig. 2). It is surprising that L-serine, the substrate of the β_2-subunit (cf. reaction 2) is as competent in eluting the complex as are the specific ligands of the α-subunit (i.e., IGP or IPP).

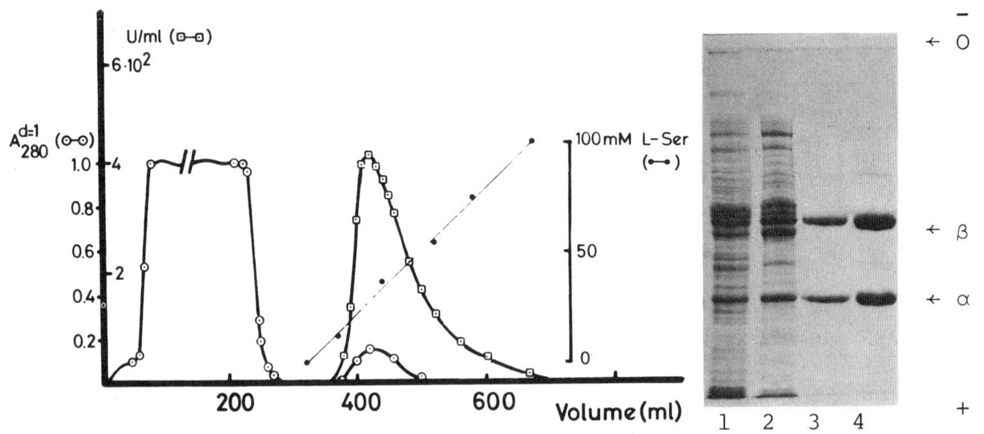

Fig. 2. Elution profile (left) and the SDS-polyacrylamide-gel electrophoresis (right) of the tryptophan synthase-complex. For details see "Methods". The yield was 65 % with five fold purification. 1 E.coli crude extract; 2 eluate from the DEAE-cellulose column; 3, 4 concentrated fractions from affinity column at 400, 420 ml eluate.

DISCUSSION

The phenomenon of phosphate induced binding of tryptophan synthase appears to be based on a superposition of hydrophobic and biospecific interactions. The unexpected finding that the α-subunit, having the lower affinity for IPP, is more strongly bound than the complex (cf. Fig. 1) can be correlated with the equally surprising fact that L-serine elutes the complex (Fig. 2). If a composite active site is formed at the subunit interface of the complex (12), the long spacer-arm will experience more steric hindrance in the complex than in the α-subunit alone and L-serine, bound to the β_2-subunit, can completely block the access of the immobilized ligand to the active site of the α-subunit (Fig. 3). This hypothesis is supported by the following observations: L-serine enhances the affinity of the complex for indolepropanol phosphate five fold (13), whereas it decreases the affinity ($1/K_i$) of N-(ε-CBZ-aminocaproyl)-TRPP (cf. VII, Tab. 1) from 0.22 μM^{-1} to 0.04 μM^{-1} (data not shown). We expect that the affinity material will be useful in purifying tryptophan synthase from other organisms with much lower levels of enzyme than the strain of E.coli used in this work.

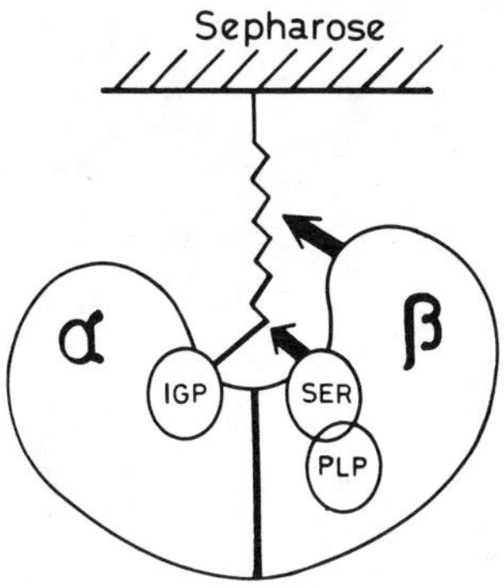

Fig. 3. Hypothetical interaction of the β-protomer and L-serine with the spacerarm of the immobilized ligand. The composite active site was suggested by Creighton (12). Only one half of the enzyme molecule is shown.

REFERENCES

(1) Yanofsky, C., Crawford, I.P. (1972) 'The Enzymes' 3rd edition (Boyer, P.D.,ed.), N.Y., Acad.Press Vol 7, pp 1-20

(2) Kirschner, K., Wiskocil, R.L., Foehn, M., Rezeau, L. (1975) Eur.J.Biochem. 60, 513-523

(3) Adachi, O., Kohn, L.D., Miles, E.W. (1974) J.Biol.Chem. 249, 7756-7763

(4) Wolf, D.H., Hoffmann, M. (1974) Eur.J.Biochem. 45, 2941-2946

(5) Owens, D., Bailey, C.J. (1974) Biochem.Soc.Trans. 2, 1331-1332

(6) Gschwind, H.-P. (1977), PhD Thesis, University of Basel

(7) Sundberg, L., Porath, J. (1974) J.Chromatography 90, 87-98

(8) Sheehan, J.C., Hess, G.P. (1965) J.Amer.Chem.Soc. 77, 1067-1068

(9) Nozaki, Y., Tanford, C. (1971) J.Biol.Chem. 246, 2211

(10) Jost, R., Miron, T., Wilchek, M. (1974) Biochim.Biophys.Acta 362, 75-82

(11) Porath, J., Erson, B. (1974) Methods.Enzymol. 34, 108-140

(12) Creighton, T.E. (1970) Eur.J.Biochem. 13, 1-10

(13) Kirschner, K., Weischet, W.O., Wiskocil, R.L. (1975) 'Protein-Ligand Interaction' pp 27-44 (H.Sund, G.Blauer, eds.), W.deGruyter, Berlin, N.Y.

AFFINITY CHROMATOGRAPHY OF THE PYRUVATE DEHYDROGENASE COMPLEX USING TPP AS A LIGAND

Jaap Visser, Walter van Dongen and Marijke Strating

Dept. of Genetics, Agricultural University, Wageningen, The Netherlands

INTRODUCTION

The development of suitable affinity chromatography systems for the pyruvate dehydrogenase multienzyme complex serves different purposes. It may provide a rapid and mild isolation procedure of the enzyme complex from various sources, enabling evolutionary studies. Moreover, one may obtain a refined tool to study mutant enzyme properties. Biochemical genetics of the pyruvate dehydrogenase complex have been studied by Henning (1,2) and Guest (3,4) in *Escherichia coli*. Structural mutants and deletions are known for all three components of the complex. Therefore, *E.coli* was chosen as a model. The enzyme complex catalyzes the following overall-reaction:

$$\text{pyruvate} + \text{CoA} + \text{NAD}^+ \rightarrow \text{acetyl CoA} + CO_2 + \text{NADH} + H^+ \quad (1)$$

The complex consists of multiple copies of three different enzymes. A normal strategy in developing biospecific matrices is to use cofactors involved in the reaction. The first partial reaction is catalyzed by pyruvate decarboxylase (E_1); the enzyme is thiaminepyrophosphate and Mg^{2+} dependent. Lipoate acetyltransferase (E_2) contains a covalently bound cofactor *viz* lipoate and converts CoA into acetyl CoA. Lipoamide dehydrogenase (E_3) is a flavoprotein and requires NAD^+ as a cofactor in the reoxidation of reduced lipoate. 5'-AMP is an effector of the *E.coli* pyruvate dehydrogenase complex. Therefore, general ligand chromatography on the basis of the following cofactors can be considered: a) thiamin pyrophosphate, b) CoA, c) AMP and NAD^+.

Matrices of the type mentioned under b) and c) are still under investigation. Preliminary data indicate that the multienzyme complex interacts both with C^8 and N^6 substituted NAD^+-Sepharose 2B matrices (van Dongen and Visser, unpublished results). The isolated E_3 component, however, shows no biospecific retardation for any NAD^+ derivative tested thus far. Here, the use of thiamin pyrophosphate as affinity ligand will be reported. Thus far, TPP or TMP affinity columns have been used to purify a thiamine binding protein of *E.coli* (5), thiamine pyrophosphokinase of rat brain (6), and pyruvate oxidase of E.coli (7).

METHODS

Preparation of Affinity Matrices

Thiamine pyrophosphate - Sepharose 2B (Matrix I). Sepharose 2B was activated by CNBr (150 mg/ml of packed gel). Ethylenediamine (0.1 M) was coupled overnight in bicarbonate buffer at pH 10.0. After washing TPP (5 mg/ml) was added in the presence of a water soluble carbodiimide (3-fold molar excess). The coupling pH was maintained at pH 6.4 (overnight, at 4°C). The gel was stored in 50% glycerol at -15°C at pH 5.0 when not in use.

Thiochrome-Sepharose 2B (matrix II). Thiochrome was obtained from thiamine by an alkaline oxidation by CNBr (8). Sepharose 2B was activated by bisoxirane (9) and reacted in the dark with 4 mM thiochrome at $4^{o}C$ at pH 9.5 during 1-2 weeks. This results in a yellow-orange fluorescent matrix which is kept in the dark.

Thiamine pyrophosphate - Sepharose 2B (matrix III) Sepharose 2B was activated with CNBr (150 mg/ml of packed gel) and subsequently 1,3 diaminopropane (0.1 M in bicarbonate buffer, pH 9.7) was coupled and used to prepare a p-aminobenzyol derivative. Then the diazo-derivative was made (10) and used to react with thiamine pyrophosphate (7)

Enzyme Isolation and Assay

E.coli K 1-1 LR 8-13, kindly given by Dr. U. Henning, was used as enzyme source. The strain was grown according to Vogel (11) and pyruvate dehydrogenase complex was partially purified by Bio gel A-50 M chromatography as described before (11) The enzyme assay was measured at 25° according to Schwartz and Reed (12).

SDS Polyacrylamide Gel Electrophoresis

Stock solutions were made according to Laemmli (13). Routinely, 10% gels were used with a 3% stacking gel. Staining of the protein occurred with Coomassie Brilliant Blue R.

RESULTS

Matrices

The matrices have the following tentative structure.

I 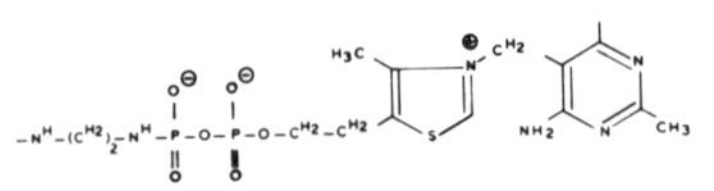

III 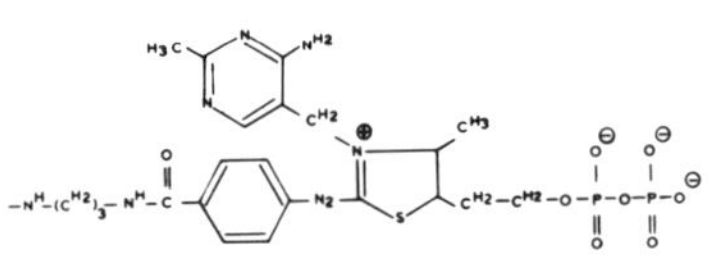

II 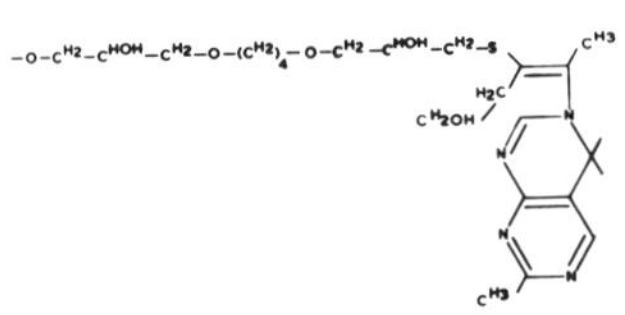

Since it is not known which cofactor modifications severely interfere with the binding of TPP to the pyruvate dehydrogenase complex, both matrix I and III were tested. Although thiochrome pyrophosphate is a better ligand in some TPP dependent enzymes like yeast pyruvate decarboxylase, the possibility to use thiochrome itself (matrix II) was explored first. A main disadvantage of matrix I is the poor stability of the ligand under operating conditions (100% release within 3 days at pH 7.0 $4^{o}C$). The lifetime is prolonged by other storage conditions (20% release at pH 5.0, $4^{o}C$)

Affinity Chromatography

Matrix I. The affinity of partially purified pyruvate dehydrogenase complex was tested on a small scale (1-2 ml bed volume). Matrix I is effective in binding the multienzyme complex. The following general observations were made testing this system.

a) The column has a capacity of 1.5 mg of enzyme per ml of gel when loaded at sufficiently low ionic strength. Normally used were 10-30 mM potassium phosphate buffers pH 7.0 which always contained 5 mM β -mercaptoethanol or 2 mM dithiothreitol.

b) The capacity is slightly reduced when magnesium ions are present in the elution buffer (2.5-5 mM). However, less aspecific protein is adsorbed.

c) The addition of the cofactor (1 mM TPP) or the $TPP\text{-}Mg^{2+}$ complex to the elution buffer did not elute the enzyme. Neither did preincubation with TPP prevent binding.

d) Elution is possible by an increase in ionic strength, either by an increase in phosphate buffer concentration or by the addition of NaCl. Recovery of the activity is almost quantitative (90-95%).

e) The matrix interacts also with the enzyme complex of various other prokaryotes e.g. Bacillus subtilis, Pseudomonas fluorescens and Azotobacter vinelandii.

The results of a larger scale experiment and the corresponding electrophoretic analysis are shown in Fig. 1 and indicate that the complex obtained is pure. The complex used did not contain contaminating α-ketoglutarate dehydrogenase activity. If a crude preparation containing both activities is applied copurification of the α-ketoglutarate dehydrogenase activity with the pyruvate dehydrogenase complex is not observed although other contaminants are now present.

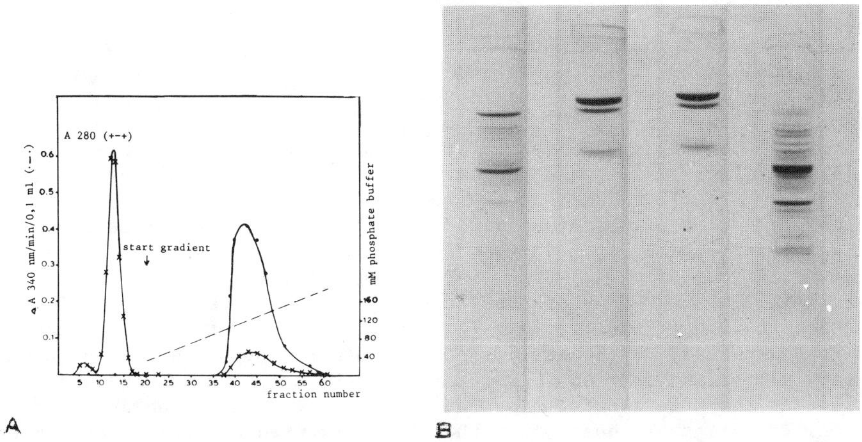

Fig. 1. A) Affinity chromatography pattern of pyruvate dehydrogenase complex on TPP-Sepharose 2B (matrix I): enzyme activity (• - •) and protein absorption (x - x). B) SDS-gel electrophoresis of starting material, enzyme fraction 42 and 45 and pooled eluate (from left to right).

Matrix II. Small scale experiments with crude extract indicate that in the absence of magnesium ions or at low concentrations of this ion (e.g. 0.1 mM) the pyruvate dehydrogenase complex and many proteins are bound to this matrix. At 5 mM $MgCl_2$ the interaction gains in specificity although the capacity is strongly decreased leading to the appearance of activity in the void volume. The retarded fraction consists mainly of pyruvate dehydrogenase complex. Larger scale purification starting with a partially purified pyruvate dehydrogenase complex is shown in Fig. 2 together with the electrophoretic data. From this it is clear that the complex can be purified to homogeneity.

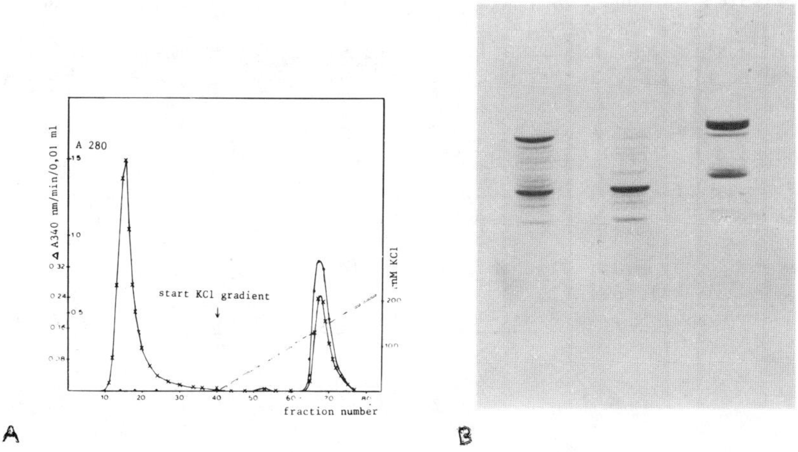

Fig. 2. A) Affinity chromatography pattern of pyruvate dehydrogenase complex on thiochrome - Sepharose 2B (matrix II): enzyme activity (. - .) and protein absorption (x - x). B) SDS gel electrophoresis of starting material, pooled eluate and active enzyme fraction (from left to right).

Matrix III. This matrix has not been found to interact with pyruvate dehydrogenase but matrices with a longer spacer arm (C_6) are more promising.

DISCUSSION

TPP, presumably bound through a phosphoamide linkage (5), interacts with the pyruvate dehydrogenase complex of E.coli and other bacteria tested. Attempts to apply this affinity adsorbent (matrix I) to the complex of eukaryotes thus far failed. Whether there is a complete lack in biospecificity remains to be seen. Reinforcement of the binding due to spacer or charge, unfavourable exchange kinetics or multiple interactions between the matrix and the large multienzyme complex (mol.wt. $8x10^6$) may all cause that may be higher TPP concentrations are required. Moreover, the matrix is biospecific for pyruvate decarboxylase apoenzyme and for a thiamin binding protein (5). TPP bound through a diazo linkage and a small spacer arm at position C_2, which is required for catalysis, does not function (matrix III). Thus, the pyrophosphate group does not seem to be the major moiety contributing to the binding. The positive result with thiochrome-Sepharose 2B is unexpected, particularly since the elution profiles of the enzyme complex are very similar to that on TPP-Sepharose (matrix I). An alternative explanation is that the free alcohol side chain has to be held responsible for the interaction. It has namely been shown that N-substituted alcohol amine Sepharose 2B (n = 2-5) resins bind the pyruvate dehydrogenase complex quite well (14). Studies with the isolated components and with mutants defective in E_1 and lacking E_1 are in progress in order to establish the nature of the interactions described for matrix I and II.

Acknowledgements: We appreciate the cooperation with Dick van Lith during part of this study, and the discussions with Dr. C.J. Bos. Mr. K. Knoop skilfully made drawings and photographs.

REFERENCES

1. Henning, U., Dennert, G., Hertel, R. and Shipp, W.S., Cold Spring Harb. Symp. quant. biol. 31, 227 (1966)
2. Henning, U., Dietrich, J., Murray, K.N. and Deppe, G. (1968) in Molecular Genetics, Springer Verlag pg. 223-236.
3. Guest, J.R. and Creaghan, I.T., J.Gen.Microbiol. 75, 197 (1973)
4. Guest, J.R. and Creaghan, I.T., J.Gen.Microbiol. 81, 237 (1974)
5. Matsuura, A., Iwashima, A. and Nose, Y., Biochem.Biophys.Res.Commun. 51, 241 (1973)
6. Wakabayashi, Y., Iwashima, A. and Nose Y. Biochim.Biophys.Acta 429, 1085 (1976)
7. O'Brien, T.A., Schrock, H.L., Russell, P., Blake, R. and Gennis, R.B., Biochim. Biophys Acta 452, 13 (1976).
8. Nishimune, T., Abe, M. and Hayashi, R. Biochim.Biophys. Acta, 279, 527 (1972)
9. Sandberg, L. and Porath, J., J.Chromatography, 90, 87 (1974)
10. Cuatrecasas, P., J.Biol.Chem., 245, 3059 (1970)
11. Vogel, O., Beikirch, H., Müller, H. and Henning, U., Eur.J.Biochem. 20, 169 (1971)
12. Schwartz, E.R. and Reed, L.J., Biochemistry, 9, 1434 (1970)
13. Laemmli, K.K., Nature 227, 681 (1970)
14. Visser, J. and Strating, M., in Proc. 4th Enzyme Engineering Conference (1977)

TEMPERATURE DEPENDENCE, ACTIVATION ENERGY AND ENTHALPY CHANGE OF THE BINDING PROCESS OF UDP-GALACTOSE 4'-EPIMERASE TO ITS IMMOBILIZED SUBSTRATE

Erich Haigis, Marie-Theres Häuptle and Richard Gitzelmann

Division of Metabolism, Department of Pediatrics, University of Zürich, Kinderspital, 8032 Zürich, Switzerland

The enzyme UDP-galactose 4'-epimerase catalyses the interconversion of UDP-galactose and UDP-glucose. In order to purify this enzyme from human red blood cells, we synthesized a biospecific affinity matrix by coupling the preferred substrate UDP-galactose to Sepharose-4B. Previous inhibitor studies with fragments and derivatives of the substrates revealed that the modification of the C-2 position of the hexose moiety did not impair recognition and binding by the enzyme (Ref. 1). The synthesized substrate-spacer-gel, shown in Fig. 1, contained 1.2 to 1.6 μmoles of ligand per ml of settled gel. Unspecifically bound enzyme was eluted from chromatography columns by adding 0.5 M KCl and the specifically bound UDP-galactose 4'-epimerase by the subsequent addition of 5 mM UDP-galactose to the buffer. A "mock affinity system" of galactosamine-spacer-Sepharose had no affinity for UDP-galactose 4'-epimerase at ionic strength of 0.3 M KCl. Nonspecific adsorption was thus ruled out.

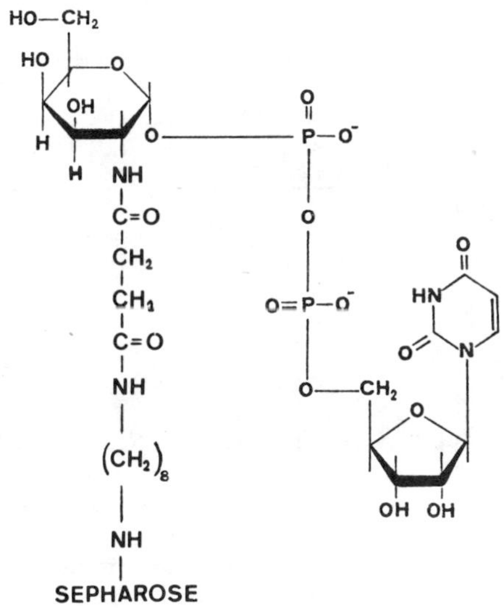

Fig. 1. The substrate-spacer-gel: UDP-galactosamine-succinyl-diaminooctyl-Sepharose

Since the immobilized UDP-galactose was available for epimerization at the C-4 of the hexose, conversion of the bound substrate to the bound product during chromatography had to be taken into account. Of the theoretically possible non-optimal conditions such as unsuitable pH or ionic strength, deforming buffer, lack of effector or coenzyme, low temperature etc. (Ref. 2), the latter was chosen to prevent the enzyme from being active. Affinity chromatography at lower temperature revealed that slightly more enzyme was bound specifically than at higher temperature. Thus, the binding of the enzyme to its immobilized substrate could be an exothermic reaction. If this were so, the equilibrium between bound and unbound enzyme would have to shift towards the enzyme-substrate complex at lower temperature.

In order to study the temperature dependence of the substrate binding reaction a new system using the substrate-spacer-gel in suspension was applied (Ref. 1). 0.5 ml of settled gel was diluted with 0.5 ml of buffer and suspended in 15 ml of enzyme solution. The concentration of bound substrate was 37.5 µM and thus close to the K_m of 20 µM for the free substrate. The rate of disappearance of free enzyme and the equilibrium were estimated by withdrawing aliquots of gel-free mixture and assaying unbound epimerase. Experiments were performed at different temperatures (-1, 2, 4, 10 °C). Figure 2 shows the calculated amount of bound enzyme as a function of time at different temperatures. The initial binding velocities followed first order kinetics. The obtained reaction constants (k) formed a straight line in the Arrhenius plot. The activation energy E_a calculated by applying the Arrhenius equation (Fig. 3) was 37 kJ/mole (9 kcal/mole).

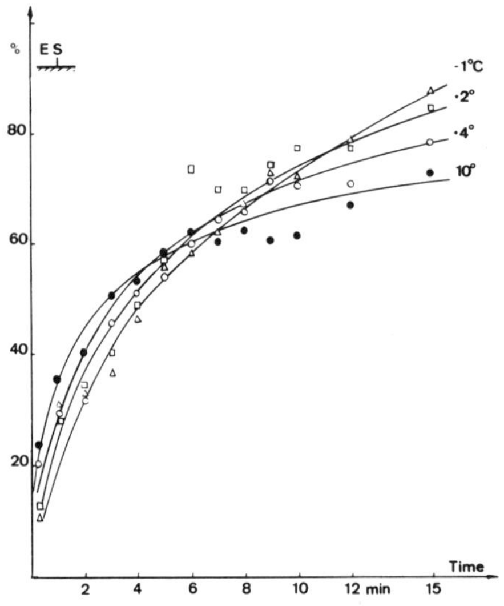

Fig. 2. Percentage of bound enzyme as a function of time at different temperatures

$$\ln \frac{k_2}{k_1} = \frac{E_a}{R} \left(\frac{T_2 - T_1}{T_2 \cdot T_1} \right)$$

Fig. 3. Arrhenius formula

The equilibria were reached after approx. 30 min. Since the concentration of the bound substrate was 10^4-fold higher than the total concentration of the enzyme in the system, equilibrium constants (K) at different temperatures were proportional to the quotient of bound to unbound enzyme. The logarithms of the quotients plotted versus the reciprocals of the absolute temperatures formed a straight line. Applying the Vant'Hoff equation revealed an enthalpy change ΔH of -42 kJ/mole (-10 kcal/mole) for the binding process. The straight line obtained for the equilibrium constants (lnK) in the Vant'Hoff plot lead us to assume that either the catalytic activity of the enzyme could be neglected or the affinity for the bound product was similar to that for the bound substrate. The calculated activation energy E_a and enthalpy change ΔH were in the range of estimations for other interactions between small molecules and enzymes (Ref. 3).

If the affinity of the enzyme for the bound substrate did not differ from that for the free substrate, a system such as the one described here could serve in elucidating the binding step of the enzyme reaction independently of the catalytic event. In the case of UDP-galactose 4'-epimerase, we found that the inhibition constant K_i for the substrate analogue N-acetyl-UDP-glucosamine was in the K_m range for UDP-galactose which indicated that the relatively large acetyl group at the C-2 position of the hexose did not reduce affinity. It could therefore be assumed that the spacer arm did not impair the affinity of the enzyme for the matrix-bound substrate, and that the calculated values for the activation energy and the enthalpy change represent reasonable approximations of those for the free substrate.

REFERENCES

1. E. Haigis, M.T. Häuptle, R. Gitzelmann, *in preparation.*

2. M.J. Harvey, D.B. Craven, C.R. Lowe, P.D.G. Dean, N^6-Immobilized 5'-AMP and NAD^+: Preparations and Applications, *Methods Enzymol. 34,* 242-253 (1974)

3. M.I. Page, Entropie, Bindungsenergie und enzymatische Katalyse, *Angew. Chem. 89,* 456-467 (1977)

AFFINITY CHROMATOGRAPHY OF CHYMOTRYPSIN ON SOYBEAN TRYPSIN INHIBITOR SEPHAROSE: APPLICATIONS IN GENETICS AND NUCLIDE LABELLING

Detlef Gabel*, Volker Kasche**, Herman Amnéus*** and Hans Lundqvist***

**Dept. of Chemistry, Universität Bremen, GFR*
***Dept. of Biology, Universität Bremen, GFR*
****Dept. of Physical Biology, The Gustav Werner Institute, Uppsala, Sweden*

Affinity chromatography of enzymes on high molecular weight inhibitors is a valuable tool to isolate and purify, in a onestep procedure, enzymes from unfractionated homogenates of organs and cells. Adsorption of the enzymes to the affinity gel is usually very strong, and by suitable gradient elutions, enzymes and isoenzymes can be recovered in electrophoretical purity.

Soybean trypsin inhibitor (STI) acts as an inhibitor for both trypsins and chymotrypsins from different sources. It can easily be bound to Sepharose by the CNBr method. When care is taken in the preparation of the adsorbant, minimizing its heterogeneity (Ref.1), two different chymotrypsins and three different trypsins can be recovered from homogenates of mouse pancreas which have been autactivated after starting activation with immobilized trypsin (Ref. 2). Under these conditions, only one enzyme is formed from each zymogen, representing the enzyme form formed under physiological conditions. By using inbred mouse strains, it was found that both chymotrypsins differed between the two strains. The following parameters were found to be changed: amino acid composition, enzymic parameters (K_M, k_{cat}), the molar extinction coefficient, the isoelectric point, the relative incorporation of ^{3}H-Leu into the different (three) chains (Ref. 2).

The differences observed between the strains were also manifested in off-springs of the cross-breeding between the two strains. Whereas the two basic enzymes were resolved by the affinity column, the two more acidic ones could be separated by subsequent isoelectric focussing in polyacrylamide gels.

Labelled, especially iodine-labelled, proteins are widely used in radioimmunoassay for specific and quantitative detection of minute quantities of proteins and other antigens (Ref. 3). Hereby, a previously purified protein is labelled with iodine by one of a variety of methods, freed from excess reagent and denatured material, and used assuming a constant specific radioactivity. We have applied a standard labelling method (labelling with ^{123}ICl (Ref. 4) to chymotrypsin in order to investigate further the properties of the labelled biologically active material. After reaction with ICl, the protein was freed from low-molecular weight material by gel chromatography on Sephadex G-100 and applied to an affinity column of STI-Sepharose. Non-adsorbed material was washed out, and desorption was carried out by a pH-gradient (1). The result of one separation is shown in Fig. 1.

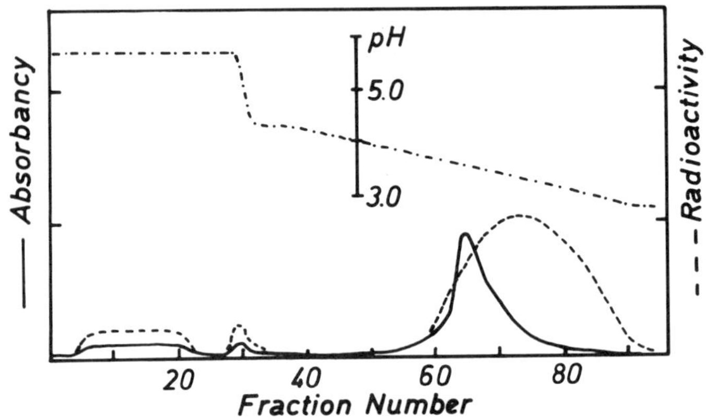

Fig. 1. Affinity chromatography of ^{123}I-labelled chymotrypsin on STI-Sepharose

As can be seen, relatively little material was destroyed during the labelling procedure. However, the desorption showed that the radioactivity did not follow the absorbancy of the protein, resulting in different specific radioactivity (expressed in radioactivity per absorbancy unit). Different fractions were further analyzed with respect to the following properties: molar extinction coefficient, K_M, and k_{cat}.

It was found that in addition to the elution profile (Fig. 1) all three parameters were affected by iodination. The molar extinction coefficient (measured by active site titration) increased with the degree of iodination, reaching, at high iodination levels, about twice the value for the native enzyme. Thus, absorbance did no longer correlate linearly with the amount of protein present. The catalytic constant k_{cat} increased by about 30 % over the native value at high iodination levels, whereas K_M increased to about three times of its original value.

Chymotrypsin was labelled in an alternative way by first preparing the soluble chymotrypsin·STI complex, labelling with ICl in the usual manner and dissociation of the complex (after gel filtration to remove unreacted reagent) by competitive affinity chromatography on trypsin-Sepharose, where STI is retained on the column. The effects of the labelling on K_M and molar extinction coefficient could be reduced considerably, but in this case k_{cat} was diminished to about half its original value. The fractionation pattern on STI-Sepharose was, however, essentially unchanged, with smaller amounts of highly labelled protein eluting after the major part of the enzyme.

These results will have the following implications on radioimmunoassay. This system (chymotrypsin·STI-Sepharose) could be used to determine, by competition between a known amount of labelled enzyme with an unknown amount of unlabelled enzyme, the concentration of chymotrypsin in a sample. The labelled enzyme was here found to consist of two fractions, one with low specific radioactivity, having about the same association constant as the unlabelled enzyme, and one with high activity with a larger association constant. Then the competition for the available binding sites on the gel will be primarily between the unlabelled enzyme (to be determined quantitatively) and the weakly labelled enzyme of the standard, leaving the radioactivity incorporated into the gel relatively constant (5). Radioactivity will then no longer correspond directly to the amount of standard enzyme bound, and because the unlabelled enzyme will compete rather infavorably with the highly radioactive material, the establishment of calibration curves will not permit a very accurate determination of the unknown protein.

As a conclusion, preparation of labelled proteins for radioimmunoassay should be followed by investigation of the heterogeneity and properties of the obtained product, in order to improve the reliability of such assay methods. Moreover, the results call for the development of more specific and selective labelling methods.

In summary, soybean trypsin inhibitor bound to Sepharose has been shown to be a valuable tool not only for the all-or-none preparation of chymotrypsins and trypsins from biological material, but also for the detection and investigation of samples which by genetic, physical or chemical factors are heterogeneous.

ACKNOWLEDGEMENTS

Technical assistance by Mss. U. Penttinen and I. Hocke is gratefully appreciated. Financial support has been received from the Swedish Atomic Research Council and the Swedish Cancer Society.

REFERENCES

(1) H. Amnéus, D. Gabel and V. Kasche, Resolution in Affinity Chromatography. The Effect of the Heterogeneity of Immobilized Soybean Trypsin Inhibitor on the Separation of Pancreatic Proteases, *J. Chromatogr.* 120, 391 (1976).

(2) V. Kasche, H. Amnéus, D. Gabel, and L. Näslund, Rapid Zymogen Activation and Isolation of Serine Proteases from Individual Mouse Pancreas by Affinity Chromatography. Genetical Heterogeneity of Chymotrypsins of *Mus musculus*, *Biochim. Biophys. Acta* 490, 1 (1977).

(3) *Radioimmunoassay Methods* (1971) (K. E. Kirkham & W. M. Hunter, eds.) Churchill Livingstone, Edinburgh & London.

(4) W. M. Hunter and F. C. Greenwood, Preparation of Iodine-131 Labelled Human Growth Hormone of High Specific Activity, *Nature* 194, 495 (1962).

(5) V. Kasche, Specific Protein-Protein Interaction and its Application in Studies on Radiation-Induced Protein Midification, *Acta Univers. Upsal.* 2, 1 (1971).

REVERSIBLE IMMOBILIZATION OF VIBRIO CHOLERAE NEURAMINIDASE BY BIOSPECIFIC SORPTION

Dietmar Ziegler, Gunda Keilich and Reinhard Brossmer

Institut für Biochemie II (Med. Fakultät), Universität Heidelberg, GFR

ABSTRACT

It has been established that N-(4-nitrophenyl)oxamic acid coupled with 1.6-diaminohexyl Sepharose 4B is a specific adsorbent for V. cholerae neuraminidase. The enzyme immobilized by non-covalent linkages retains its full activity. Natural and synthetic substrates were hydrolyzed in 50 mM sodium acetate buffer pH 5.0, containing 2 mM calcium ions; however, with none of the substrates (up to 50 µmol/ml gel) any enzyme activity was eluted from the adsorbent. The adhesion of neuraminidase is reversible; it may be desorbed by sodium acetate buffer pH 5.0, containing 1.0 M sodium chloride.
A simple method for the preparation of asialo-glycoproteins is presented, whereby the immobilized neuraminidase may be used continuously without loosing its activity.

INTRODUCTION

Neuraminidases have been known for over thirty years but their precise biological function is yet unknown. In order to elucidate more specifically what part this important enzyme takes in metabolism and in all aspects of cell behaviour an enzyme preparation free of contaminating activities was needed. In addition, this enzyme has become an increasingly useful tool for glycoprotein research and for cell biology (Ref. 1). The use of a pure as well as immobilized neuraminidase eliminates the interference due to other impurities and the technical problem of product and enzyme separation.
While investigating the relationship between the structure and the function of V. cholerae neuraminidase a series of N-phenyl oxamic acid derivatives have been

---

Enzyme: neuraminidase, acylneuraminyl hydrolase (EC 3.2.1.18)
Abbreviations: NNPOA-AHS 4B, N-(4-nitrophenyl)oxamic acid coupled with 1.6-diaminohexyl Sepharose 4B; EDC, 1-(3-dimethylaminopropyl)-3-ethyl-carbodiimide hydrochloride; NeuAc, N-acetyl-D-neuraminic acid.

found to be inhibitory whereby those containing a nitro group have been most effective (Ref. 2).

Following these observations N-(4-nitrophenyl)oxamic acid was coupled with 1.6-diaminohexyl Sepharose 4B to produce NNPOA-AHS 4B, a highly specific adsorbent for neuraminidase (Ref. 3). In this report we show that the biospecific sorption of this enzyme may not only be used for its purification but also for its reversible immobilization.

MATERIALS AND METHODS

The preparation of NNPOA-AHS 4B has been described in a previous report (Ref. 3). Crude neuraminidase was a gift from Behringwerke AG, Marburg, BRD; the enzyme was purified following the procedure described earlier (Ref. 3). The enzyme activity was measured by using a standard assay procedure; 1 milli-unit (mU) expresses the release of 1nM NeuAc/min. The release of NeuAc was determined by the method of Warren (4) with crystalline NeuAc as standard.

The following substrates were used: Benzyl α-ketoside of NeuAc; N-acetyl-neuraminosyllactose; α_1-acid glycoprotein as prepared by K. Schmid et al. (5); collocalia mucin I and II prepared according to G. Keilich et al. (6).

Preparation of immobilized neuraminidase: a sample of enzyme was adsorbed on a column of NNPOA-AHS 4B (1-5 ml) (3000-3500 mU/ml adsorbent) in 50 mM sodium acetate buffer pH 5.0, 2 mM $CaCl_2$, at 4°C; if a crude enzyme preparation was used, contaminating proteins were separated by washing the column repeatedly with the same buffer until the absorbance at 280 nm was approaching zero.

Cleavage of the substrates: the substrates dissolved in buffer (20 mg/ml) were applied to the column containing the immobilized enzyme (up to 40 µmol ketosidically bound NeuAc/ml gel) in 50 mM sodium acetate buffer pH 5.0, 2 mM $CaCl_2$; the substrate was left at 37°C up to 30 min. Subsequently, the products were eluted and the released NeuAc was measured.

Preparation of asialo-glycoproteins: the glycoproteins were incubated under the conditions given above except the time was extended up to 2 hrs.; the eluate was fractioned and those fractions showing an absorbance at 280 nm were pooled. After separation of NeuAc by dialysis or ultrafiltration the asialo-glycoproteins were concentrated by lyophilisation.

RESULTS AND DISCUSSION

NNPOA-AHS 4B was obtained by coupling N-(4-nitrophenyl)oxamic acid with 1.6-diaminohexyl Sepharose 4B using EDC (Ref. 3) as shown in Fig. 1.

The effectiveness of the affinity adsorbent in the purification of neuraminidase was pointed out. With NNPOA-AHS 4B the enzyme activity was completely retained and in addition, it was separated from the bulk of the protein when washed with 50 mM sodium acetate buffer, pH 5.0. The neuraminidase activity was eluted as a sharp band using the same buffer containing 1.0 M sodium chloride. The purification was found to be 80-fold and the recovery was more than 9o%.

Due to the fact, that neuraminidase is only weakly retained on a column of 1.6-diaminohexyl Sepharose 4B and it is not adsorbed at all by Sepharose 4B (Ref. 3) the adsorption is effected particularly by the ligand N-(4-nitrophenyl)oxamic acid. In addition, N-acetylneuraminate pyruvate-lyase, an enzyme that also accepts NeuAc as substrate, was found to pass through a column of NNPOA-AHS 4B without being significantly retained. This result provides further evidence for the assumption that this material is a specific adsorbent for V. cholerae neuraminidase.

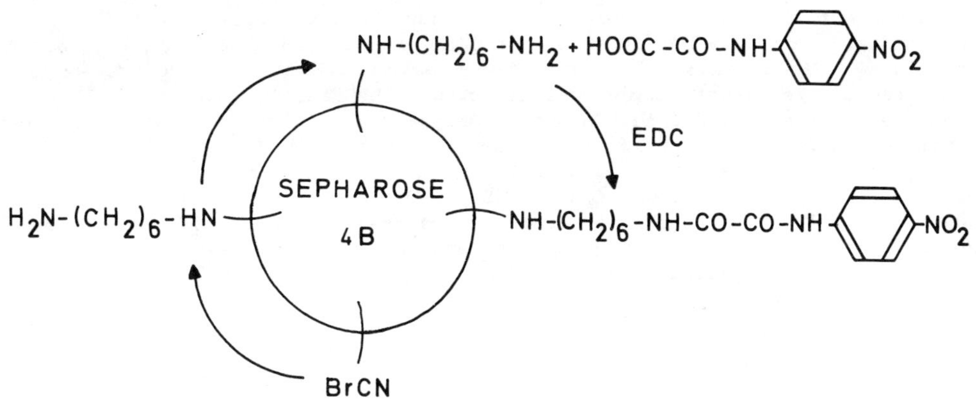

Fig. 1 Schematic presentation of NNPOA-AHS 4B and its preparation

Furthermore, two questions evolved from the tight adhesion of the enzyme: First, whether neuraminidase remains active when it is immobilized by non-covalent linkages; secondly, whether the immobilized enzyme is stripped off the column by incubation with other substrates. Therefore, to settle these questions natural and synthetic substrates (up to 50 µmol ketosidically bound NeuAc/ml gel) were incubated with the immobilized enzyme in 50 mM sodium acetate buffer pH 5.0 containing 2 mM calcium ions for 2 hrs at 37°C (see Table 1).

TABLE 1 Cleavage of Substrates by Immobilized Neuraminidase

Substrates	NeuAc content (%)	Release of bound NeuAc by neuraminidase[+] immobilized[++] 2 hrs	in solution[++] 24 hrs
N-acetylneuraminosyl lactose	48.0	87.9	95.0
α_1-acid glycoprotein	11.8	94.9	100.0
collocalia mucin I	16.5	100.0	100.0
collocalia mucin II	17.8	100.0	100.0
benzyl α-ketoside of NeuAc	77.0	93.8	98.3

+ for detail see Material and Methods
++ values expressed in per cent of NeuAc content

All the substrates tested so far were almost completely hydrolyzed. Even with higher concentration of the substrates no neuraminidase activity was eluted from the adsorbent. The release of bound NeuAc by the immobilized enzyme seems to be about 10-times faster than by the soluble neuraminidase. This fact is probably due to the high capacity of the affinity adsorbent. When NNPOA-AHS 4B was saturated with the enzyme, about 3500 mU of neuraminidase were adsorbed on 1 ml gel, i.e. about 400 mU per mg dry weight.

The immobilized enzyme was found to be stable for weeks at 4°C. In addition, it retains its activity during several runs of enzymatic substrate hydrolysis. Neuraminidase could only be eluted by using 50 mM sodium acetate buffer, pH 5.0, containing 1.0 M sodium chloride as mentioned above. After employing the enzyme 6 times the recovered activity was still found to be 85%.

Immobilized neuraminidase allows a simple preparation of asialo-compounds without the difficult problem of enzyme and product separation. Soluble asialo-glycoproteins may best be prepared by the procedure indicated in "Material and Methods". The incubation up to 2 hrs. was found to be sufficient. With the glycoproteins shown in Table 1 the yield of the asialo-derivatives was about 90%. Therefore, there is no adsorption of either the substrates or the products.

The availability of a neuraminidase immobilized by non-covalent linkages offers an attractive tool for the studies on glycoproteins and on cell systems.

REFERENCES

(1) A. Gottschalk and R. Drzeniek, Neuraminidase as a tool in structural analysis; in Gottschalk, A. (1972) Glycoproteins, Elsevier Publishing Company, Amsterdam.

(2) R. Brossmer, G. Keilich, and D. Ziegler, Inhibition studies on V. cholerae neuraminidase, Hoppe-Seyler's Z. Physiol. Chem., 358, 391 (1977).

(3) R. Brossmer, D. Ziegler, and G. Keilich, Selective isolation of V. cholerae neuraminidase using an immobilized 4-(nitrophenyl)oxamic acid, Hoppe-Seyler's Z. Physiol. Chem., 358, 397 (1977).

(4) L. Warren, The thiobarbituric assay of sialic acids, J. Biol. Chem., 234, 1971 (1959).

(5) K. Schmid, M. B. Mac Nair, and A. F. Bürgi, The chromatographic separation and purification of acidic proteins on carboxylated ion exchange resins, J. Biol. Chem., 230, 853 (1958).

(6) G. Keilich, E.-Ch. Kopp, and R. Brossmer, Charakterisierung von Collocalia mucoid, Hoppe-Seyler's Z. Physiol. Chem., 358, 257 (1977).

AFFINITY CHROMATOGRAPHY OF CARBOHYDRATE BINDING PROTEINS. PREPARATION AND USE OF SOME POTENTIAL LIGANDS

Marc Claeyssens, Francois Deleyn, Eric Saman and Clement K. De Bruyne

Laboratorium voor Algemene en Biologische, Scheikunde, Rijksuniversiteit Gent, K. L. Ledeganckstraat 35, B-9000 Ghent, Belgium

ABSTRACT

The preparation of several ligands derived from carbohydrates (glycosides, 1-thio-glycosides and glycosylamines) is described. Some novel synthetic methods are introduced. The use of these ligands in affinity chromatography experiments of ß-xylosidases from fungal or bacterial origin is discussed.

METHODS

Synthesis of Carbohydrate Ligands

1. ω-Amino-alkyl ß-D-xylopyranosides (A) were prepared starting from the corresponding ω-chloro-derivatives (1), substituting azide for the halogen (in dimethylforamide as solvent, 2h at 80°) and catalytic reduction ($Pd/BaSO_4$ in methanol solution, 12h, room temperature and H_2 at 2 kg/cm^2).

2. p-Aminophenylalkyl 1-thio ß-D-xylopyranosides (B) were prepared as described (2) starting from 2,3,4-tri-O-acetyl ß-D-xylopyranosylmercaptan and p-nitrobenzylbromide or p-nitrophenylethylbromide.

3. 2-Imino-2-methoxyethyl 1-thio ß-D-xylopyranoside (D) was prepared from 2,3,4-tri-O-acetyl xylopyranosylmercaptan and chloroacetonitrile, yielding cyanomethyl 2,3,4-tri-O-acetyl 1-thio ß-D-xylopyranoside (C) and subsequent sodium methoxide catalysed formation of compound D as described by Lee et al. (3).

4. 5'-Amino-pentyl 1-thio ß-D-xylopyranoside (E) was prepared from 2,3,4-tri-O-acetyl xylopyranosyl mercaptan and carbobenzoxy 5-amino-pentyliodide.

5. ß-D-xylopyranosylamine (F) and 2' amino-acetyl ß-D-xylopyranosylamine (G) Ligand F was prepared as described (4) and the N-chloroacetyl ß-D-xylopyranosylamine was obtained by acylation (DMF, 20°, 12h) of the amine F with chloroacetylanhydride (Fluka). Substitution by azide and subsequent reduction as under 1.

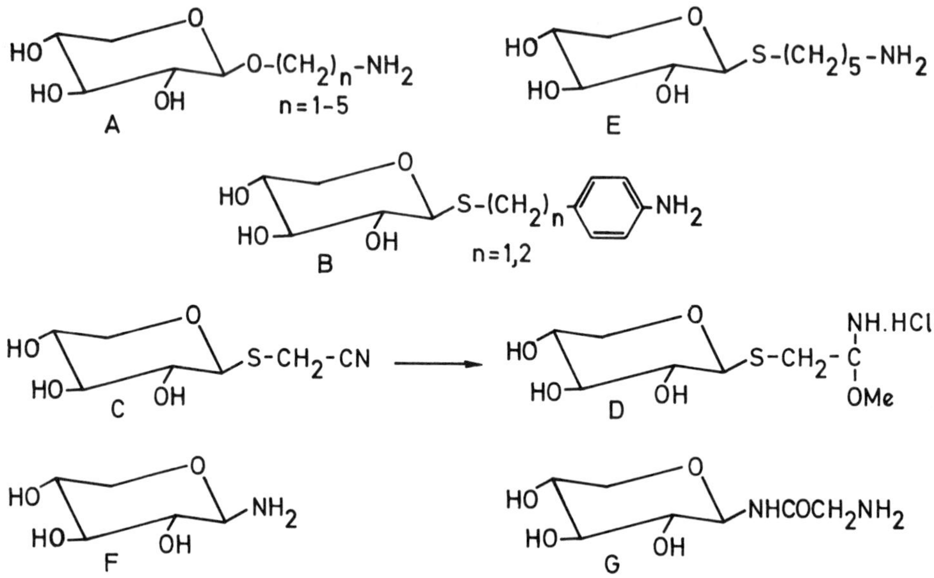

Coupling to Matrix

Compounds A, B, E, F and G were coupled directly to Sepharose 4B (Pharmacia) as described (5). Compound D was reacted with AH-Sepharose (Pharmacia) essentially in the same manner as Lee et al. (3) attach 2-imino-2-methoxy ethyl glycosides to proteins. The imidate/AH-Sepharose-ratio was 400 (AH-Sepharose expressed in amino content). No free amine groups could be detected after reaction (in borate buffer, pH 9.0, 20°, 24h) by the trinitrobenzenesulfonic acid reagent.

ß-Xylosidase Preparations (E.C. 3.2.1.21)

Crude preparations of ß-D-xylosidase from Bacillus Pumilus and enzymic assays were performed as reported (6). ß-D-Xylosidase from Penicillium wortmanni was obtained as described (7) and contains a ß-D-glucosidase as major contaminant.
The enzyme activity was measured using the p-nitrophenyl glycosides at pH 7.2 for the bacterial enzyme and pH 3.3 for the fungal enzyme. The molecular weight (~ 100 000 daltons) and isoelectric point (4.6-5.0) for both enzymes are similar.

RESULTS

1. ß-D-Xylosidase from Bacillus pumilus

The results of binding and desorption experiments at 4°C with different immobilized ligands are shown in table 1.

TABLE 1 Purification of β-D-Xylosidase from *Bacillus pumilus*

covalently bound ligand	association constant in solution (M^{-1})	matrix	binding	elution: 1 M xylose	elution: 0.1 M KCl
1-thio-xylosides					
amidine	100-150	AH-Sepharose	+	-	+
ω-amino-alkyl		Sepharose-4B	-	-	-
p-aminophenyl	160	"	-	-	-
p-aminobenzyl	370	"	+	+	-
p-aminophenylethyl	2000	"	+	+	-
N-Xyloside					
ω-amino-acyl	260	"	-	-	-
O-Xyloside					
ω-amino-alkyl	50-900	"	+	+	-

2. ß-D-Xylosidase from Penicillium wortmani

Only adsorbents with B were effective, desorption with high ionic strength buffers does not separate the enzymes present (ß-xylosidase and ß-glucosidase). These activities could be resolved by hydroxyl apatide chromatography (7).

DISCUSSION

Purification by affinity chromatography of the ß-xylosidase from Bacillus pumilus is successful when ω-amino alkyl ß-D-xylopyranosides (A) or p-aminophenyl alkyl 1-thio ß-D-xylopyranosides (B) are used as ligands. The corresponding ω-amino alkyl 1-thio derivative (E) is ineffective and this could be explained by the observed aglycon specificity of the enzyme, which is distinctly different for these 1-oxygen and 1-thio-xylosides (8).
The results obtained with the amidine derivative (D) point to ion-exchange phenomena when the crude enzyme preparation is applied to columns containing compound D covalently bound to AH-Sepharose (Table 1). Similar effects seem to be predominant when the fungal ß-xylosidase is used in affinity chromatography experiments.
This illustrates the fact that it is difficult to devise, for this class of enzymes, a general ligand affinity gel, and corroborates similar observations for ß-galactosidases (9).

REFERENCES

1. C.K. De Bruyne and F.G. Loontiens, Nature 204, 396 (1966).
2. M. Claeyssens and C.K. De Bruyne, Carbohydrate Res. 22, 460 (1972).
3. Y.C. Lee, C.P. Stowell and M, Kranth, Biochemistry 15, 3956 (1976).
4. H.S. Isbell and H.L. Frush, J. Org. Chem. 23, 1309 (1958).
5. M. Claeyssens, H. Kersters-Hilderson, J.P. Van Wauwe and C.K. De Bruyne, FEBS-Letters 11, 336 (1970).
6. H. Kersters-Hilderson, F.G. Loontiens, M. Claeyssens and C.K. De Bruyne, Eur. J. Biochem. 7, 434 (1969).
7. F. Deleyn, M. Claeyssens, C.K. De Bruyne and J. Van Beeumen, Can. J. Biochem., in press.
8. H. Kersters-Hilderson, M. Claeyssens, F.G. Loontiens, A. Krynski and C.K. De Bruyne, Eur. J. Biochem. 12, 403 (1970).
9. P. O'Carra, S. Barry and T. Griffin, Biochem. Soc. Transactions, 1973, 289.

DIRECT COUPLING OF REDUCING OLIGOSACCHARIDES TO AMINOHEXYL-SEPHAROSE: PURIFICATION OF α,α-TREHALASE FROM *ARTEMIA SALINA*

Mario Bergami and Marcello G. Cacace

Institute of General Physiology, Chair of Molecular Biology, University of Rome and Laboratory of Molecular Embryology, 80072 Arco Felice, Naples, Italy

ABSTRACT

Reducing oligosaccharides can be covalently linked to the spacer arm of aminohexyl Sepharose by reductive amination The reaction is carried out in the presence of sodium borohydride in aqueous medium and under mild conditions. An affinity column containing lactose-coupled Sepharose was used for the purification of α,α-trehalase from *Artemia salina* embryos.

INTRODUCTION

Sugars may be covalently attached to an insoluble matrix by i) previous conversion to aldonic acid (1) or amino (2-4) derivatives and ii) reaction with epoxy-activated Sepharose (5). The latter method, although not requiring any prior modification of the carbohydrate, allows non-specific reaction of hydroxyl groups which could hinder correct recognition of the ligand. Direct coupling of reducing sugars to aminoethyl–polyacrylamide gels by reductive amination with cyanoborohydride has been reported (6). The need for both specific linkage of the ligand and the interposition of a suitable spacer group, prompted us to search for a method of covalently coupling reducing oligosaccharides to a gel containing a spacer arm. Lactose-coupled Sepharose was successfully used for the purification of α,α-trehalase (α,α-trehalose 1-D-glycohydrolase, EC 3.2.1.28) from *Artemia salina* embryos.

METHODS

Aminohexyl Sepharose 4B was obtained from Pharmacia Fine Chemicals, sodium borohydride from Aldrich Chemical Co., lactose monohydrate and trehalose dihydrate from Fluka AG, D-glucose-1-^{14}C lactose (16 mCi/mmol) from the Radiochemical Centre, Amersham. *Artemia salina* cysts were from San Francisco Bay Brand Ind. All other chemicals were reagent grade. Labelled lactose, covalently bound to Sepharose, was measured after solubilizing the gel in 2M HCl at 100 °C for 2h; 10 μl aliquots were spotted onto glass filter disks and the latter counted in a toluene-based scintillation fluid. α,α-trehalase activity was determined by incubating the enzyme for 60 min at 40 °C in 0.5 ml of 0.1 citrate buffer, 0.5 mM $MgCl_2$, 30 mM trehalose, pH 5.6. After stopping the reaction in a boiling bath for 5 min, glucose was determined fluorometrically according to Lowry and Passonneau (7). One trehalase unit is the amount of enzyme which hydrolizes 1 μmol of trehalose per min under the conditions described above. Protein was determined by the method of Lowry et al.(8). Galactose, the enzymatic hydrolysis product of covalently-bound lactose, was searched for in the eluates from columns containing absorbed trehalase by paper chromatography as described by Gray (6).

RESULTS

Coupling of Sugar to Aminohexyl Sepharose

All operations were carried out at 2-4 °C. The gel (6-10 μmol $-NH_2$/ml swollen gel) was washed and equilibrated with 0.2 M phosphate buffer, pH 7.0. 200 mg of lactose were added to 20 ml of loosely packed gel in a stoppered flask and the slurry gently shaken for 30 min; 0.1 g of sodium borohydride were then added. After 6 h a second addition was made and the reaction allowed to proceed for 22h. During the reaction, the slurry was continuously shaken. The extent of the re-

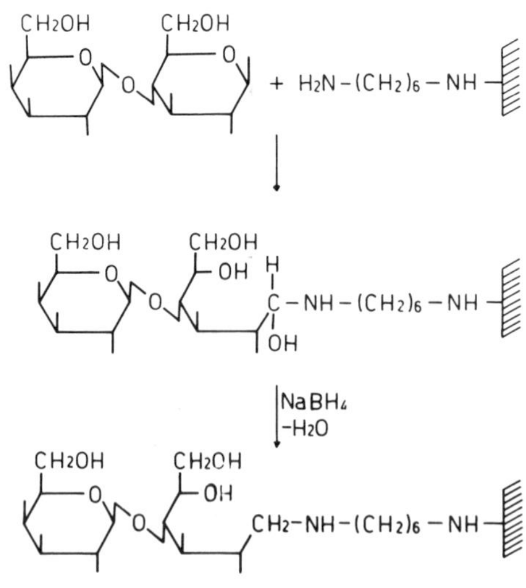

Fig.1. Coupling of lactose to aminohexyl Sepharose

action was measured by coupling ^{14}C-lactose (5 mCi/mmol) to 1 ml of aminohexyl Sepharose under the same conditions and by measuring the incorporated radioactivity in the thoroughly washed gel. 1.6 μmol of lactose per ml of swollen gel were found to be covalently bound. Figure 1 shows a scheme of the coupling reaction Aminohexyl Sepharose treated with sodium borohydride in the absence of lactose was used as a control for affinity chromatography experiments.

Affinity Chromatography of α,α-Trehalase

The control and the lactose-coupled gels were packed in 1.5 x 4 cm columns cooled at 2 °C and equilibrated with 25 mM phosphate buffer, 5 mM dithiotreitol, 1 mM EDTA, 5 mM $MgCl_2$, 20% glycerol (v/v), pH 6.0 (buffer A). The column used in the experiment reported in Fig.1c was equilibrated in buffer A, pH 7. Ten ml of a partially purified (34-fold over the crude extract, specific activity 3.65 U/mg)

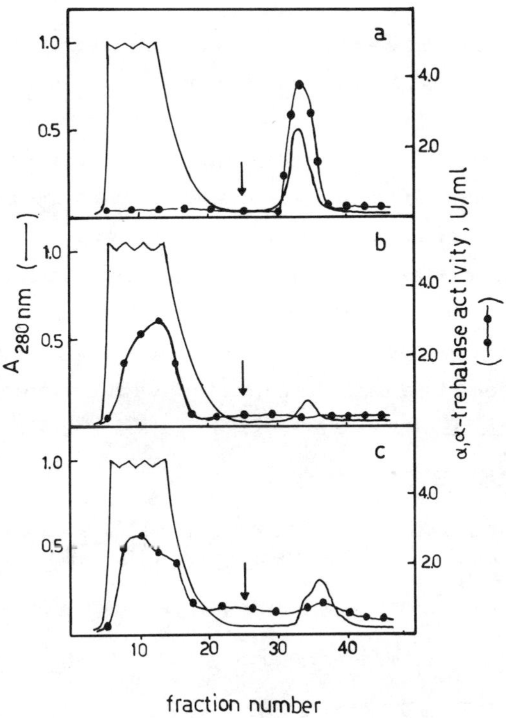

Fig.2. Affinity chromatography of α,α-trehalase

Columns (1.5 x 4 cm) containing lactose-coupled aminohexyl Sepharose 4B (panels a and c) and aminohexyl Sepharose 4B (panel b) were loaded with 10 ml of a partially purified (3.65 U/mg protein) α,α-trehalase preparation from Artemia salina cysts. Equilibrating buffer was buffer A, pH 6.0, in panels a and b, and buffer A, pH 7.0, in panel c. Elution was carried out with buffer A, 0.5 M NaCl, pH 6.0, at the position indicated by the arrow. Flow rate was 30 ml/hr and fraction volume was 0.9 ml. Jacketed columns cooled at 2 °C were used.

trehalase preparation were loaded onto the columns. Elution was carried out with buffer A, 0.5M NaCl as indicated in Fig.2. The specific activity of α,α-trehalase after the affinity chromatography step was 20.5 U/mg with a 5-fold purification. This value was comparable to that obtained after two steps of the alternative purification procedure formerly employed, which included gel permeation and ionic exchange chromatography.

DISCUSSION

Trehalase is highly specific for its natural substrate, however it can also utilize lactose at 20% efficiency(9). In fact, the enzyme was absorbed by the lactose-coupled Sepharose at a pH value close to the optimum (pH 5.6) for enzymic activity(Fig.2a). Nevertheless, under these conditions, no lactose hydrolysis was observed even when the enzyme was left on the column for as long as 2h. As shown in Fig.2b, the enzyme was not retained by the aminohexyl Sepharose column. It can therefore be excluded that a non-specific interaction with the hydrophobic spacer arm is primarily involved in the absorption of trehalase, at least under our experimental conditions. No binding was observed with lactose-coupled Sepharose at pH 7.0 (Fig.2c). The reduced affinity for the derivatized gel at a pH significantly different from the optimum value for enzymic activity indicates that trehalase biosorption results mainly from specific substrate-enzyme interaction. The use of epoxy-activated Sepharose, although permitting the preparation of derivatized gel with a spacer arm, suffers from the drawback of high pH for the coupling reaction; under these conditions many labile ligands could give undesirable side reactions. The present method could represent a valid alternative for coupling reducing sugars to an insoluble matrix with an interposed spacer arm and under mild conditions.

REFERENCES

(1) Hayes, C.E. and Goldstein, E J.(1974) J.Biol.Chem. 249, 1904-1914.

(2) Jeffrey, A.M., Zopf, D.A. and Ginsburg, V. (1975) Biochem.Biophys.Res. Comm. 62, 608-613.

(3) Rafestin, M.E., Obrenovitch, A., Oblin, A. and Monsigny, M. (1974) FEBS Letters 40, 62-66.

(4) Harpaz, N., Flowers, H.M. and Sharon, N. (1974) Biochem.Biophys.Acta 341, 213-221.

(5) Vretblad, P. (1974) FEBS Letters, 47, 86-89.

(6) Gray, G.R. (1974) Arch.Biochem.Biophys. 163, 426-428.

(7) Lowry, O.H. and Passonneau, J.V. (1972). A Flexible System of Enzymatic Analysis pp.174-175. Academic Press, New York.

(8) Lowry, O.H., Rosebrough, N J., Farr, A L. and Randall, R J. (1951) J.Biol.Chem. 193, 265-275.

(9) Ballario, P., Bergami, M., Cacace, M.G., Scala, F. and Silvestri, L. Submitted for publication.

PURIFICATION OF PECTINESTERASE FROM *ASPERGILLUS NIGER* BY CHROMATOGRAPHY ON AN AMIDE DERIVATIVE OF THE CROSS-LINKED PECTATE

M. A. Vijayalakshmi, Christine Bonaventure, Daniel Picque and Emile Segard

Institut de Technologie des Surfaces Actives 60206 Compiègne

We, here, report a preliminary work on the purification of Pectinesterase (PE), (3.1.1.11) activity from *Aspergillus Niger* commercial enzymes. PE saponifies the methyl ester groups of the esterified polygalacturonic acid (pectin) present in the fruits and fruit juices. This enzyme plays an important role in initiating the degradation of pectin and hence of interest in the clarification of fruit juices.

The classical methods of protein purification such as precipitation with Ammonium sulfate, gel filtration and ion exchange chromatography on Sephadex ion exchangers do not give satisfactory results and are rather tedious.

Actually supports for a selective purification of PE are not available, though REXOVA-BENKOVA et al (1972) have reported the specific adsorption of Polygalacturonase (PG) (pectate-hydrolase) (3.2.1.15) on the cross-linked pectate. We now report a derivative of this cross-linked pectate which adsorbs rather specifically the PE activity of the fungal origin (*Aspergillus Niger*).

It is well known that PE is quite specific for the galacturonide methyl esters and it has been suggested that the enzyme acts on the ester groups next to free carboxyl groups (Solms and Deuel 1955).
Lee and Mc Millan (1970) have reported that at least 50 % of the tomato PE activity is initiated near the reducing ends of the highly esterified pectin molecules. Based on these hypothesis and our preliminary studies of PE inhibition, we have tried to prepare derivatives of the cross-linked pectate which could selectively adsorb PE.

MATERIALS AND METHOD

Cross-linked pectate is prepared according to a modification of the method described by REXOVA-BENKOVA et al (1972) from Genu Pectin-rapid set (Kobenhavns Pektinfabrik, Denmark). Crude enzyme preparations are gifts from Rapidase SA (Seclin, France) and from Rohm GmbH (Darmstadt, W.Germany). The substrate Pectin (methyl galacturonic acid) is obtained from Sigma Chemical Co. All other reagents are from Merck Chem. Co. (reagent grade).

Enzyme assays

The PE activity was measured by the titration of the liberated COOH groups with 0.02 M NaOH (Metrohm pH stat) at pH 4,5 and at 25°C, from a 0.5 % pectin solution in 0.1 M NaCl. One unit PE = 1 m.eq of NaOH per minute. The protein content is monitored by the absorption at 280 nm.

Preparation of amide derivative

One gram of dry cross-linked pectate prepared according to Vijayalakshmi et al (1977) is suspended in 10 ml of absolute Methanol saturated with ammonia. After 4 days of stirring at 4°C, the suspension is decanted and washed on glass filter with a mixture of isopropanol-Hcl (5 % Hcl) and then with 70 % isopropanol and then with 96 % alcohol. The product is dried at room temperature and the ligand concentration is expressed as the nitrogen content per 100 g of the dry gel. The nitrogen content is measured by the Kjeldahl method (Vogel 1975).

Elution of the enzyme

The Rapidase C_{10} -pectic enzyme- non-retained peak-on a cross-linked pectate column, ie. containing the PE activity and a little of PG activity and other proteins, or the Pectinase Rohm previously treated with acetone to reduce the PG activity to a minimum are used. The enzymes are dissolved in 0.02 M Sodium acetate buffer at pH 4.0 and applied to a 2 x 10 cm column of the amide derivative of the cross-linked pectate, previously equilibrated with the starting buffer at 4°C. The elution is continued at the rate of 300 ml hr^{-1} until no protein is eluted. Then 0.2 M NaCl is added to the elution buffer and the elution continued. The PE activity as well as the protein content of the fractions are monitored. Finally the fractions containing PE activity are pooled, concentrated and tested for the absence of PG activity according to Somogyi and Nelson (1952). Purity was tested with polyacrylamide gel 7.5 % according to Davis (1964) with Tris-Glycine buffer at pH 8.5.

RESULTS

The elution at pH 4.5 of the crude commercial pectic enzymes containing PG, PE and other proteins on the cross-linked pectate retains only the PG activity. The PE and other proteins get eluted almost at the total volume of the column. When these fractions free from most of the PG activity, but containing the total PE and other proteins is eluted on the amide-derivative of the cross-linked pectate column, about 90 % of the PE activity and only 20 % of protein are retained and can be eluted with 0.2 M NaCl in the same buffer. (Fig. 1). The same picture is obtained when the acetone treated Pectinase from Rohm is used.

Purity of the PE

The electrophoretic patterns on polyacrylamide gel (7.5 %) at pH 8.9 and at pH 4.3 of the purified Pectinesterase show only a faint band of relatively slow migration is seen after coloration with coomasie blue. This result is in accordance with Lee and Mc Millan (1968) in the case of tomato PE.

Capacity of the column

Upto about 500 units of PE/g of dry matrix can be eluted with the same degree of purification and yield.

Stability

A decrease in the Nitrogen content of the matrix is observed when the column was run for about 10 times with 500 u PE/g each time. This decrease in the nitrogen content results in a decrease in the amount of PE retained on the column and an increase in the non-specific adsorption is observed. There is a direct relation between the nitrogen content of the matrix and the amount of PE retained/g of the matrix. We observe that the capacity decreases rather rapidly. But a regeneration of the matrix seems possible to restore the selective adsorption of PE.

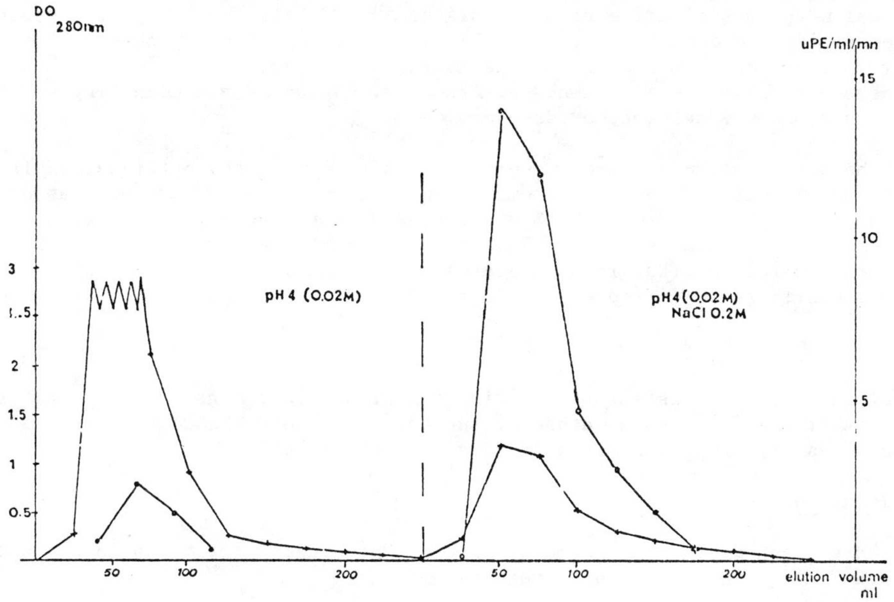

Fig. 1. Elution pattern of acetone treated Pectinase-Rohm on the Amide derivative of cross-linked pectate. (+——+) Absorption at 280 nm (o——o) Pectinesterase activity.

DISCUSSION

In the case of the cross-linked pectate alone no PE was retained, contrary to the observations of Rexova-Benkova et al. (1977) with tomato PE.

The PE activity can be inhibited by (1) Poly-or oligo-galacturonic acid with a Ki = 7 x 10^{-3}M (expressed as monomer units) (2) Glycol esters of pectate and (3) amide derivative of pectate in solution. The inhibitors (1) and (2) were found not useful as insoluble supports for PE retention.

An attempt to prepare the methyl esters of the cross-linked pectate results in the hydrolysis of the matrix therely rendering it unsuitable. Only the amide derivative (treatment with Methanol-ammonia) of the cross-linked pectate, gives an adsorbent for PE.

However, the exact nature of the derivative is not yet well defined. According to Pilnik and Rombouts (personal communications) the same treatment of the soluble pectin gives substitution products both at the reducing end as well as at the carboxyl end of the chain, depending upon the degree of esterification of the pectin. Moreover, Lee and Mc Millan (1970) have indicated that in the case of tomato PE, the activity is initiated at both ends of the polysaccharide chain.

The fact that no change in the swelling volume or flow rate of the gel is observed, after this treatment, rules out the possibilities of gel hydrolysis and hence an adsorption of PE due to gel effects only. A similar treatment of the cross-linked dextran hydrolyses the gel.

After prolonged usage of the support, considerable non-specific adsorption is noticed, along with loss of amide groups. This could indicate that the amide groups provide specific interaction sites for PE. Moreover, if we block these sites by previously loading the column with PG, the matrix can no more retain PE, though it does not undergo any hydrolysis or leakage. Probably, PG and PE are competing for the same sites along the polysaccharide chain.

Though the purification phenomenon seems partially ionic, the matrix also plays a role for the specific retention as our attempts to retain the same enzymes on the Sephadex ion exchangers or on synthetic resins do not give any purification inspite of non specific protein retention. It has been reported by Mark Sheiman (1974) that the PE acts on its substrat in an electrostatic manner. So it seems that the ionic interaction with its modified substract could be a specific one.

CONCLUSION

This method gives a satisfactory purification with a very high capacity. But it remains to define better the nature of the derivative and the mode of interaction. These studies are being further carried out in our laboratory.

Acknowledgements

The authors thank Prof. Pilnik and his group (Wageningen) for valuable discussions and D.G.R.S.T. (France) for the financial aid.

REFERENCES

(1) Rexova-Benkova, L. and Tibensky, V. Biochimica Biophysica Acta. 268, 187 (1972).

(2) Solms, J. and Deuel, H. Helv. Chim. Acta. 38, 321 (1955).

(3) Lee, M. and Mac Millan, J.D. Biochemistry, 9, 1930 (1970).

(4) Vijayalakshmi, M. ; Segard, E. ; Rombouts, F. and Pilnik, W. Food Science and Technology (submitted for publication) (1977)

(5) Lee, M. and Mac Millan, J.D. Biochemistry, 7, 4005 (1968).

(6) Rexova-Benkova, L. ; Markovic, O. and Foglietti, M.J. Collection of Czechoslovak Chemical Communications, 42, 1736 (1977).

(7) Mark Sheiman, Doctoral Thesis, Rutgers University, New Jersey, USA (1974).

(8) Vogel, A.I. Quantitative Inorganic Analysis, 256, (1975).

A CHYMOTRYPSIN INHIBITOR FROM ADULT *CULEX PIPIENS* ISOLATED BY AFFINITY CHROMATOGRAPHY

A. Spiro-Kern and P. S. Chen

Institute of Zoology, University of Zürich, 8006 Zürich, Switzerland

Inhibitors of proteolytic enzymes are of wide natural occurrence. But information about their presence in insects is very limited. Crude homogenates from adult Aedes aegypti were used to examine the inhibitory phenomenon (Yang and Davis, 1973), whereas a chymotrypsin (CHT) inhibitor has been isolated from Drosophila melanogaster which contains no detectable activity of this enzyme (Kang and Fuchs, 1974). Since these inhibitors differ greatly in their action against chymotrypsin-like proteases from various organisms, their species-specificity appears questionable.

In a previous study we found that chymotrypsin activity in the larval midgut of the mosquito Culex pipiens amounted to 187% of that in the whole animal (Spiro-Kern, 1974). This over-recovery suggests the occurrence of a CHT-inhibitor. By employing affinity chromatography we succeeded to isolate such an inhibitor from adult animals. Our isolation procedure has two great advantages: First, work with small insects usually suffers from shortage of material. As the inhibitor can be isolated in a single step, the loss of material is much reduced. Second, we used purified Culex CHT as ligand (see below). Thus, the purified sample must be species-specific. In this report our experimental conditions and some biochemical properties of the Culex CHT-inhibitor are described.

Isolation of the inhibitor

In the first step chymotrypsin from Culex pipiens larvae was purified as follows: Sepharose 4B was activated with CNBr (Porath et al., 1973) and then coupled with the competitive inhibitor 4-phenylbutylamine (PBA) (Stevenson and Landman, 1971). The inhibitory effect of PBA against mosquito chymotrypsin has been ascertained in a preliminary experiment. The enzyme absorbed on the column was eluted with 0,1 m acetic acid in the reversed direction. Fractions which contained CHT activity were pooled, desalted on a Sephadex G-15 column and concentraded with Amicon filter.

The mosquito CHT purified as above was subsequently coupled to activated Sepharose. Crude homogenate from adult animals could be directly applied to the column without pretreatment. A maximal inhibitory effect was located in the peak immediately after buffer change (Fig. 1). At the concentration level indicated in Fig. 2, the enzyme activity was inhibited to about 82% by the purified sample.

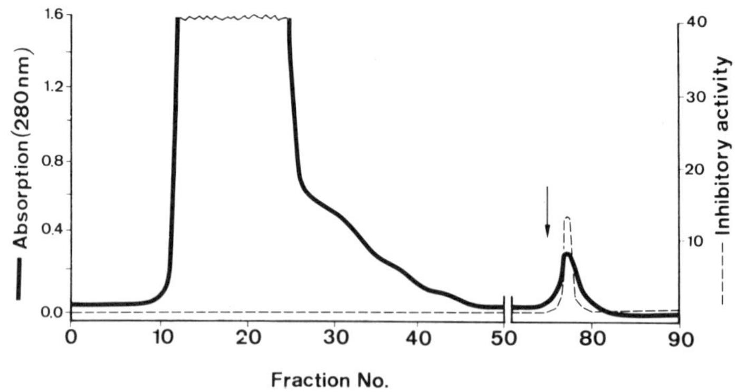

Fig. 1. Isolation of a chymotrypsin inhibitor from adult Culex on Sepharose to which purified larval chymotrypsin was covalently bound. The sample (homogenate from 3 g mosquitos) was applied to the column (1.5 x 5 cm) and eluted with 0.1 M borate buffer, pH 8.0. The arrow indicates replacement of the borate buffer with 0.2 M KCl-HCl, pH 2.0. One unit of inhibitory activity was defined as the amount of inhibitor which reduced the hydrolysis of 1 μMole GPNA/min.

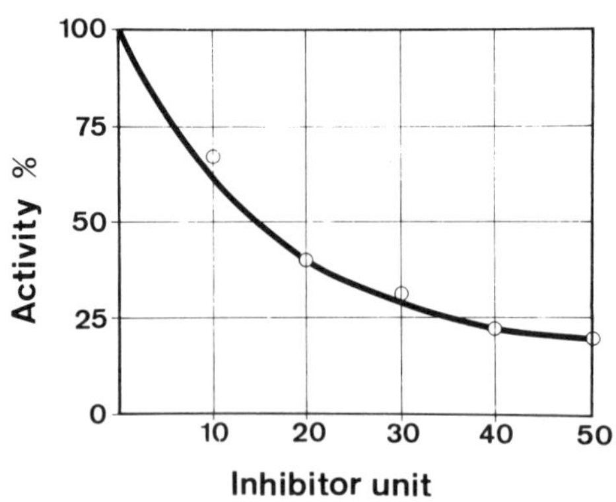

Fig. 2. Inhibition of chymotrypsin from larval Culex by purified CHT-inhibitor from adult Culex. For unit of inhibitory activity, see legend to Fig. 1.

The Culex chymotrypsin covalently bound to Sepharose exhibits a high stability. The column prepared in our laboratory has been repeatedly used for about one and half years. Following warming the column to 37°C and the application of an adequate amount of the synthetic substrate N-glutaryl-L-phenylalanine-nitroanilide, the location of CHT activity can be visualized by the yellow colour of p-nitroanilide set free through the enzyme action (modified method given by Erlanger et al., 1966).

Electrophoretic separation of the purified inhibitor on polyacrylamide gel (Davis, 1964) yielded a strong and a faint band. Owing to its low concentration we do not yet know whether the faint band has any inhibitory effect, or if it is simply due to slight contamination.

Some properties of the inhibitor

Molecular weight: The elution volumes of the inhibitor and several reduced markers were estimated on a 6% Agarose column (0.8 x 70 cm) by using 6 m guanidine/HCl as the eluting solvent (Fisch et al., 1969). The calculated value of 6800 indicates that the compound must be a peptide of rather low molecular weight.
Stability: At pH above 5 the inhibitor remains fully active for about 6 days. It shows a high stability against thermal denaturation. There is no detectable loss of activity after being heated for 20 min at 60°C or for 5 min at 97°C. In fact, we observed a 32% increase in its inhibition following the temperature treatment.
Specificity: The isolated sample yielded a 82 and 80% inhibition of chymotrypsin from *Culex pipiens* and *Aedes aegypti* respectively, whereas its effect against bovine CHT was only 32%.
Complex formation: The formation of the enzyme-inhibitor-complex has been indicated by gel electrophoresis (Reisfeld *et al.*, 1962). The mobility of the complex differs distinctly from that of either chymotrypsin or the inhibitor (Fig. 3).

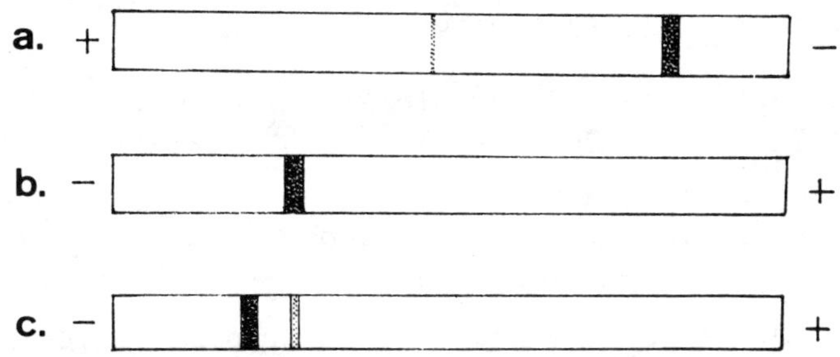

Fig. 3. Polyacrylamide gel elektrophoresis. a, pattern of the purified CHT-inhibitor from adult Culex (Davis, 1964). b, pattern of the purified chymotrypsin from larvae (Reisfeld et al., 1962). c, pattern of the enzyme-inhibitor-complex after preincubation for 10 min at room temperature (Reisfeld et al., 1962).

Our hitherto results demonstrate that at least in Culex pipiens a protease inhibitor can be isolated by coupling the corresponding enzyme to an activated Sepharose. In the mosquito larvae there is a high CHT activity. But in the adults no CHT activity could be detected even after the uptake of a blood meal (Spiro-Kern and Chen, 1972; Spiro-Kern, 1974). Our *in vivo* and *in vitro* experiments designed to answer the question, as to whether the absence of CHT activity in adult Culex is causally related to a high level of the inhibitor, are now underway.

References

B. J. Davis, Disc electrophoresis - II. Method and application to human serum proteins, Ann. N. Y. Acad. Sci., U. S. 121, 404-427 (1964).

B. Erlanger, F. Edel, A. Cooper, The action of chymotrypsin on two new chromogenic substrates, Arch. Biochem. Biophys. 115, 206-210 (1966).

W. Fish, K. Mann, C. Tarrford, The estimation of polypeptide chain molecular weights by gel filtration in 6 M guanidine hydrochloride, J. biol. Chem. 244, 4989-4994 (1969).

S. H. Kang, M. S. Fuchs, Isolation of a chymotrypsin inhibitor from Drosophila melanogaster, Insect Biochem. 4, 1-8 (1974).

J. Porath, K. Asperg, H. Drevin, R. Axen, Preparation of cyanogen bromide-activated agarose gels, J. Chromatography 86, 53-56 (1973).

R. A. Reisfeld, U. J. Lewis, D. E. Williams, Disk electrophoresis of basic proteins and peptides on polyacrylamide gels, Nature 195, 281-283 (1962).

A. Spiro-Kern, Untersuchungen über die Proteasen bei Culex pipiens, J. comp. Physiol. 90, 53-70 (1974).

A. Spiro-Kern, P. S. Chen, Ueber die Proteasen der Stechmücke Culex pipiens, Rev. suisse Zool. 79, 1151-1159 (1972).

K. J. Stevenson, A. Landman, The isolation of chymotrypsin-like enzymes by affinity chromatography using Sepharose -4-phenylbutylamine, Can. J. Biochem. 49, 119-126 (1971).

Y. J. Yang, D. M. Davis, A mosquito chymotrypsin inhibitor in tissues of adult Aedes aegypty, Comp. Biochem. Physiol. 43B, 137-141 (1972).

SEPARATION OF PROTEINASES AND IN PARTICULAR OF CLOSTRIPAIN AND COLLAGENASE BY AFFINITY CHROMATOGRAPHY

Istvan Emöd, Isabelle Trocheris and Borivoj Keil

Chimie des Protéins, Institut Pasteur, 75015 Paris, France

ABSTRACT

Selective separations of proteinases can be achieved by affinity chromatography using different Sepharose-bound amino acids. Trypsin, chymotrypsin, papain, clostripain, pepsin, collagenases of Cl. histolyticum and A. iophagus, carboxypeptidase A and B were assayed on columns of Sepharose-bound arginine, histidine and phenylalanine. Highly active preparations of clostripain and collagenase were isolated from crude collagenase of Cl. histolyticum using five different Sepharose-bound ligands.

EXPERIMENTAL AND RESULTS

This short communication is concerned with a methodical study of the application of affinity chromatography to the isolation of proteolytic enzymes. First, we present a general approach to the isolation of enzymes by affinity systems which is based on the differences of the binding sites in proteolytic enzymes. Subsequently, we discuss its application in the particular case of a mixture of collagenase and clostripain.

The differences exhibited by proteolytic enzymes in their capacity to bind selectively certain amino acid residues are well established. According to the data on the inhibition of different proteinases by free amino acids or their derivatives pepsin, chymotrypsin and carboxypeptidase A are inhibited by aromatic amino acids (1-4), and clostripain and carboxypeptidase B by basic amino acids (5-6). Collagenase and carboxypeptidase A bind histidine (3, 7). Affinity chromatography based on this specific interaction has been described for particular proteinases, such as the binding of chymotrypsin on Sepharose-bound tryptophan ester (2) and of several proteinases on the phenylalanyl-derivative of Sepharose (8).

In the present study we have attached three different amino acids to Sepharose, arginine, histidine and phenylalanine, respectively, via their amino groups, the carboxyl group being left free. We have chosen chymotrypsin and trypsin as representatives of serine proteinases, papain and clostripain as representatives of thiol proteinases, pepsin as a carboxylproteinase, two collagenases and carboxypeptidase A and B as representatives of metalloproteinases.

Using a column of Sepharose-L-arginine, a clear difference was found between trypsin and clostripain: both degrade arginine substrates but only clostripain binds to the column. Carboxypeptidase B was retained by the same column under the conditions used, a result in complete accord with its specificity. A high selectivity was also observed in the case of a column of Sepharose-L-histidine: the collagenases from *Achromobacter iophagus* and from *Clostridium histolyticum* were retained, whereas chymotrypsin or carboxypeptidase A emerged with the void volume.

	Arg	His	Phe
TRYPSIN	±	−	
CHYMOTRYPSIN		−	−
PAPAIN	+	+	
CLOSTRIPAIN	+	+	
PEPSIN			+
COLLAGENASE *Clostridium*	−	+	
COLLAGENASE *Achromobacter*		+	
CARBOXYPEPTIDASE A		−	±
CARBOXYPEPTIDASE B	+	−	

Table 1: Binding of proteinases to Arg-, His- and Phe-Sepharoses

The results are summarized in Table 1, showing that the binding to immobilized amino acids reflects, in most of the cases, interactions in solutions. This study was of course not aimed at the elaboration of new purifications of such well known proteinases as chymotrypsin or pepsin for which very efficient and specific methods have already been described. On the other hand, when dealing with complex unknown mixtures of proteinases in biological fluids and culture media, the retention and elution of different constituents on a series of small columns to which different amino acids are attached will indicate the possible nature of an inhibitor, or the relationship to known groups of proteinases; at the same time it can help in the choice of the whole purification procedure.

This preliminary study helped us to elaborate separation procedures for two proteolytic enzymes, collagenase and clostripain, which have been well known for years. Nevertheless, very little is known about the relationship between their structure and function, in comparison with other proteinases.

The bacterial collagenase from *Clostridium histolyticum* has been purified in several laboratories using different methods (9-16) but the conclusions as to its molecular weight, subunits, homogeneity and specific activity differ considerably. The highest specific activity of one of five active fractions was obtained by Kono (14), but the yield was very low. Clostripain, which is produced by Clostridium together with collagenase, is an extremely interesting thiol-proteinase which cleaves proteins exclusively next to arginine residues and which is reversibly converted to the highly active form by cleavage of disulphide bonds (5, 17-19). We have worked out a system of successive affinity chromatographies which yield these two enzymes in a highly active and homogeneous state (20).

Among the ligands, bound to Sepharose or cellulose compounds which are known to be competitive inhibitors of clostripain, arginine, polyarginine and benzamidine were tried. As clostripain is a thiol proteinase, we used mercurial p-amino-phenylmercuric acetate too, as we did in the purification of papain. The use of soy bean trypsin inhibitor was suggested by our earlier studies on the selective inhibition of clostripain by different polypeptidic trypsin inhibitors. The last ligand of choice was Rivanol which interacts with proteins sometimes in an unpredictable manner and which was found to inhibit collagenase (21).

Good separations of a mixture of clostripain and collagenase were obtained on a column of Sepharose-bound arginine and on a column of soy bean trypsin inhibitor (20). In the case of clostripain the effect which contributes to the separation is primarily inhibitor-enzyme interaction, but it is probable that some secondary effects, such as hydrophobic interactions contribute to the retention of collagenase. In the case of Sepharose-bound Rivanol, the composition of the elution buffer has a pronounced effect: in tris buffer clostripain emerges with the void volume whereas collagenase is fully retained, whereas with distilled water both enzymes are retained.

Using a combination of columns of mercuri- and arginine-Sepharose it is possible to purify clostripain in a single run (Fig. 1).

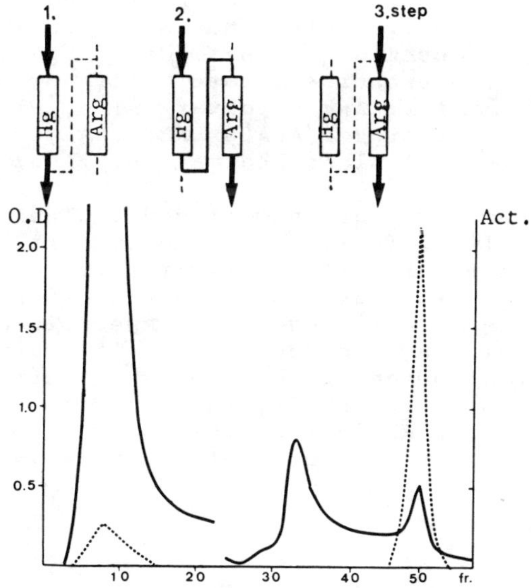

Fig. 1: Purification of clostripain by a three-step passage through two different columns (see text) —— OD_{280}; ..... clostripain activity (BAE)

First, clostripain is selectively retained by the mercurial and the other components pass through into the collector. Then by addition of dithiothreitol, clostripain is desorbed from the first column and is quantitatively adsorbed on the arginine column; the other thiol-bound proteins are eluted. Finally, pure clostripain is eluted by higher salt concentration (20).

Column	Crude COLLAGENASE *Cl. histolyticum*			
	CLOSTRIPAIN		COLLAGENASE	
	Purif. x	Yield %	Purif. x	Yield %
Arg	55	85	16	90
Benz-NH_2	14	84		
HgAc	14	89		
STI	1.5	19	12	39
HgAc → Arg_n	73	79		
RIV → G75 → Arg → Arg	109	16	117	n.d.
AcA44 → DEAE → STI			600	25
	4.94 µ kat/mg		0.655 nkat/mg	

Table 2: Purifications of clostripain and collagenase from crude Clostridium collagenase by use of different Sepharose-bound ligands. Benz-NH_2, benzamidine; HgAc, p-amino-phenyl-mercuric-acetate; STI, soy bean trypsin inhibitor; Arg_n, polyarginine; RIV, Rivanol; AcA44, Ultrogel AcA44 LKB; DEAE, diethylaminoethyl-cellulose Whatman; n.d. not determined.

The different procedures are summarized in Table 2. The best results are underlined:for clostripain, the five µkatals per mg represent the highest specific activity ever obtained; in comparison with 66 units/mg previously described this value represents 300 units. In the case of collagenase the specific activity is also higher than the highest recorded in the literature. These examples underline the great advantage of the use of affinity chromatography in a proteolytic system, namely, the rapidity of the whole procedure which allows the isolation of intact enzymes which otherwise risk autolysis or degradation by other proteolytic components of the mixture.

REFERENCES

(1) W.T. Jackson, M. Schlamowith and A. Shaw, Kinetics of the Pepsin catalysed hydrolysis of N-acetyl dipeptides, Biochemistry 5, 4105 (1966).

(2) P. Cuatrecasas, M. Wilchek, C.B. Anfinsen, Selective enzyme purification by affinity chromatography, Proc. Nat. Acad. Sci. USA 61, 636 (1968).

(3) E. Elkins-Kaufman, H. Neurath, Structural requirements for specific inhibitors of Carboxyl-peptidase, J. Biol. Chem. 178, 645 (1949).

(4) J.R. Whitaker, F. Menger, M.L. Bender, The kinetics of some Carboxypeptidase A and Acetyl carboxypeptidase A catalysed hydrolys s, Biochemistry 5, 386 (1966).

(5) W.M. Mitchell and W.F. Harrington, Clostripain (1971) Enzymes, 3rd Ed., Boyer, Academic Press, London and New York.

(6) E.C. Wolff, E.W. Shirmer, J.E. Folk, The kinetics of Corboxy-peptidase B activity. I. Kinetic parameters, J. Biol. Chem. 237, 3094 (1962).

(7) E. Harper, Mechanism of action of Collagenase. Irreversible inhibition by cysteine; reversible inhibition by histidine or imidazole, Federation Proc. 25, 790 (1966).

(8) K. Fujiwara, D. Tsuru, Affinity chromatography of several proteiolytic enzymes on carbobenzoxy-D-phenylalanyl-triethyl-enetetramine-Sepharose, Int.J. Peptide Protein Res. 9, 18 (1977).

(9) S. Seifter, P.M. Gallop, L. Klein and E. Meilman, Studies on Collagen. II. Properties of purified collagenase and its inhibit on, J. Biol. Chem. 234, 285 (1959).

(10) I. Mandl, S. Keller and J. Manahan, Multiplicity of *Clostridium histolyticum* Collagenases, Biochemistry 3, 1737 (1964).

(11) E. Yoshida and H. Noda, Isolation and characterization of collagenases I and II from *Clostridium histolyticum*, Biochem. Biophys. Acta 105, 562 (1965).

(12) E. Harper, S. Seifter and V. Hospelhorn, Evidence for subunits in bacterial collagenase, Biochem. Biophys. Res. Commun. 18, 627 (1965).

(13) S. Takashashi and S. Seifter, New culture conditions for *Clostridium histolyticum* leading to production of Collagenase of high specific activity, J. Appl. Bacteriol. 35, 647 (1972).

(14) T. Kono, Purification and partial characterization of collagen-olytic enzymes from *Clostridium histolyticum*, Biochemistry 7, 1106 (1968).

(15) W.M. Mitchell, Pseudocollagenase A protease from *Clostridium histolyticum*, Biochem. Biophys. Acta 159, 554 (1968).

(16) M.R. Kula, D. Hatef-Haghi, M. Tauber-Finkelstein and S. Shaltiel, Consecutive use of ω-amino-alkylagaroses. Resolution and purification of clostripain and collagenase from *Clostridium histolyticum*, Biochem. Biophys. Res. Commun. 69, 389 (1976)

(17) W.H. Porter, W.M. Mitchell, Anomalous lectrochemical behaviour of clostripain: evidence for disulfide isolation conformers, Mol. Cell. Biochem. 1, 95 (1973).

(18) B. Labouesse, L'hydrolyse du glucagon par la clostripaine, Bull. Soc. Chim. Biol. 42, 1293 (1960).

(19) J.W. Shih and J.J. Hash, The N,O-diacetylmuramidase of Chalaropsis species, J. Biol. Chem. 246, 994 (1971).

(20) I. Emöd and B. Keil, Five Sepharose-bound ligands for the chromatographic purification of Clostridium collagenase and clostripain, FEBS-Letters 77, 51 (1977).

(21) F.K. Cowey, M.W. Whitehouse, Biochemical properties of anti-inflammatory drugs. VII Inhibition of proteolytic enzymes in connective tissue by chloroquine (resochin) and related anti-malarial/antirheumatic drugs, Biochem. Pharmacol. 15. 1071 (1966).

SYNTHESIS OF IMIDAZOLE CONTAINING MATRICES (POLYHYDROXYETHYLMETHACRYLATES, POLYSACCHARIDES) AND ITS APPLICATION IN AFFINITY CHROMATOGRAPHY

P. Mohr*, K. Pommerening*, M. Kühn*, J. Stamberg** and M. Benes**

**Department of Biocatalysis, Central Institute of Molecular Biology of Academy of Sciences of the GDR, GDR-1115 Berlin-Buch, Lindenberger Weg 70*
***Institute of Macromolecular Chemistry, Czechoslovak Academy of Science, Prague 6, 1888 Petriny, Czechoslovakia*

ABSTRACT

Methods for the covalent binding of imidazole to poly-(hydroxyethylmethacrylates) and polysaccharides are described. These matrices are suitable for affinity chromatography of hemoproteins (hemoglobin/myoglobin, cytochrome P-450/cytochrome P-420 etc.). The separation effectivity depends on the spacer length of the matrix, pH and temperature. Ligand specific contacts (coordinative binding between hemin iron and imidazole) are essential in case of a short spacer. Long spacers are responsible for unspecific interactions. Separation of hemoglobin/myoglobin is possible also by means of matrices containing ω-aminoalkyl groups.

INTRODUCTION

The functional centre of hemoproteins is formed by an iron porphyrin and amino acid residues of the protein component. Anions and neutral bases (CN^-, F^-, imidazole etc.) can be bound coordinatively to the porphyrin iron (1-3). Therefore the application of imidazole containing matrices for the affinity chromatography of hemoproteins was studied.

SYNTHESIS OF IMIDAZOLE CONTAINING MATRICES

Three types of imidazole containing matrices have been synthesized (Table 1).
In type I matrices the imidazole is linked via the annular nitrogen. Three different spacer lengths were employed. The macroporous poly-(ethylenmethacrylate-arylsulfonates) were coupled with imidazole and its derivatives in the form of alkaline salts (Fig. 1). The modified matrices are chemically stable even under extreme conditions. This new method has been developed recently (4).

Abbreviations: Hb = hemoglobin, MetHb = methemoglobin, MetMb = metmyoglobin, cyt = cytochrome

Table 1 Synthesized Imidazole Containing Matrices

Type	Matrix	Side chain Chemical structure	length (atoms)
I	Poly-(hydroxyethyl-methacrylate)	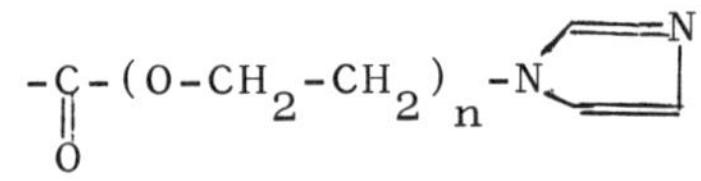	4, 7, 10
II	Spheron	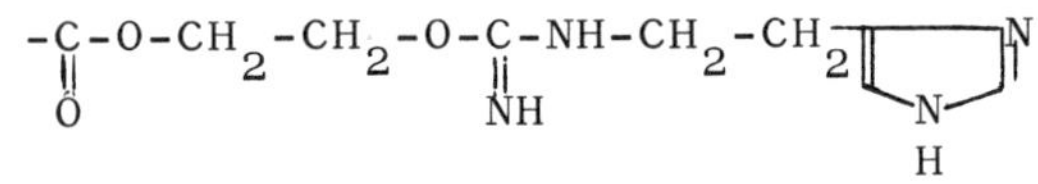	9
III	Cellulose, Sepharose	-R	18
III	Spheron	$-\overset{\overset{O}{\|}}{C}-O-CH_2-CH_2-R$	22

n = 1,2,3

R = 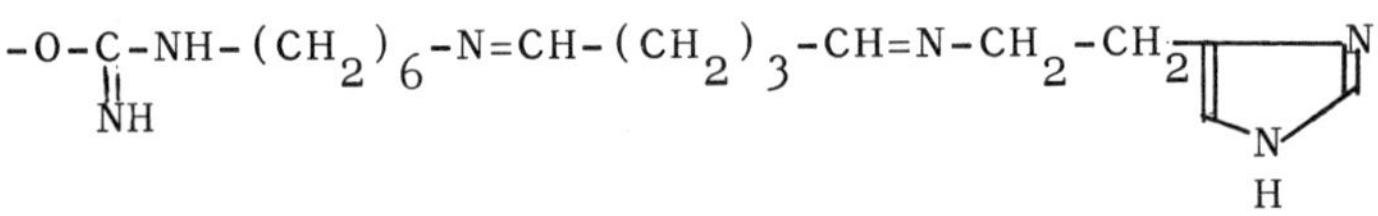

Type II matrices were synthesized by reaction with histamine after CNBr activation of Spheron and type III matrices by hexamethylendiamine and glutaraldehyde treatment preceding the histamine treatment.

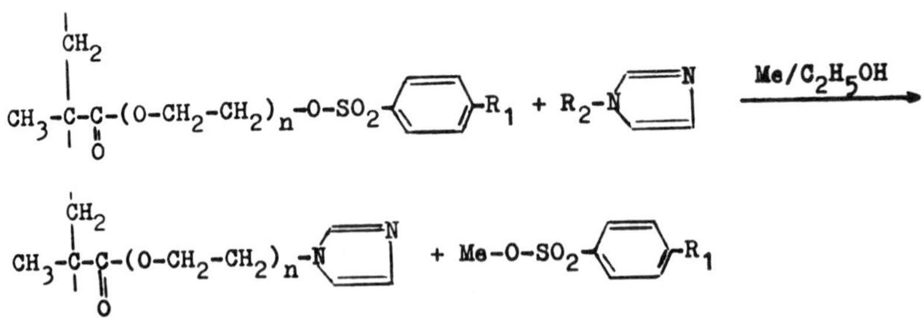

n = 1, 2, 3; R_1 = H, CH_3; R_2 = H, CH_2-CH_2-OH; Me = Na, Li

Fig. 1. Scheme of synthesis for type I matrices

ADSORPTION OF HEMOGLOBIN

Matrix bound imidazole retained its bioaffinity for the heme pocket as was tested in experiments using Hb (human and bovine). The three types of matrices adsorbed MetHb. Two different Hb-matrix-interactions were observed.

In type I and II matrices the imidazole is responsible for the adsorption. The adsorption is dependent on the imidazole content and decreases if the imidazole is replaced by an OH group or if Hb derivatives are being used which do not form an imidazole complex (HbO_2, HbCO). The coordinative attachment of the imidazole to the iron of the hemin is evidenced by EPR spectroscopy (data not shown).

In type III matrices unspecific interactions between the matrix and the protein should be effective: the Hb adsorption is independent of the oxidation state of the iron and, furthermore, Hb is still bound when imidazole is replaced by an OH group on the gel. This mechanism is also corroborated by the fact that Hb is adsorbed on poly-methacrylate, containing the following side chains:

$$\underset{\underset{O}{\|}}{C}-O-CH_2-CH_2-NH-R; \quad R = C_2H_5,\ CH_2-CH_2-NH_2,\ (CH_2)_6-NH_2.$$

APPLICATION OF IMIDAZOLE CONTAINING MATRICES

In addition to Hb various other hemoproteins are adsorbed on the synthesized matrices. In Table 2 some applications are summarized.

Table 2 Applications of Imidazole Containing Matrices

Type	Matrix	Separation
I	Poly-(hydroxyethylmethacrylates)	MetHb/MetMb (bovine); n = 2, pH 8, r.t.[x)] Cyt P-420/P-450 (rabbit); pH 7,25, Na-cholate; r.t.,4°C
III	Sepharose	MetHb/MetMb (bovine); pH 8, r.t.
III	Sphcron	MetHb/MetMb (bovine); pH 8, r.t., 4°C

x) r.t. = room temperature

Typ I matrices proved to be useful in selective MetHb adsorption from a mixture with MetMb. Cyt P-450 could be purified by selective adsorption of the biologically inactive cyt P-420 fraction.

The MetHb/MetMb separation could be performed with type III matrices too. With Spheron the adsorption of MetHb is temperature dependent. At room temperature the MetHb is totally adsorbed, at 4°C, however, it is only retarded with respect to MetMb.

The separation is also possible by means of poly-methacrylate containing the following side chain: $-\underset{\underset{O}{\|}}{C}-O-CH_2-CH_2-NH-(CH_2)_6-NH_2$. This indicates that unspecific interactions with type III matrices which contained imidazole are rather effective in separating different heme proteins.

Further applications for the isolation or purification of peroxidase, catalase, etc. will be investigated.

REFERENCES

(1) Antonini, E. and M. Brunori (1971), Hemoglobin and myoglobin in their reaction with ligands, North-Holland, Amsterdam-London.

(2) P. Mohr, H. Rein, W. Pfeil, O. Ristau, K. Müller and W. Scheler, Untersuchungen zur Natur der Ligand-Rezeptor-Beziehungen im Methämoglobinkomplexen mit Imidazol, 1,2,4-Triazol und Pyridin, Acta biol. med. germ., 28, 221 (1972).

(3) G.-R. Jänig, R. Misselwith, D. Zirwer, E. Buder, H. Rein and K. Ruckpaul, Differentiation between Type I and Type II Substrate Binding to Cytochrome P-450 by Temperature Studies, Croat. Chem. Acta 49, 263 (1977).

(4) M. Benes, M. Kühn and K. Pommerening, Polymethakrylsäureester mit Imidazolseitengruppen und Verfahren ihrer Herstellung. DDR-Patent 124405 (1977).

PURIFICATION OF NEUTRAL PROTEASES FROM LEUCOCYTES ON PROTAMINE-SEPHAROSE

Manfred Mattausch

Department of Experimental Surgery, 1st Surgical University Clinic, Vienna, Austria

It is well known that leucocytes have the capacity to split fibrin (1). Examinations of plasma clots have shown that the capacity of lysis will increase if leucocytes will disintegrate in the thrombi. These findings show that leucocytes have proteases which are able to split fibrin in the absence of plasminogen.

MATERIALS AND METHODS

To obtain these proteases, leucocytes from whole blood must be concentrated. Fresh ACD* blood from healthy donors was collected into tubes. To 15 ml of blood 3 ml of 0.5 % human fibrinogen were admixed. The erythrocytes were sedimented at room temperature. The leucocyte-rich supernatant was centrifuged at 240 x g for 30 min. and the pellet was washed 2 x with ACD solution and 3 x with isotonic saline. The remaining erythrocytes were destroyed by hypotonic shock. After further washing procedures the leucocytes were counted and adjusted to a concentration of 80 000/μl.

The concentrated leucocytes were destroyed by ultrasonication and dissolved with different agents. Solubilization and activation was optimal with cationic detergents and proteins.

TABLE 1: Solubilization and Activation

Cationic detergents		Cationic proteins	
Cetylpyridinechloride	1 %	Protamine sulfate	1 %
Cetavlon	1 %	Lysozyme	1 %
Cephirol	1 %		

After incubation for 4 hours at room temperature homogenates were centrifuged for 1 hour at 100 000 x g. The active supernatant was dialysed for 16 hours against tap water and then against physiological saline. This raw material was tested for its fibrinolytic activity on heated fibrin tubes (2), a modification of the heated fibrin plate technique of Astrup and coworkers (3). The method of

*ACD = Acid Citrate Dextrose

heated fibrin tubes gives a much better quantitation of the lytic activity than the fibrin plate technique.

Fibrin agar electrophoresis was used according to the method of Heimburger and Schwick (4) as modified by Gottlob et al. (2).
The coupling of protamine sulfate to CNBr activated Sepharose 4B from Pharmacia Fine Chemicals was performed according to the manufacturer's recommendation.
Laurell antigen-antibody crossed electrophoresis (5) was performed according to Clark and Freeman (6).
Anti human FDP* D and FDP E serum from rabbit was obtained from Behringwerke Marburg/Lahn.
Raw material used later was frozen at -30° C.

RESULTS

After solubilization and activation the greatest activity of the supernatant was obtained by using cationic detergents or proteins. After removal of the detergent by dialysis the activity decreases. Addition of cationic detergent or protamine sulfate can restore the fibrinolytic activity of the preparation.

Cetavlon activated and dialysed raw material was submitted to a fibrin-agar-electrophoresis. Fig. 1 shows that the protease migrates to the cathode as does plasmin. After incubation in a moist chamber at 37° C for 24 hours the slides were covered by a layer of agarose containing protamine sulfate or Cetavlon and incubated at 37° C for 24 hours again (Fig. 2).

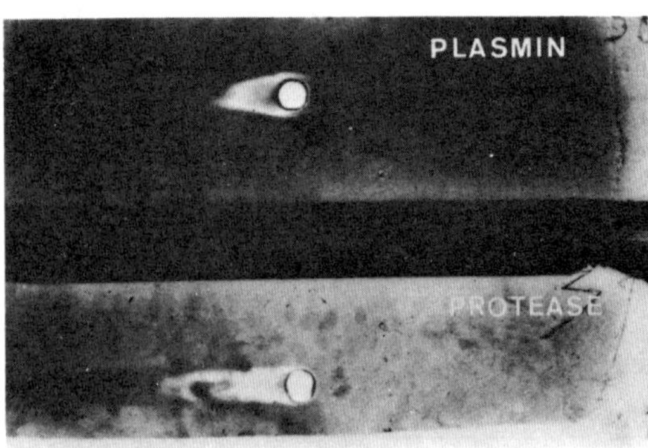

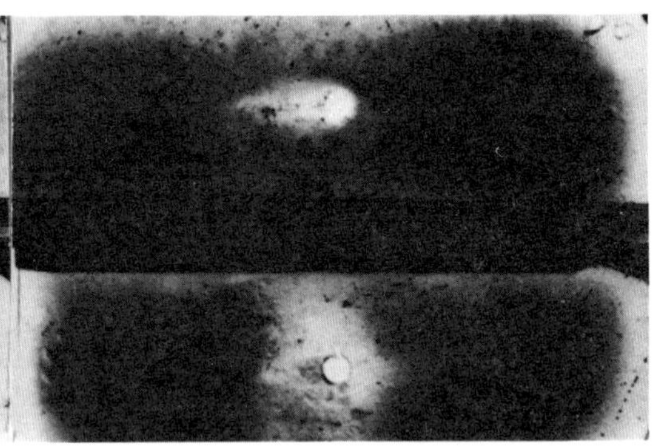

Fig. 1 Electrophoresis slides without "sandwich"

Fig. 2 Electrophoresis slides with "sandwich"

The lytic zone of plasmin remained unchanged. However, the lytic zone produced by the protease was extended to form an almost circular lytic zone surrounding the central well. This phenomenon can be tentatively interpreted as a protease - inhibitor complex without electrophoretic migration. It seems that the inhibitor has a higher affinity to the cationic detergent or protamine sulfate than to the protease.

---

*FDP = fibrin degradation products

To study the fibrinolytic properties of the neutral protease it was necessary to free them of the inhibitor. In order to remove the inhibitor the homogenate of leucocytes was chromatographed on a Sepharose 4B-Protamine column (Fig. 3).

The protease could be eluted with 0.1 M $NaHCO_3$ and 0.5 M NaCl, pH 8.0. However, the activity of this material could not be further enhanced by addition of cationic detergents. The purity of the protease was tested by gel chromatography as well as by PAA*gel electrophoresis. The pattern in gel chromatography was the same as in PAA gel electrophoresis (only two bands, both enzymatically active).

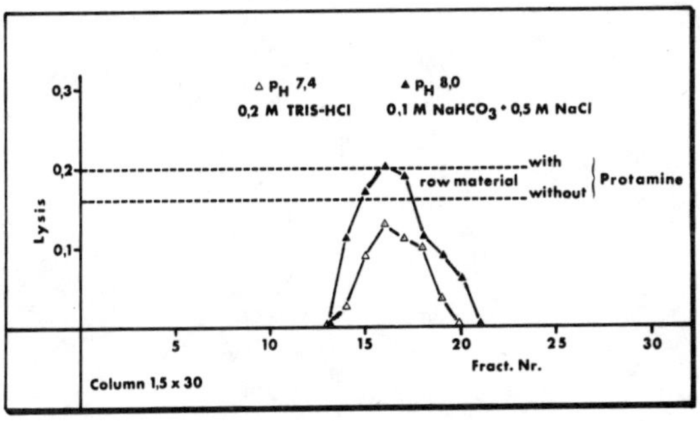

Fig. 3 Chromatography on a Sepharose 4B-Protamine column (Sample 2 ml, fraction volume 3 ml)

The active fractions were lyophilized and tested with several inhibitors. The protease was inhibited by human serum as well as by heparin and α1-antitrypsin. The inhibition by serum can be abolished by adding Cetavlon. Examinations about the inhibition of the protease and its molecular weight were carried out by Tschesche (7).

TABLE 2: Differences between the neutral protease and plasmin and trypsin by activators and inhibitors

	Triton X 100	Cation. Deterg.	Cation. Protein	Trasylol	Soy Bean Inhibitor
Protease	i	+	+	0	0
Plasmin	+	i	0	i	i
Trypsin	+	+	0	i	i

i : Inhibition + : Activation 0 : No change

Leucocyte protease with Antithrombin III and C_1-Inactivator show no inhibition at different concentrations in contrast to plasmin (5 CU/ml).

* PAA = polyacrylamide

No inhibition was obtained by p-chloromercuribenzoic acid (0.357 %), by EDTA solution (1 %) or ε-aminocaproic acid (1 %).

Heating neutral protease for 20 minutes at 80° C reduced the activity slightly, heating at 50° C for the same time did not reduce the activity.

The protease was resistent to acids down to pH 4 and to alkaline solution up to pH 9 when kept for 1 hour at room temperature under these conditions.

Within 24 hours the neutral protease is capable of splitting fibrin to the fibrin degradation product (FDP) D and E (Fig. 4). Plasmin is splitted considerably faster, the process is finished after 2 hours.

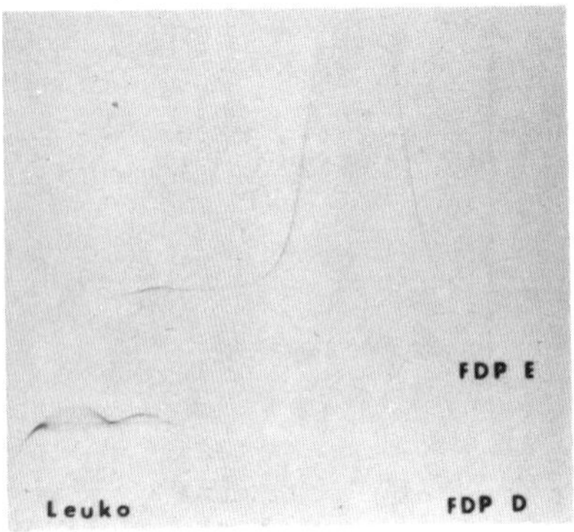

Fig. 4 Crossed immunoelectrophoresis of FDP D and E

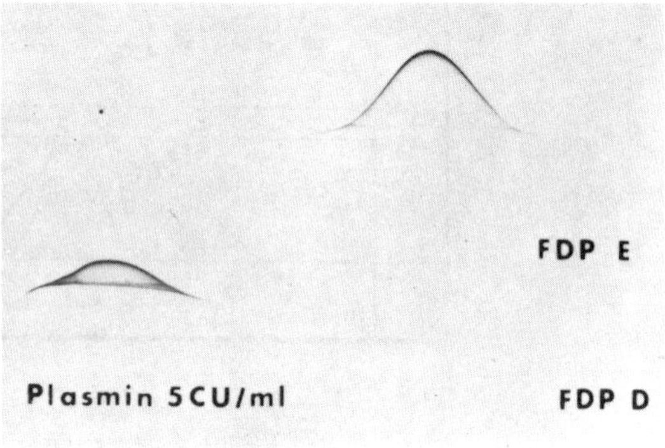

Fig. 5 Crossed immunoelectrophoresis of FDP D and E

DISCUSSION

The activation of the dialysed protease by cationic detergents or proteins may be explained by a competition of these cationic agents with the cationic protease. This assumption also explains why after passing the Sepharose 4B-Protamine column the protease obtained is no longer activable by these cationic compounds. Most conveniently the protease may be obtained by affinity chromatography if ultrasonicated concentrate of leucocytes is passed through a Sepharose 4B Protamine column.

REFERENCES

(1) T. ASTRUP et al.: Protease content and fibrinolytic activity of human leucocytes, Blood 29, 134 (1967).

(2) R. GOTTLOB et al.: The possible role of leucocytes for spontaneous and induced thrombolysis and some properties of leucocyte protease, in "Progress in Chemical Fibrinolysis and Thrombolysis", Vol. 3, Raven Press (in press).

(3) T. ASTRUP and S. MÜLLERTZ: The fibrin plate method for estimating fibrinolytic activity, Arch. Biochem. 11, 149 (1952).

(4) N. HEIMBURGER and G. SCHWICK: Die Fibrinagar-Elektrophorese, Thromb. Diath. Haemorrh. 7, 432 (1962).

(5) C.-B. LAURELL: Antigen-antibody crossed electrophoresis, Anal. Biochem. 10, 358 (1965).

(6) H.G.M. CLARK and T. FREEMAN: A quantitative immunoelectrophoresis method (Laurell electrophoresis), Pp. 503-509 in H. Peeters (ed.) Prot. Biol. Fluids, 14 (1967), Elsevier, Amsterdam.

(7) H. TSCHESCHE: Molekulare Aspekte zellulärer Proteinasen und ihrer Inhibitoren, Munich 1977 (in press).

PART III

HYDROPHOBIC INTERACTION CHROMATOGRAPHY

HOMOLOGOUS SERIES OF HYDROCARBON-COATED AGAROSE IN HYDROPHOBIC CHROMATOGRAPHY

S. Shaltiel, G. Halperin, Z. Er-el, M. Tauber-Finkelstein and A. Amsterdam

The Weizmann Institute of Science, Rehovot, Israel

FROM BIOSPECIFIC AFFINITY CHROMATOGRAPHY TO HYDROPHOBIC CHROMATOGRAPHY

The development of hydrophobic chromatography (1,2) was the result of a reluctance to overlook some unexplained observations while attempting to prepare glycogen-coated agarose for biospecific affinity chromatography and for studying the assembly of glycogen-recognizing enzymes. The procedure used for the preparation of the columns consisted of activation of the agarose with CNBr (3) and reaction with an α,ω-diaminoalkane to obtain an ω-aminoalkyl agarose, then binding of CNBr-activated glycogen to the ω-aminoalkyl agarose (Fig. 1).

Two glycogen-coated agarose preparations which differed only in the length of the hydrocarbon chains bridging the ligand and the bead (Seph-C_8-glycogen and Seph-C_4-glycogen) displayed an unexpected behaviour: whereas the column with 8-carbon-atom bridges adsorbed the enzyme, the column with 4-carbon-atom bridges did not even retard it (1,4). Additional experiments showed that the same behaviour could be reproduced with columns that did not contain any glycogen at all (Seph-C_8-NH_2 and Seph-C_4-NH_2) and that in fact, the binding of n-alkylamines to CNBr-activated agarose results in materials with new adsorption properties. Homologous series of such materials, similar in all respects except for the length of the attached hydrocarbon chains, form the basis of "hydrophobic-chromatography" (1,2,4-13) a new approach to the resolution, purification and probing of proteins and other biomolecules or biological particles.

In principle, it is possible to visualize several types of series of such columns (Fig. 2). Such series will derive their adsorption and discriminating power not only from hydrophobic interactions but also from other types of interactions (electrostatic, hydrogen bonding, donor-acceptor, etc.). However, with appropriately designed homologous series it is possible to gradually increase the contribution of the hydrophobic interactions and tailor them in each case so as to optimize resolution and improve the purification of a desired biomolecule.

Abbreviations: Seph-C_n-X, denotes Sepharose 4B activated with CNBr then reacted with an n-aminoalkane derivative NH_2-$(CH_2)_n$-X (X = H, NH_2, COOH, etc. and n = 1,2, 3...); Seph-C_o, represents unmodified Sepharose 4B; Seph-C_o* denotes Sepharose 4B activated with CNBr and hydrolyzed in alkali.

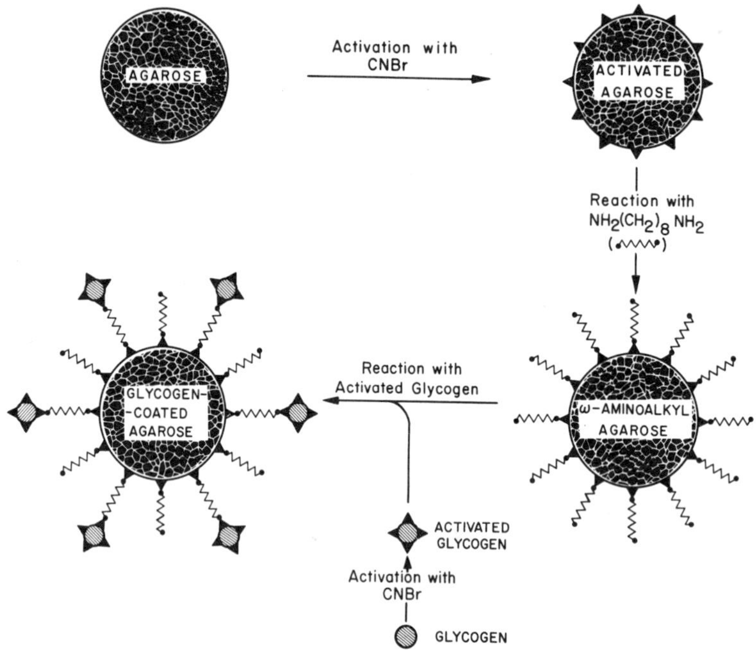

Fig. 1. Schematic representation of the preparation of Seph-C_8-glycogen (▶, represents a reactive functional group introduced by CNBr-activation and •∿∿∿• represents 1,8-diaminooctane).

USEFUL FEATURES AND POTENTIAL OF THIS APPROACH

1) Adjustability of adsorption forces.

Using homologous series of hydrocarbon-coated agaroses it is possible to delicately adjust the adsorption forces of a given protein and thus avoid an overly strong retention, which often dictates rather drastic conditions at the elution step. This delicate adjustment is particularly important in the purification of labile enzymes, receptors, etc. whose native conformation, activity, or ability to recognize specific biological signals may be lost in the course of purification. Such adjustability of adsorption forces provides not only enzyme preparations with high specific activity but also results in high yields.

2) Additional criterion for resolution

Protein mixtures which have already been resolved by procedures based on differences in stability, solubility, charge, size or biospecific ligand recognition, can be further resolved by hydrocarbon-coated agaroses which discriminate among proteins mainly through differences in hydrophobic interactions between the hydrocarbon chains of the alkyl agarose and appropriately-sized hydrophobic "pockets" or "patches" in the various proteins.

3) Probes for protein surfaces

Columns of hydrocarbon-coated agaroses may be used as probes for the surface of

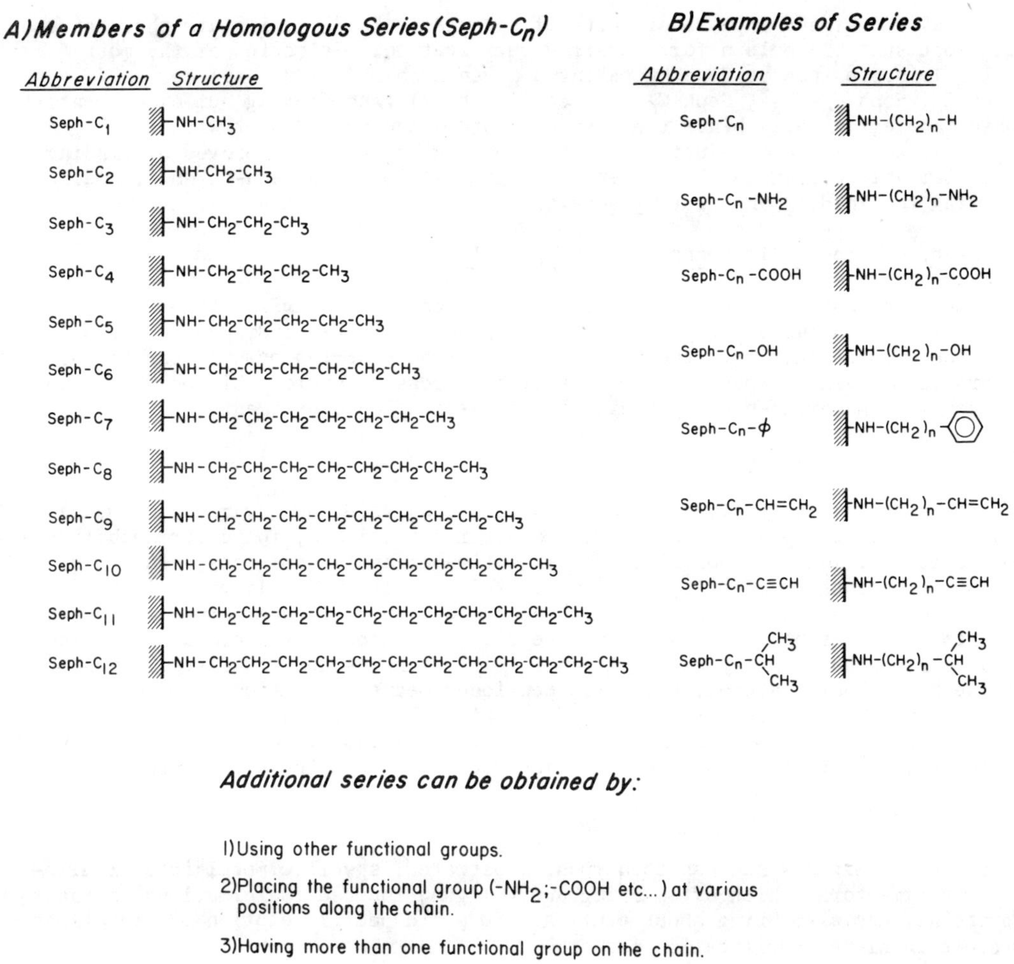

Fig. 2. Variety of homologous series of alkyl agaroses which may be used for chromatography of proteins and other biomolecules or biological particles.

protein molecules and for the detection of structural changes in proteins, if these are reflected in the size and distribution (or density) of their available hydrophobic "pockets" or "patches" (cf. reference (14)).

4) Basis for consecutive fractionators.

Homologous series of hydrocarbon-coated agaroses may form the basis for the development of automatic consecutive fractionators for the purification of proteins (15), e.g. by passage of a crude extract through Seph-C_1 then Seph-C_2, Seph-C_3 ... Seph-C_n. Each column in such fractionators will retain a few proteins out of the mixture, which can then be further resolved by consecutive detachment (13,16). Consecutive extraction of proteins is potentially very useful, as it allows maximal utilization of expensive, scarcely available tissues, or a simultaneous identification and determination of several components in a given biological fluid or tissue.

5) Rapid experimental optimalization of resolution

Especially designed exploratory kits of columns (4,5,13) facilitate the choice of the most suitable column for a desired purification. Selection of the most suitable series can be done by taking a given member from several series (e.g. Seph-C_6, Seph-C_6-NH_3^+, Seph-C_6-COO^- etc., Fig. 2) and finding out experimentally which one efficiently binds the desired protein and excludes it with the highest specific activity upon elution. Resolution can be further improved by finding the most suitable column in the chosen homologous series, say Seph-C_n-NH_3^+, and by varying the loading and eluting conditions (see below, section 7).

6) Study of lipophilic membrane-bound proteins.

An important potential use of these columns lies in the purification and study of lipophilic membrane-bound proteins, which have accessible hydrophobic regions used for their localization in the membrane. Homologous series of hydrocarbon-coated agaroses may also be used as carriers with increasing hydrophobic character for assembly of membrane-bound proteins in reconstitution experiments.

7) Variety of means for elution

Proteins can be eluted from hydrocarbon-coated agaroses by a variety of means. These include inexpensive means such as alterations in pH, ionic strength (increase or decrease), ionic composition, concentration of apolar solvents, additions of deforming agents or of mild, fully reversible denaturants. In view of the fact that the availability of hydrophobic patches on the surface of a protein depends on its conformation, and since the retention of proteins by hydrocarbon-coated agaroses depends largely on the lipophilicity, size, shape and distribution of these hydrophobic patches, the above-mentioned means of elution may function either by directly disrupting the hydrophobic bonds between the protein and the column material or by changing the conformation of the protein (or the column material). In fact, both mechanisms may sometimes function simultaneously.

8) Biospecific elution

Biospecific ligands such as coenzymes, substrates, specific metal ions or allosteric effectors, which often bring about ligand-induced conformational changes in proteins, may also bring about elution. Such biospecific elution may sometimes result in higher purification factors.

9) Stability, capacity, flow-rates and repeated use

When stored at neutral or acidic pH and 4°, the column materials are quite stable. With three arbitrarily chosen columns from three different series (Seph-C_4, Seph-C_4-NH_2 and Seph-C_1-COOH) it was shown that after 21 days less than 5% of the chains were detached from the matrix (17). Appropriate protection against microbial or fungal growth prolongs, of course, the shelf life of the columns. The columns have a good capacity (in the case of glycogen phosphorylase b, Seph-C_4 binds up to 10 mg protein per ml settled volume of the column) and on a laboratory scale they can be run at fairly high flow rates (usually 1-2 ml/min). The columns can be used repeatedly - in one case, a column of Seph-C_5-NH_2 was used over 20 times during a period of about 4 months without significant loss in capacity or resolution power.

10) Resolution of cells and probing of their surface

Homologous series of hydrocarbon-coated agaroses seem to provide a useful tool in cytology. Erythrocytes, for example, can be efficiently adsorbed onto and desorbed from Seph-C_n columns under physiological, isotonic conditions (18). This reversible adsorption (*cf.* also Halperin and Shaltiel, this volume) can be carried out with essentially no physical entrapment or lysis of the cells, as indicated by the fact that over 95% of the cells applied on the columns can be recovered. The erythrocytes eluted from such columns and those yet to be applied were found to be morphologically indistinguishable and to exhibit an identical osmotic fragility profile, suggesting that they are not significantly damaged by the adsorption-desorption procedure. Furthermore, erythrocytes from different sources exhibit different adsorption profiles on such column series, illustrating their potential use in discriminating between closely-related cells and in probing cell surfaces (18).

STRUCTURE OF HYDROCARBON-COATED AGAROSES

Several structural studies were carried out, in order to establish whether the differences observed in the resolution power of the various members of a given column series should be attributed mainly to differences in the length of their attached hydrocarbon chains or whether they might be associated (wholly or in part) with changes in the ultrastructure of the column materials. Morphological studies under a phase contrast light microscope show that activation of agarose with CNBr and subsequent reaction with various alkylamines does not affect significantly the size or shape of the beads (19). Furthermore, ultrastructural studies under the electron microscope show that untreated agarose (Seph-C_o) (20) and arbitrarily chosen modified agaroses (Seph-C_4, Seph-C_4-NH_2, Seph-C_8-NH_2) are indistinguishable in their ultrastructure. Thin sections of all the above agaroses embedded in Epon revealed a network skeleton with variable pore size (Fig. 3). These pores or channels are distributed at random in different sections of the bead.

The hydrocarbon chains in these columns are distributed at random both on the surface and within the beads, as shown by autoradiography of Seph-C_4-NH_2 with tritiated hydrocarbon chains (Fig. 4) (19). Examination of the several bead slices (1 μ sections) revealed that the autoradiographic grain density, and thus the density of the hydrocarbon chains, does not fluctuate by more than ± 15% throughout the bead.

Activation of agarose with CNBr and coupling with amines results in several types of functional groups and of ligand-to-matrix bridges (21), some of which (the N-substituted isourea linkages) are positively charged at neutral pH (Fig. 5). The possibility was considered that since the reaction between activated agarose and alkylamines is carried out at quite a high concentration of the ligand (1,4), the alkylamine itself (occupying a considerable volume of the reaction mixture, especially as the molecular weight of the alkylamine increases) might affect to a different extent the properties of the medium in which the reaction takes place and thus lead to a different array of functional groups introduced into the agarose as a result of the coupling reaction. Also, one should not exclude *a priori* the possibility that hydrocarbon chains of different lengths might alter to a different extent the dielectric constant in the microenvironment of positively charged groups linking some of the chains to the column and thus lead to differences in the apparent pK_a values of these groups. This could result in column materials with a different net charge (say at neutral pH) as the length of the hydrocarbon chain increases.

In an attempt to investigate whether this is the case, we have carried out careful potentiometric titrations of arbitrarily chosen members of the Seph-C_n series (from Seph-C_1 to Seph-C_{12}) (18). Figure 6 represents several examples of such titrations and Fig. 7 depicts the determination of the apparent pK_a values from such titrations. These pK_a values were found to be 9.7 in all cases (18).

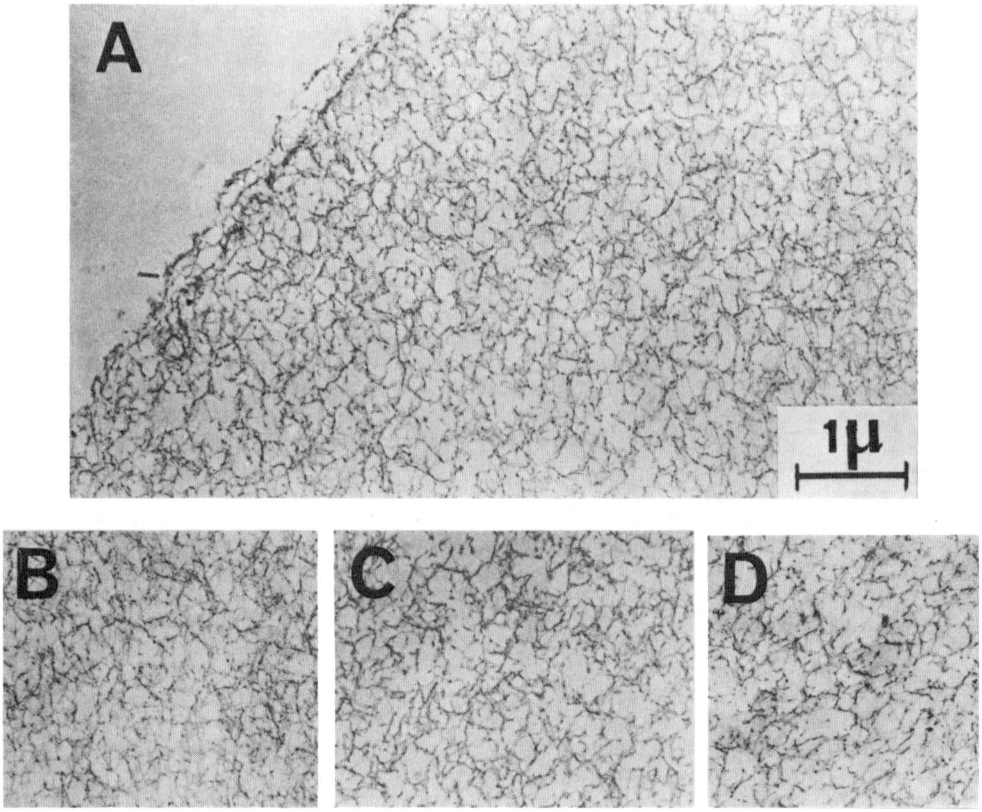

Fig. 3. Electron micrographs showing the ultrastructure of (A), Seph-C_4; (B), Seph-C_0; (C), Seph-C_4-NH_2; (D), Seph-C_8-NH_2. Thin sections (∿ 1000 Å) were prepared from Epon-embedded beads, then doubly stained with uranyl acetate and lead citrate. The stained sections were examined and photographed under a Jeolco 100-B electron microscope operated at 80 kV.

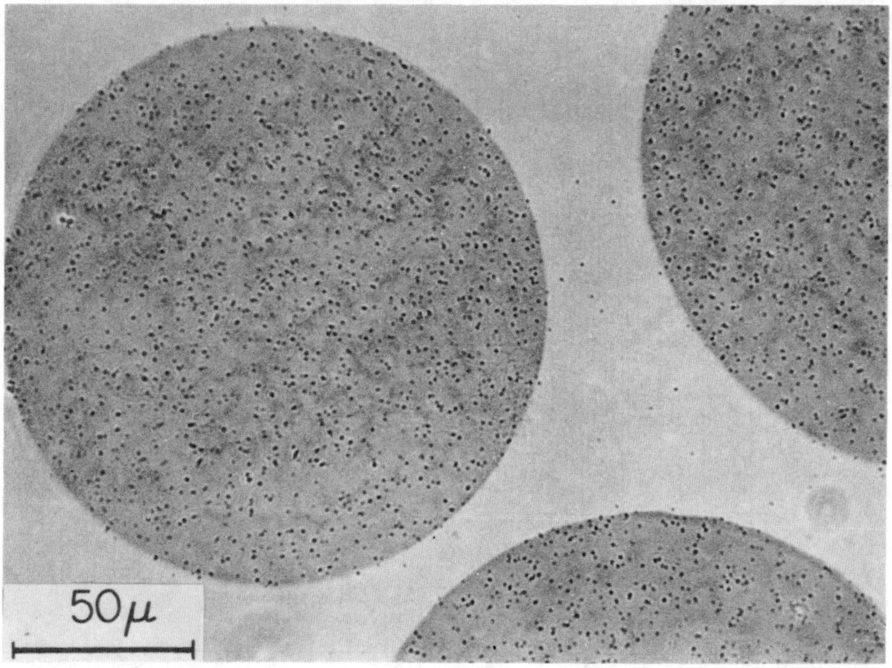

Fig. 4. Autoradiogram showing the random distribution of radioactively labeled ^{3}H-hydrocarbon chains in Seph-C_4-NH_2.

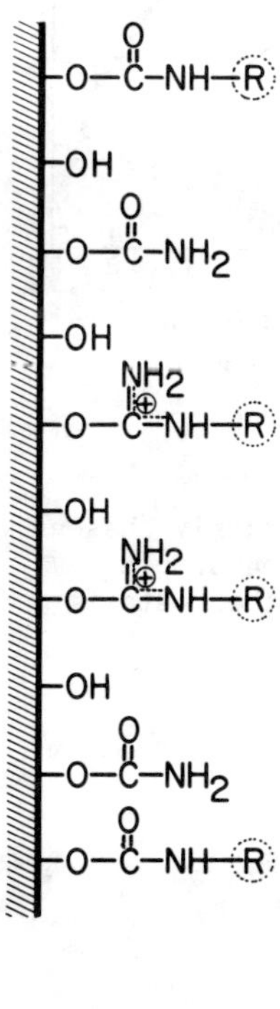

Fig. 5. Schematic representation of the product formed after activation of agarose with CNBr and subsequent reaction with an alkylamine (cf. (21)).

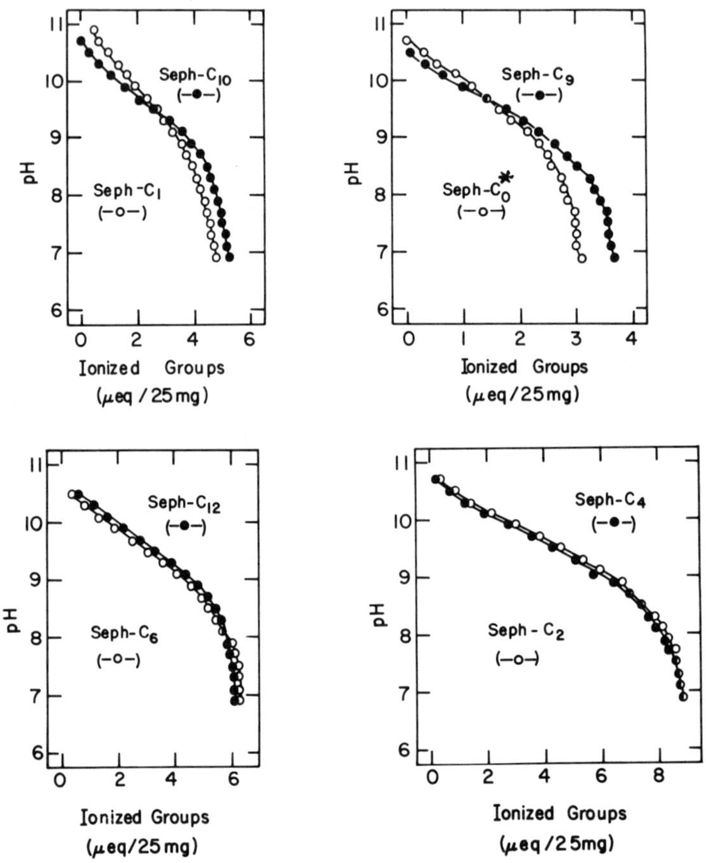

Fig. 6. Potentiometric titrations of arbitrarily chosen members of the Seph-C_n series, carried out on samples (25 mg) of dried column materials. For further details see reference (18).

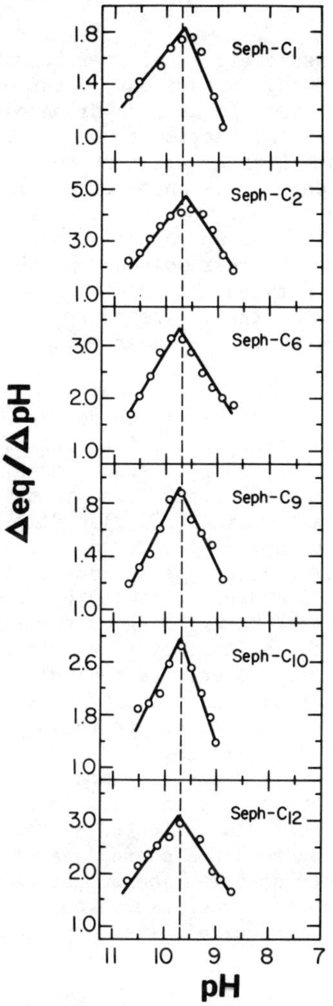

Fig. 7. Determination of pK_a values of arbitrarily chosen members of the Seph-C_n series by plotting $\Delta eq/\Delta pH$ vs. pH from the potentiometric titrations depicted in Fig. 6.

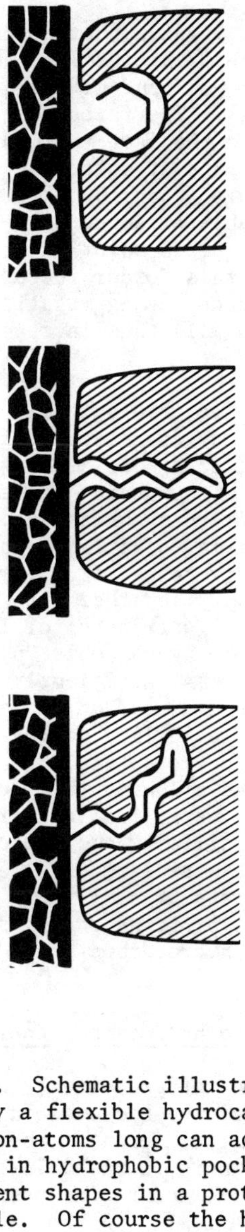

Fig. 8. Schematic illustration of the way a flexible hydrocarbon chain 6-carbon-atoms long can accomodate itself in hydrophobic pockets of different shapes in a protein molecule. Of course the hydrophobic regions may be very superficial patches on the molecule.

ON THE MECHANISM OF ACTION OF HYDROCARBON-COATED AGAROSES

Nature of hydrophobic interactions and their involvement in hydrophobic chromatography.

It is commonly known that nonpolar molecules tend to associate or cluster together in water. This interaction, which cannot be accounted for by covalent, electrostatic, hydrogen bond or charge transfer forces is referred to as a "hydrophobic bonding" (22). The favourable net free energy for this interaction seems to be mainly due to the fact that water molecules have a much greater affinity for each other than for most solutes (22-24). Attractive van der Waals-London dispersion forces may also contribute to the free energy of interaction. Furthermore, there is considerable evidence suggesting that the water molecules surrounding nonpolar solutes in aqueous solution are more "ordered" than in the bulk solvent and that a decrease in this "order" occurs when the contact area between the solutes and water is reduced by aggregation of the solutes with each other. The entropy of the solution will thus increase with the enhanced freedom of the water molecules (25-27).

In their native, biologically functional conformation, proteins are folded so as to bury as many as possible of their hydrophobic side chains in the interior of the molecule and to expose as many as possible of their polar, charged side chains to interaction with water. In fact, this seems to be a major driving force in the folding of protein chains into their 3-dimensional conformation (28,29). Still, complete burying of all hydrophobic groups is generally not achieved, so that in many native proteins some hydrophobic groups remain exposed at the surface of the molecule (29). Together with hydrophobic components of charged amino acids, such as the $(-CH_2-)_n$ stretches of lysines and arginines or the phenyl ring of tyrosines, these may form hydrophobic "patches" or "pockets" at the surface or in crevices of the molecule. A sufficiently large hydrophobic patch may constitute a binding site for the hydrocarbon chains implanted on the relatively inert agarose matrix. Due to the flexibility of the n-hydrocarbon chains they can accomodate themselves within such pockets (Fig. 8) and form "hydrophobic bonds", freeing "ordered" water molecules and allowing them to interact with each other. It seems reasonable to assume that the available hydrophobic patches of different proteins vary in size, shape, lipophilicity and number, and that these variations are reflected in the relative affinities of different proteins to a given hydrocarbon-coated agarose. It is the properties of these patches and perhaps their distribution on the surface of different proteins that most likely play a major role in the resolution of proteins on hydrocarbon-coated agaroses, as suggested by the following experimental evidence.

Effect of the hydrocarbon chain length

The retention of proteins by hydrocarbon-coated agaroses is greatly affected by the length of the hydrocarbon chains attached to the matrix. This is indicated by the fact that different members of a given homologous series of alkyl agaroses (which are indistinguishable in the size of the beads, their shape and their ultrastructure and having a similar charge density and hydrocarbon chain density) differ dramatically in their capacity to bind a given protein under a given set of conditions. For example, both the Seph-C_2 and Seph-C_4 columns used in the experiment depicted in Fig. 9 had at pH 7 a charge density of $\sim$ 8.8 μ eq/ml of settled column material (cf. also Fig. 6) and both had a hydrocarbon chain density of $\sim$ 32 μmoles/ml (settled volume), yet, under the same reaction conditions, Seph-C_2 excluded glycogen phosphorylase b while Seph-C_4 retained it. The Seph-C_2 column can be regarded in this case as a very close control for the Seph-C_4 column (Fig. 10), and the dramatic difference in retention power has to be associated with the fact that each hydrocarbon chain in Seph-C_4 is two carbon

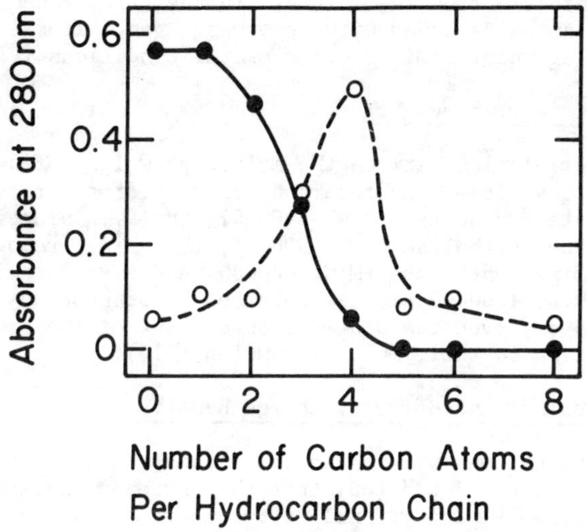

Fig. 9. Adsorption profile (-●-) of glycogen phosphorylase b on a Seph-C_n kit equilibrated at 22° with a buffer composed of sodium β-glycerophosphate (50 mM) and EDTA (1 mM), pH 7. Elution (--o--) is attempted with a "deforming buffer" composed of imidazole (0.4 M) adjusted to pH 7 with citric acid. Aliquots of an AMP-free enzyme (1 mg in 1 ml) were applied on each of the columns. Fractions of 2 ml were collected first with the glycerophosphate buffer and then with the "deformer", and the absorbance of these fractions at 280 nm was monitored.

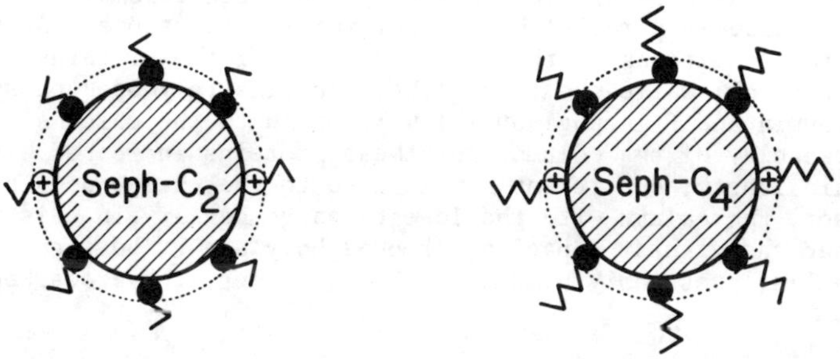

Fig. 10. Schematic representation of Seph-C_2 and Seph-C_4 illustrating their similarity in charge density and in their hydrocarbon chain density and the fact that under the conditions of preparation of these columns (4) the hydrocarbon chain density > charge density, i.e., that many of the ligand-to-matrix bonds are most likely not positively charged at neutral pH. Of course the schematic figure distorts the relative size of the beads and the chains. It should also be emphasized that the hycrocarbon chains are not limited to the outer surface and are evenly distributed throughout the bead (*cf*. Fig. 4).

atoms longer than the hydrocarbon chains of Seph-C_2. A further illustration of the importance of the length of the hydrocarbon chain for retention is found in the fact that the tightness of binding of a given protein usually (though not always) increases upon increasing the number of carbon atoms in the hydrocarbon chains, as judged from the fact that one has to apply more and more drastic conditions in order to detach the protein from the column. In the case of phosphorylase b, for example, the enzyme can be detached with deforming agents from Seph-C_4, but not from Seph-C_6 or Seph-C_8 under similar experimental conditions (Fig. 9).

No correlation with molecular weight

There is no direct correlation between the molecular weight of a protein and the member in a homologous series of hydrocarbon-coated agaroses required to retain it (4,16). Thus glutamate synthetase (M.W. 800000), the P_{IID} protein involved in the regulation of glutamine metabolism (M.W. 50000), the uridyl removing enzyme (M.W. 130000), glutamine synthetase (M.W. 600000) and ATP:glutamine synthetase adenyl transferase (M.W. 130000) are all retained by Seph-C_5-NH_2 under a given set of conditions. Moreover, even the order of detachment of the proteins from the column bears no relation to their molecular weight (16).

Increased capacity not accounted for by charge density

In a series of Seph-C_4 columns varying in their charge density and in their hydrocarbon chain density, it was found (30) that the increase in capacity for glycogen phosphorylase b was roughly proportional to the increase in chain density, rather than to the increase in charge density. For example - upon raising the charge density about 2.5 fold and the chain density 11.1 fold, the capacity of the column for the enzyme increased ∿ 10-fold. Similarly, while a Seph-C_4 column retained glycogen phosphorylase b (∿ 10 mg/ml of settled column material), a DEAE Sephadex column with a charge density about 5 fold higher retained, under the same conditions, a negligible amount of the enzyme (< 0.3 mg/ml of settled column material).

The lack of correlation between the charge density of Seph-C_n columns and their capacity to bind proteins can also be seen in the experiment depicted in Fig. 11. Four different proteins were applied on a series of such columns, whose charge density was determined in parallel by potentiometric titrations. As seen in Fig. 11, under the conditions of the experiment one of the proteins (cytochrome c) was not adsorbed at all while the three other proteins (glycogen phosphorylase b, bovine serum albumin and β-lactoglobulin) were bound to the columns. Generally speaking, the capacity of the columns for these proteins increased with increasing hydrocarbon chain length, but it was not related to the charge density of the columns. In fact, the column with the lowest charge density in this particular set (Seph-C_8) had the highest capacity for phosphorylase b, bovine serum albumin and β-lactoglobulin, presumably because it had the longest hydrocarbon chains in this set.

It should also be pointed out that column materials with a very similar charge density (e.g. Seph-C_2, Seph-C_4 and Seph-C_5 all having charge density of 8-9 μeq of positively charged groups (at pH 7.4) per ml of settled gel) do differ in their capacity for the above mentioned proteins. In all these cases Seph-C_5 adsorbed more than Seph-C_4 which, in turn, adsorbed more than Seph-C_2 (Fig. 11A,B,D). However, the adsorption ratio of various proteins was also found to depend on the length of the hydrocarbon chains, illustrating the fact that the discrimination power of these columns is hydrocarbon-chain-length dependent (Fig. 12).

Negative net-charge of the protein - not a prerequisite for binding

Both negatively and positively charged proteins may (but need not) be retained on hydrocarbon-coated agaroses, which are known to be positively charged at neutral pH. For example, lysozyme (isoelectric pH 10.5 - 11.0 (31)) is more and more retarded on the Seph-C_n series ($n > 5$) as the pH of the eluent is lowered (Fig. 13).

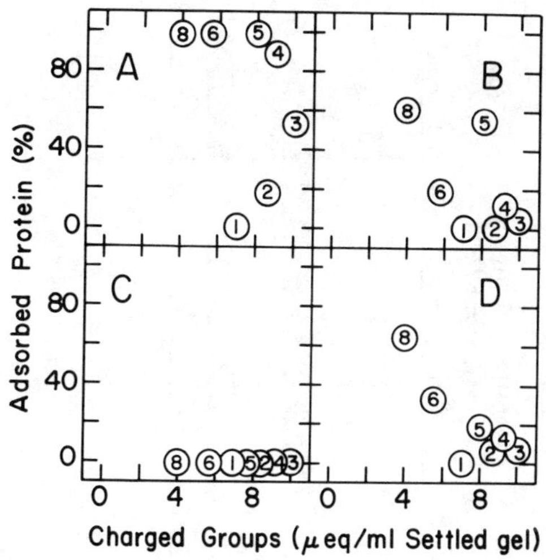

Fig. 11. Lack of correlation between the adsorption of proteins on a series of Seph-C_n columns and their charge density. Samples (1 mg) of four proteins, glycogen phosphorylase b (A); bovine serum albumin (B); cytochrome c (C); and β-lactoglobulin (D)) were applied on each of a series of Seph-C_n columns (5 x 0.5cm) equilibrated and run with a 0.1 M phosphate buffer, 7.4 (22°). The first 2 ml were collected and their absorbance at 280 nm was measured and used to calculate the percent of adsorbed protein in each case. The charge density of the columns at pH 7.4 was determined by potentiometric titrations as described previously (18,30). Numbers within the circles stand for "n" in Seph-C_n.

Similarly, the calf thymus histone H3 (isoelectric pH > 9.0 cf. reference (32)) is retained on Seph-C_9 at pH 4 and low salt concentration (Fig. 14) (cf. reference 33). On the other hand, the histidine binding protein J, which is negatively charged at neutral pH (isoelectric point 5.5) is excluded at pH 7.2 by all Seph-C_n-NH_2 columns tested (n = 4-12) even at low ionic strength (34). It seems therefore that in order to achieve retention of a protein onto alkylagaroses, conditions should be found under which hydrophobic regions in the protein become available for interaction with the hydrocarbon side chains of these columns.

Increasing the number of charged groups per chain often decreases retention of an oppositely-charged protein

The binding of proteins to hydrocarbon-coated agaroses is affected by ionic interactions. However, the ionic interactions will sometimes promote binding, sometimes they will have no effect and sometimes they will oppose the interactions. For example, side chains 4-carbon-atoms long suffice to retain glycogen phosphorylase b in the alkylagarose series whereas 6-carbon-atom chains are needed for retention of this enzyme (under the same conditions) in the ω-aminoalkyl series, which (having an additional -NH_3^+ group at the tip of the chains) should have been more effective in binding a protein with a negative net charge at neutral pH. In fact, these results are not surprising if we keep in mind that by implanting a charged group at the tip of the hydrocarbon chains we do not only allow the column to participate more efficiently in ionic interactions with the negatively charged protein, but also reduce the hydrophobic character around this ionized group.

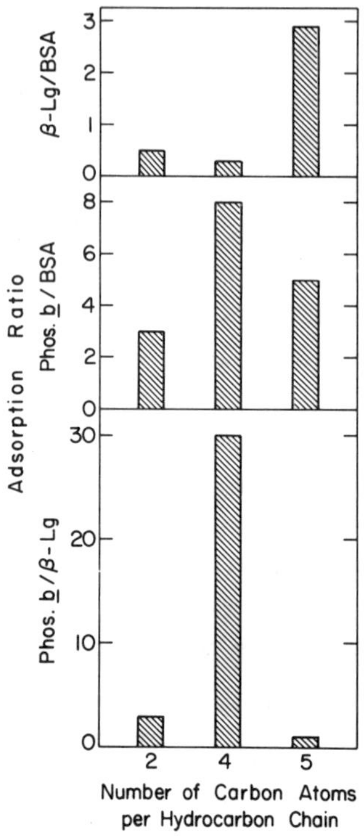

Fig. 12. Adsorption ratios of proteins bound onto three columns differing in the number of carbon atoms in their hydrocarbon chains but having a similar charge density (see text). Results calculated from Fig. 11. Phos. b, glycogen phosphorylase b, β-Lg, β-lactoglobulin, BSA, bovine serum albumin.

Abolishing the positive charge of the columns by acetylation does not cancel the chain-length-dependent adsorption of phosphorylase to a homologous series of Seph-C_n columns

The relative importance of ionic and hydrophobic interactions in the binding of proteins to Seph-C_n columns could also be assessed by chemically modifying the columns in a way that would cancel their charge and comparing the adsorption profile on the modified agaroses against the profile on the original columns. Of course, in doing so we assume that the chemical modification affects mainly the charged loci and either has no other effect on the columns, or that such effect is negligible.

Wilchek and Miron have recently shown (35) that acetylation of alkyl agaroses with acetic anhydride abolishes their positive charge at neutral pH by modifying the N-substituted isourea linkages formed upon activation of agarose with CNBr and subsequent coupling with alkylamines (Fig. 15).

The experiment depicted in Fig. 16 shows that at least in the case of glycogen phosphorylase b, acetylation of Seph-C_n columns does not significantly affect the adsorption profile of the enzyme and that there is still a clear cut hydrocarbon-chain-length dependent adsorption of the protein (cf. also Halperin and Shaltiel, this volume).

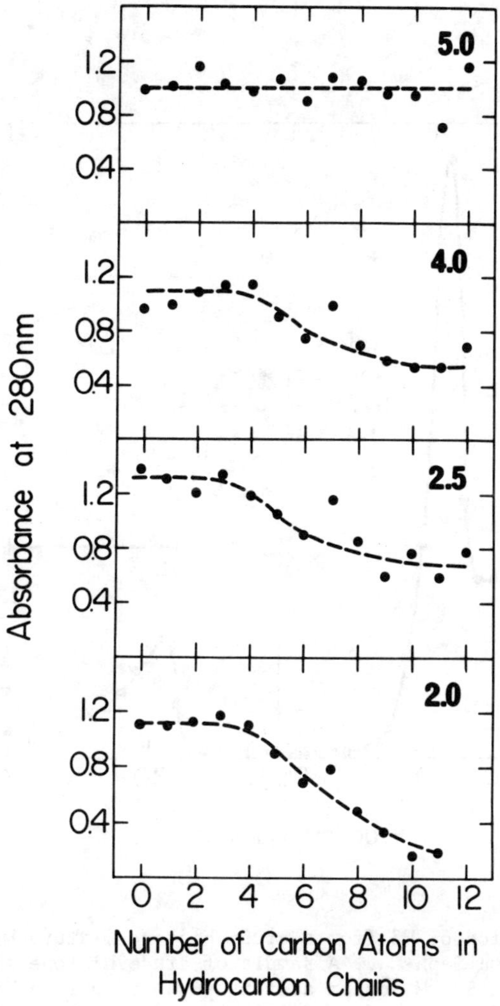

Fig. 13. Effect of pH on the retardation of lysozyme on a kit of Seph-C_n columns (n = 0-12). Samples (1 mg protein dissolved in 0.1 ml of various buffers) were applied on each of the columns (5 x 0.4 cm), equilibrated in the same buffer at 20°. Fractions of 2 ml were collected and their absorbance at 280 nm was monitored. The buffers used(from the top panel downwards) were 0.1 M citrate phosphate, pH 5.0, pH 4.0 and pH 2.5, then a KCl-HCl buffer pH 2.0.

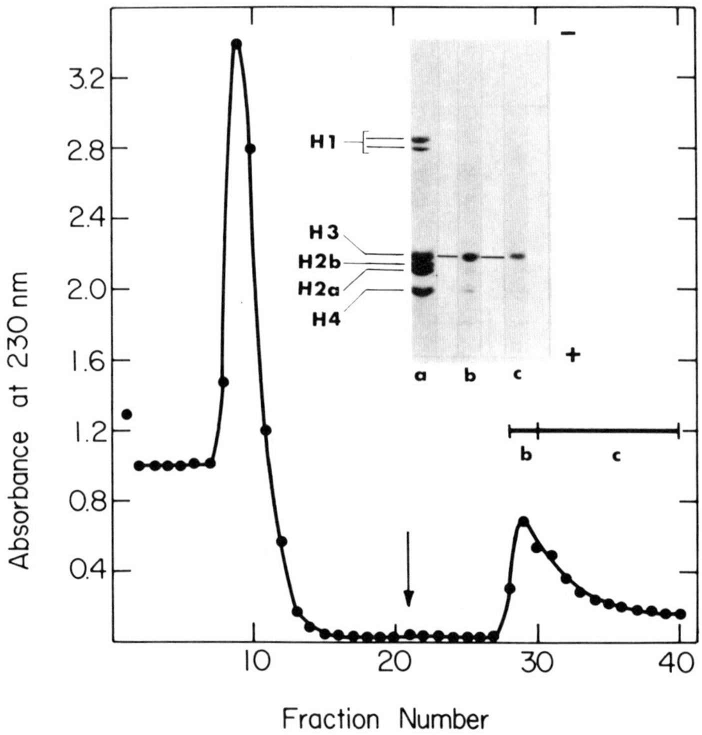

Fig. 14. Isolation of H3 from a crude histone mixture by chromatography on Seph-C_9. A sample of crude histone mixture was dissolved in 50 mM sodium acetate pH 4, to give a histone concentration of ∿ 5 mg/ml. After passage through a millipore filter, 1 ml of the clarified solution (4.3 mg/ml) was applied on a Seph-C_9 column, equilibrated at 20° with the same buffer. The column was eluted first with 10 mM HCl and then (arrow) with 6 M guanidine·HCl. Fractions of 1.3 ml were collected and their absorbance at 230 (-●-) was monitored. Inset: samples of the crude histone mixture (a), and of pools "b" and "c" analyzed by polyacrylamide gel electrophoresis in the presence of sodium dodecylsulfate. For further details see reference (33).

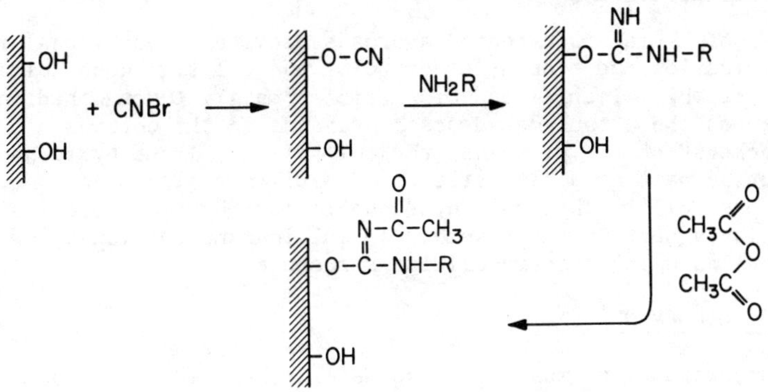

Fig. 15. Acetylation of Seph-C_n columns by acetic anhydride, following the procedure proposed by Wilchek and Miron (35).

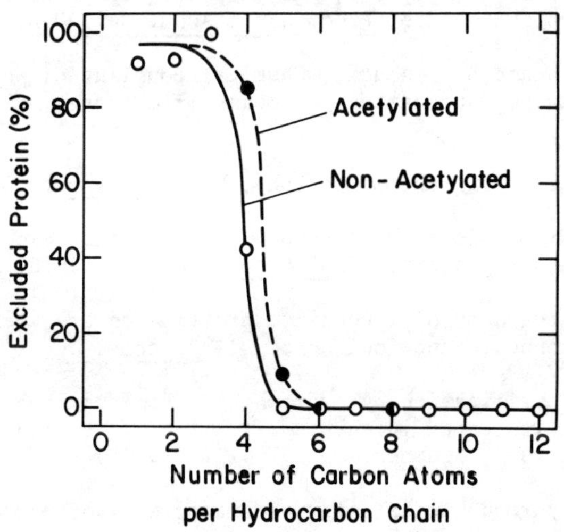

Fig. 16. Adsorption profile of glycogen phosphorylase b on homologous series of Seph-C_n columns before and after acetylation with acetic anhydride. The columns were equilibrated at 22° with a buffer composed of sodium β-glycerophosphate (50 mM), pH 7. Aliquots of the enzyme (1 mg in 0.1 ml) were applied on each of the columns. The first 12 ml were collected and their absorbance at 280 nm was monitored and used to calculate the percentage of excluded protein.

CONCLUDING REMARKS

Homologous series of hydrocarbon-coated agaroses provide an additional route to the resolution, purification and probing of proteins and cells. These columns seem to discriminate between biomolecules and bioparticles mainly through hydrophobic interactions between the attached hydrocarbon chains on the columns and hydrophobic "patches" or "pockets" on the proteins, the cells, etc. These hydrophobic pockets may, but need not,be part of a catalytic or a regulatory site. However, this does not exclude the possibility that such hydrophobic loci have a biological assignment since they may play a part in the assembly of multienzyme complexes, or in the anchoring of proteins onto an intracellular structure.

ACKNOWLEDGEMENT

This work was supported by a grant from the Gesellschaft für Strahlen und Umweltforschung mbH, München, Germany, and the National Council for Research and Development, Israel.

REFERENCES

1. Z. Er-el, Y. Zaidenzaig and S. Shaltiel, Hydrocarbon-coated Sepharoses. Use in the purification of glycogen phosphorylase, Biochem. Biophys. Res. Commun., 49, 383 (1972).

2. S. Shaltiel and Z. Er-el, Hydrophobic chromatography: Use for purification of glycogen synthetase, Proc. Natl. Acad. Sci. USA, 70, 778 (1973).

3. R. Axén, J. Porath and S. Ernback, Chemical coupling of peptides and proteins to polysaccharides by means of cyanogen halides, Nature, 214, 1302 (1967).

4. S. Shaltiel, Hydrophobic chromatography, Meth. Enzymol., 34, 126 (1974).

5. S. Shaltiel, Hydrophobic chromatography. Use in the resolution, purification and probing of proteins, FEBS Proc., 40, 117 (1975).

6. R.J. Yon, Chromatography of lypophilic proteins on adsorbents containing mixed hydrophobic and ionic groups, Biochem. J., 126, 765 (1972).

7. R.J. Yon, Enzyme purification by hydrophobic chromatography: an alternative approach illustrated in the purification of aspartate transcarbamylase from wheat germ, Biochem. J. 137, 127 (1974).

8. B.H.J. Hofstee, Hydrophobic affinity chromatography of proteins, Anal. Biochem., 52, 430 (1973).

9. B.H.J. Hjertén and N.F. Otilio, Protein binding by agarose carrying hydrophobic groups in conjunction with charges, Biochem. Biophys. Res. Commun., 50, 751 (1973).

10. S. Hgerten, Some general aspects of hydrophobic interaction chromatography, J. Chromatogr., 87, 325 (1973).

11. J. Porath, L. Sundbey, N. Fornstedt and I. Olsen, Salting-out in ampiphilic gels as a new approach to hydrophobic adsorption, Nature, 245, 465 (1973).

12. S. Shaltiel, Hydrophobic chromatography in the study of regulatory enzymes,

in "Metabolic interconversion of enzymes - 1973", E.H. Fisher, E.G. Krebs, H. Neurath and E.R. Stadtman eds., pp. 379-392, Springer-Verlag, Berlin (1974).

13. S. Shaltiel, Hydrophobic chromatography, in "Chromatography of synthetic and bio-polymers, Vol. II, Hydrophobic, affinity and ion exchange methods", R. Epton, ed. pp. 1-29, Ellis Horwood for The Chemical Society (1978).

14. Z. Er-el and S. Shaltiel, Hydrophobic chromatography in the resolution of the interconvertible forms of glycogen phosphorylase, FEBS Lett., 40, 142 (1974).

15. M.R. Kula, D. Hatef-Haghi, M. Tauber-Finkelstein and S. Shaltiel, Consecutive use of ω-aminoalkylagaroses. Resolution and purification of clostripain and collagenase from *Clostridium histolyticum,* Biochem. Biophys. Res. Commun., 69, 389 (1976).

16. S. Shaltiel, S.P. Adler, D. Purich, C. Caban, P. Senior and E.R. Stadtman, ω-Aminoalkyl agaroses in the resolution of enzymes involved in regulation of glutamine metabolism, Proc. Natl. Acad. Sci. USA, 72, 3397 (1975).

17. S. Shaltiel, Hydrophobic chromatography and its potential use in biotechnology, in "Enzyme engineering", Vol. III, E.K. Pye and L.B. Wingard, eds., pp. 321-330, Plenum, New York (1977).

18. G. Halperin and S. Shaltiel, Homologous series of alkylagaroses discriminate between erythrocytes of different sources, Biochem. Biophys. Res. Commun., 72, 1497 (1976).

19. A. Amsterdam, Z. Er-el and S. Shaltiel, Ultrastructural identity of hydrocarbon-coated agaroses, Israel J. Med. Sci. 10, 1580 (1974).

20. A. Amsterdam, Z. Er-el and S. Shaltiel, Ultrastructure of beaded agarose, Arch. Biochem. Biophys., 171, 673 (1975).

21. J. Porath, J.C. Janson and T. Låås, Agar derivatives for chromatography, electrophoresis and gel bound enzymes. I. Desulfated and reduced cross linked agar and agarose in spherical bead form, J. Chromatogr., 60, 167 (1971).

22. W.P. Jencks, "Catalysis in chemistry and enzymology", pp. 393-436 McGraw-Hill, New York (1969).

23. J.A.V. Butler, The energy and entropy of hydration of organic compounds, Trans. Faraday Soc., 33, 229 (1937).

24. D.D. Eley, On the solubility of gases, Part I - The inert gases in water, Trans. Faraday Soc., 35, 1281 (1939).

25. H.S. Frank and M.W. Evans, Free volume entropy in condensed systems. III. Entropy in binary liquid mixtures, partial molal entropy in dilute solutions: Structure and thermodynamics in aqueous electrolytes, J. Chem. Phys., 13, 507 (1945).

26. G. Némethy and H.A. Scheraga, Structure of water and hydrophobic bonding in proteins. II. Model for thermodynamic properties of aqueous solutions of hydrocarbons, J. Chem. Phys., 36, 3401 (1962).

27. W. Kauzmann, Some factors in the interpretation of protein denaturation, Adv. Prot. Chem., 14, 1 (1959).

28. R.E. Dickerson and I. Geis, "The structure and action of proteins" Harper and Row, New York (1973).

29. C. Tanford, "The hydrophobic effect", pp. 120-142, Wiley, New York (1973).

30. Z. Er-el, "Hydrophobic chromatography in the study of enzymes involved in glycogen metabolism", Ph.D. Thesis submitted to The Weizmann Institute of Science, Rehovot (1975).

31. G. Alderton, W.H. Ward and H.L. Fevold, Isolation of lysozyme from egg white, J. Biol. Chem., 157, 43 (1945).

32. L.S. Hnilica, "The structure and biological functions of histones" p. 5, CRC Press, Cleveland (1972).

33. H.A. Arfmann and S. Shaltiel, Resolution and purification of histones on homologous series of hydrocarbon-coated agaroses, Eur. J. Biochem., 70, 269 (1976).

34. S. Shaltiel, G. Ferro-Luzzi Ames and K.D. Noel, Hydrophobic chromatography in the purification of the histidine binding proteins J from *Salmonella typhimurium*, Arch. Biochem. Biophys. 159, 174 (1973).

35. M. Wilchek and T. Miron, On the mode of adsorption of proteins to "hydrophobic columns", Biochem. Biophys. Res. Commun., 72, 108 (1976).

ADSORPTION OF PROTEINS AT HIGH SALT CONCENTRATIONS ON HYDROPHOBICALLY INTERACTING MATRICES

Sven Påhlman

Institute of Biochemistry, University of Uppsala, Box 576, S-751 23 Uppsala, Sweden

INTRODUCTION

Soon after Cuatrecasas *et al.* (1) had introduced affinity chromatography, Cuatrecasas (2) showed that, in many cases, a spacer between the gel matrix and the affinity group had to be introduced in order to achieve biospecific adsorption. The spacer, normally a hydrocarbon chain, was said to circumvent sterical hindrance of the gel, and indeed, the chain had to be of a certain minimum length.

However, in 1972 Yon (3), Er-el *et al.* (4) and Hofstee (5) independently found that water-soluble proteins could bind to agarose gels substituted with such hydrocarbon spacers frequently used in affinity chromatography. Shaltiel and co-workers (4,6) showed that increasing the lengths of the attached carbon chains increased the strength of the adsorption. Contrary to Cuatrecasas, they suggested that the adsorption was due to hydrophobic interaction. Since proteins differed in affinity for the hydrocarbon-coated gels, they were suggested for use in protein fractionation (3-5). The new separation method was termed "Hydrophobic Chromatography" by Shaltiel *et al.* (6).

During the past five years, different separation techniques based on the interaction between biomolecules and hydrophobic matrices substituted with non-polar groups have been developed, and the adsorption mechanisms in hydrophobic chromatography have been studied in more detail. The agarose derivatives used by Yon (3), Shaltiel *et al.* (4,6) and Hofstee (5) were synthetized from cyanogen-bromide activated agarose (7) and alkyl mono- or di-amines. The proteins were adsorbed to these gels at low ionic strength ($I \sim 0.01$ M)* and desorbed by an increase in salt concentration ($I \sim 0.1$-0.5 M). The adsorption and desorption conditions were thus similar to those used in ion exchange chromatography. When Hofstee (8) demonstrated in 1973 that almost exclusively negatively charged proteins were adsorbed to these columns, he concluded that the gels, besides non-polar groups, also contained positive charges. Hjertén (9) came to the same conclusion, but from other considerations. He pointed out that hydrophobic interaction should increase with ionic strength and not *vice versa.* In addition, he was not able to show an increased capacity of these gel derivatives when the temperature was raised, indicating that hydrophobic interaction was not the dominant type of interaction. He also showed that proteins could be adsorbed to alkylamino agarose at high ionic strength ($I=4$ M)

*I = ionic strength

and desorbed by lowering the salt concentration.

The effect of the charges on the gel was further investigated by Jost *et al.* (10), who compared non-charged alkyl hydrazide agarose with charged alkylamino agarose. Ovalbumin and α-lactalbumin were found to bind tightly to the charged gels but not to the non-charged ones. Bovine serum albumin adsorbed, however, to both types, which was explained by the presence of fatty acid binding sites on serum albumin. The mechanism by which amines are coupled to CNBr-activated dextrane or agarose as well as the nature of the resulting derivatives have been studied in detail (11). Several different derivatives are possible, many of which are positively charged at neutral pH (12), confirming the results cited above.

However, separation methods based on the interaction of biomolecules with both charged and non-polar groups, as in "hydrophobic chromatography", have proven to be very efficient, which is obvious from the extensive number of publications in the field. Besides the introduction of a new separation technique, the work by the groups of Shaltiel, Yon, Hofstee and many others has also been of importance in affinity chromatography. In view of their results, the non-polar spacers used in affinity chromatography have been replaced by hydrophilic ones, and this modification together with a proper choice of coupling procedures has allowed non-specific adsorption to be eliminated (10,12,14). It should be pointed out that not only the non-polar spacers discussed above, but also the charges introduced with the CNBr-method (10) can give non-specific adsorption.

To avoid the influence of charges, several types of neutral alkylated and arylated agarose derivatives have been synthetized (10,15,16,17). Adsorption of proteins to these derivatives increases with increasing ionic strength and *vice versa.* In these systems, hydrophobic interaction is believed to be the dominant driving force.

As shown by Hjertén (9), Hofstee (18), Hammar *et al.* (19), Rimerman and Hatfield (20) and Doellgast and Fishman (21), agarose gels containing both charged and non-polar groups (as the alkylamino agarose discussed above) also bind proteins at high salt concentrations. The effect of the charges is then quenched, and the adsorption-desorption mechanism is probably the same as for the neutral gels. For reasons given earlier (22), we call the salt-induced type of chromatography, "hydrophobic interaction chromatography", which will be the subject of this contribution.

HYDROPHOBIC INTERACTION

The tendency of non-polar molecules to self-associate in an aqueous medium is called hydrophobic interaction. The association is not primarily due to interactions between the non-polar solutes, but rather to the lack of affinity between water and non-polar or "hydrophobic" molecules (23). It is generally accepted that hydrophobic interactions play a major role in biological systems *e.g.,* lipid bilayer formation and protein stabilization. However, these systems are normally too complex for study of hydrophobic interaction and simpler systems have been chosen. In a classical paper in 1945, Frank and Evans (24) estimated the separate contributions of enthalpy and entropy to the free energy of transfer of simple hydrocarbons from a non-aqueous medium to water. The characteristics of these transfer reactions are a negative enthalpy change, a large negative entropy change and an extraordinary large positive change in heat capacity. Frank and Evans interpreted the large negative changes in entropy (*i.e.,* increasing order) as an ordering of water molecules around the hydrocarbons. Although hydrocarbons do not form hydrogen bonds with water, the net change in hydrogen bonding is believed to be small (23) when non-polar solutes are accomodated in water, whereas the freedom of the water molecules close to the solutes is restricted (25). The interactions between hydrocarbons in water are thus entropy-driven, and the induced order of the system is also reflected in the large change in heat capacity.

A question which immediately arises is whether the thermodynamic quantities for the hydrophobic effect in biological systems are the same as for the transfer of hydrocarbons from a non-aqueous medium to water. From binding studies with proteins and hydrocarbons (26), it appears that the transfer of hydrocarbons from the interior of a protein into water is thermodynamically similar to the more simple systems discussed above. The quantities are given in Table 1. Both the ΔH and ΔS of the

TABLE 1 Thermodynamic quantities associated with the dissociation protein-alkane → protein+alkane (H_2O)

Protein	Ligand	ΔG^o kcal/mole	ΔH^o kcal/mole	ΔS^o cal/mol-deg	ΔC_p^o cal/mol-deg
β-Lactoalbumin A dimer	Pentane	7.8	3.3	- 15	140
	Butane	6.5	2.9	- 12	71
ferrihemoglobin	Pentane	7.0	3.4	- 12	106
	Butane	6.1	1.0	- 17	79
ferrimyoglobin	Pentane	6.7	0.9	- 19	114
	Butane	5.5	0.1	- 18	65

All quantities refer to 25°C using a standard state of unit mole fraction. Data taken from refs. 25 and 26.

reactions contribute to the ΔG, although the entropy term dominates. It should be stressed that ΔC_p is positive for these reactions, which seems to be the best indication of hydrophobic interaction (23,27). In an extensive review on the thermodynamic basis of the stability of biological systems, Edelhoch and Osborne (28) make the point that from the point of view of thermodynamics the stabilization of micelles, membranes and proteins can be compared with the transfer of hydrocarbons from water to an organic solvent. The general conclusions drawn by Frank and Evans (24) are thus still valid, and the more recent experiments with complex biological systems just support the original view (23,25).

As suggested as early as 1959 by Kauzmann (29), the unique three-dimensional structure of globular proteins is to a great extent favoured by hydrophobic interaction between non-polar amino acids. He proposed that these amino acids should be found in the interior of the proteins. This hypothesis has been supported by X-ray crystallographic data from a number of proteins. However, Klotz (30) has shown that there are non-polar amino acid residues at the exterior of globular proteins. The existence of "hydrophobic sites" in water-soluble proteins has also been demonstrated with hydrophobic probes (31). It seems likely, therefore, that these hydrophobic areas can interact with non-polar groups attached to hydrophobic matrices.

The thermodynamic functions ΔG, ΔH, ΔS, and ΔC_p have not been measured for the adsorption of proteins to neutral alkylated and arylated agarose gels. If it is possible to determine these functions, the final answer may be given whether the adsorption process is due mainly to hydrophobic interaction or not.

Despite the lack of thermodynamic data, we believe that the main driving force in the adsorption of proteins to neutral alkyl agarose derivatives, is hydrophobic interaction. The reasons are as follows: The protein-gel interaction increases with 1) hydrophobicity of the substituent 2) the ambient temperature and 3) the ionic strength of the medium. Moreover, at constant ionic strength the interaction varies with the type of salt, in the same way as the salting-out behaviour of proteins

(Hofmeister series). Figure 1 shows schematically the interaction between a solute (*e.g.*, a protein) and a gel substituted with non-polar groups.

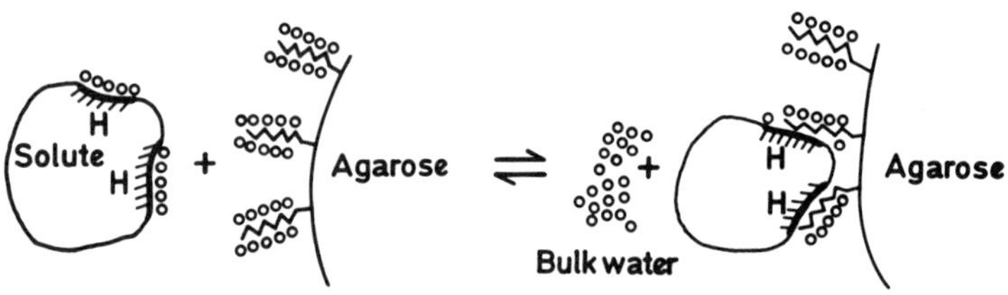

Fig. 1. Schematic representation of hydrophobic interaction between a solute and alkyl-agarose. Water molecules (o) are ordered at hydrophobic regions (H) of the solute and around the alkyl groups. The decreased order (*i.e.*, gain in entropy) of the water molecules upon interaction, is believed to be the driving force in hydrophobic interaction.

HYDROPHOBIC INTERACTION CHROMATOGRAPHY

Gel derivatives

The coupling of amines with different types of non-polar groups to cyanogen bromide-activated agarose (7), has been the most common method for synthetizing amphiphilic gels. Besides the positive charges introduced by the cyanogen bromide method, gel derivatives with additional charges on the ligands have been synthetized (4,32). The extra charges decrease the effect of the non-polar groups (32) and differences in both hydrophobic and ionic properties of proteins, will govern the separation pattern on these gels.

The bisoxirane method introduced by Sundberg *et al.* (33) is a convenient way to couple different types of ligands to agarose without the introduction of charges. The low coupling yields and the long reaction times are probably the reasons why this method has not been used for preparing amphiphilic gels. Coupling alkyl- and aryl-glycidyl ethers to agarose in dioxane, however, gives good yields, high reproducibility and a stable gel derivative as shown by Hjertén *et al.* (16) and Rosengren *et al.* (34). The coupling reaction is shown in Fig. 2. The advantages of synthetizing these gels in dioxane are the short reaction time and the ease in solubilizing extremely hydrophobic ligands such as dodecylglycidyl ether. In aqueous medium the reaction must be performed at high temperatures and pH for several hours (Jan Rosengren, personal communication) with cross-linked Sepharose. Under similar conditions aryl and alkyl halides have been coupled to cross-linked agarose (15), thus giving a neutral amphiphilic gel matrix without the hydrophilic spacer introduced with the glycidyl ether method.

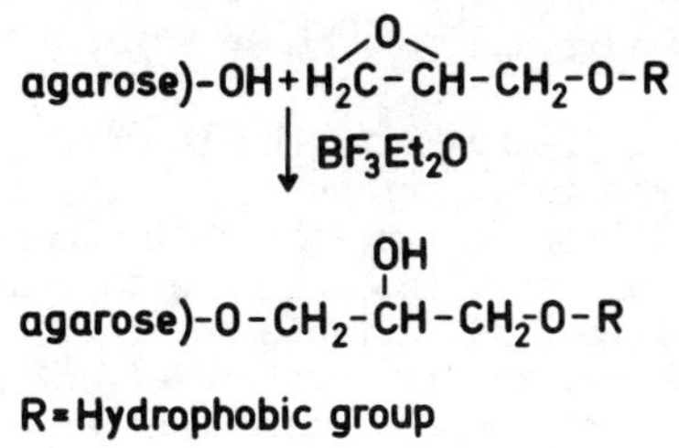

Fig. 2. Coupling of glycidyl ethers to agarose in an organic solvent (*e.g.*, dioxane). The agarose derivatives are named after the R-groups, *e.g.*, in octyl-agarose, R is C_8H_{17}.

A third possibility for synthetizing neutral gels has been introduced by Jost *et al.* (10). They couple alkylhydrazides in dimethyl formamide-bicarbonate, giving neutral gels above pH 4 (10). Wilchek and Miron have also acetylated cyanogen bromide coupled amines, which reduced the number of charges significantly (35). The leakage of the ligands from these gels is, however, much greater than from non-acetylated ones.

Degree of substitution

The ligand concentration of the gel is often of great importance in order to achieve a good yield and an effective purification, as will be discussed below. We therefore need accurate and simple methods to determine the degree of substitution. One such method is to use radioactive ligands as described by Jennissen and Heilmeyer (36). Hofstee (37) uses a dye which interacts with the positive charges of the alkyl amino derivatives, and the degree of substitution of the gels is determined spectrophotometrically. Other workers titrate these gels with a result similar to that of Hofstee. The methods based on determination of the charge density on alkyl amino agarose are especially dubious since only a non-constant fraction of the alkyl groups is linked *via* a charged group (12).

Gels prepared with the glycidyl ether method presented above do not contain any charges or hetero elements. A method based on NMR for determining the degree of substitution has therefore been developed (34). The alkyl agarose is partially hydrolyzed in formic acid, and the hydrolysate is transferred to a deuterated medium and finally dissolved in deuterated dimethyl sulphoxide. The area of the NMR signals from the ligand protons (signal positions determined for the free glycidyl ether in a separate recording) is divided by the area of the signals from the galactose protons, giving the degree of substitution in moles ligand per mole galactose. For details see Ref. (34). The accuracy of the method increases with the degree of substitution and the number of protons in the substituent. For butyl and pentyl agarose the lower limit for determining the ligand concentration lies around 20-30 mmol ligand/mol galactose. However, gels suitable for protein fractionation normally have a degree of substitution between 40-100 mmol/mol galactose. The ligand concentration is also of importance in affinity chromatography, especially when non-specific interactions also contribute to the adsorption process. In such cases when the ligand concentration must be controlled, the NMR-method might be very useful, for it provides the possibility of determining the concentration of ligands lacking, for instance, charges or hetero elements.

Adsorption and desorption conditions

The protein-gel interaction increases with the temperature, the hydrophobicity of the ligand and the solute, the ionic strength, and the degree of substitution, and furthermore it varies with different types of salt according to the Hofmeister series. Normally it is not necessary in each case to examine all parameters affecting the interaction in order to find suitable adsorption and desorption conditions. In an actual separation problem, the temperature effect and the salt dependence according to the Hofmeister series are of more theoretical than of practical value. In the following paragraphs, however, the different parameters and how they effect the adsorption and desorption of proteins to alkyl agarose will be discussed. The discussion is limited to neutral gels and applies to charged amphiphilic gels only under conditions when the effect of the charges is quenched by salt.

From theory one can expect that the protein-gel interaction will increase with increasing hydrophobicity of the ligand and increasing salt concentrations (23). In practice, at 1 to 4 M sodium chloride most water-soluble proteins will bind to alkyl agarose gels with carbon chains varying from butyl to octyl. The capacity of these gels increases drastically with the length of the carbon chain (19,34,38), and, for instance, in most cases dodecyl agarose will not give an efficient purification because of low yields. The capacity (34) and the resolution (39) of the gels are also affected by the ligand concentration. Rosengren *et al.* (34) and Jennissen (40) have shown that more than one alkyl group must interact with each protein molecule in order for adsorption to occur. This limits the useful range of the degree of substitution; excessively high ligand concentrations lead to irreversible binding even at low salt concentrations, and low degrees of substitution give low capacities. An intermediate degree of substitution of 40-100 mmol substituent/mol galactose has been found to be suitable in most column experiments (34). In cases when amphiphilic agarose gels have been used as matrices for enzyme reactors (16,41) the ligand concentrations were 5 to 10 times higher than for the gels used for protein fractionation (41). These highly substituted gels could adsorb up to 35 mg β-amylase per ml gel with at most 50 % of the total activity recovered in immobilized form (41). The enzyme could not be completely desorbed from these gels unless detergents such as Triton X-100 were used (42).

The interaction of proteins with neutral alkyl agarose gels is entropy-driven, and the capacities of the gels will hence increase with increasing temperature. The effects are not drastic (16), and since purification of proteins must often be performed at low temperature, changes in the ambient temperature have not come into frequent use for either desorption or adsorption of proteins to amphiphilic gels.

Several groups have investigated the effect of different types of neutral salts on the adsorption-desorption of proteins to both non-charged and charged alkyl derivatives of agarose (17,22,36,43). The trend seems to be similar irrespective of whether alkyl amino agarose (adsorption at low ionic strength) (36,43) or neutral alkyl agarose (adsorption at high ionic strength) (17,22) are used. For the adsorption of human serum albumin (HSA) to neutral pentyl agarose the following salt dependence is shown at constant ionic strength (22): the capacity was highest in Na_2SO_4 and decreased as follows NaCl > $(NH_4)_2SO_4$ > NH_4Cl > NaBr > NaSCN. Circular dichroism spectra of HSA in some of these salts recorded at the same ionic strength revealed that the low binding capacities in NaSCN and NaBr could be explained by structural changes in the proteins. However, no structural changes could be detected in Na_2SO_4 and NaCl, indicating that the differences in binding capacities in the presence of these salts are not due to changes in protein structure (22). The salt dependence is thus in accordance with the Hofmeister series for salting-out of proteins (22). Schrier and Schrier (44) have shown that this effect can be attributed to hydrophobic interaction, which thus implies that hydrophobic interaction really contributes to the adsorption of proteins to alkyl derivatives of agarose. A point to emphasize in this context is that so-called chaotropic ions (*e.g.*, NaSCN, NaBr) can not be used for desorption of protein from neutral alkyl gels. In the example related above 3 M NaSCN promotes the adsorption of HSA to pentyl agarose (∿2 mg HSA/ml gel) whereas no detectable amount of HSA is bound to the gel without any salt present (22). Chaotropic salts might, however, be very useful for desorption purposes when charged alkyl gels are used since the ionic interaction can be counteracted, whereas the hydrophobic interaction will be only slightly enhanced.

Material adsorbed to amphiphilic agarose derivatives at high salt concentrations can be desorbed by a stepwise or continuous decrease in ionic strength. An illustrative example is given in Fig. 3, where human plasma proteins have been fractionated on pentyl agarose by a stepwise decrease of the sodium chloride concentra-

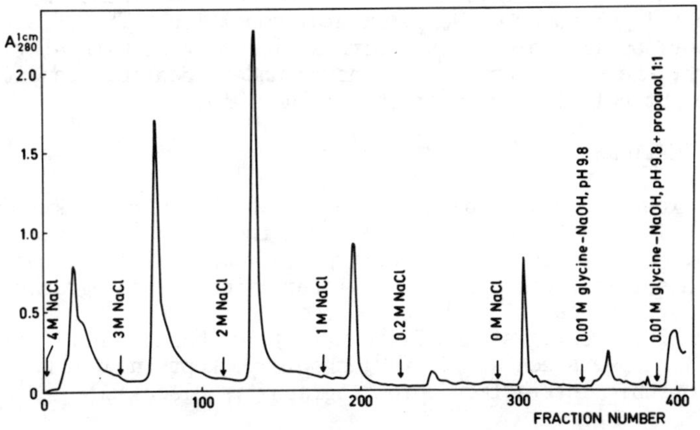

Fig. 3. Fractionation of human plasma on pentyl agarose. Sample: 2 ml plasma in 10 mM sodium phosphate pH 6.8 + 4 M NaCl. Temperature: 4° and column dimensions: 36 x 1.5 cm I.D. Degree of substitution not determined.

tion. Further desorption was achieved by raising the pH, and finally by addition of a polarity-reducing solvent such as propanol or ethylene glycol. Salt gradients normally give rise to broad elution peaks with extensive dilution of the material as a result (38). A superimposed gradient in increasing ethylene glycol concentration gives a more effective desorption as demonstrated by, for instance, Janson and Låås (39). The example in Fig. 3 also illustrates the general way to proceed in this type of chromatography. A gel with short alkyl chains is chosen, and the low intrinsic capacity of the gel can then be compensated for by using high salt concentrations in the adsorption step. The bound material can be desorbed in high yield without using too drastic desorption conditions. As mentioned earlier, detergents are very efficient for desorbing tightly bound material, but they are difficult to remove from the gel. Organic solvents partially remove the detergents, but the fraction retained on the gel changes the protein-binding capacity (42). Irreversibly bound protein material seems not to change the capacity of the alkyl agarose significantly, and after rinsing at pH 10 at low ionic strength, the gel can be reused. Cross-linked agarose can be further treated with high concentrations of sodium hydroxide at high temperatures, which will hydrolyze bound proteins.

Von der Haar (45) and Wilchek and Miron (35) have suggested that the adsorption mechanism in hydrophobic interaction chromatography is not an interaction between the hydrophobic ligands and the proteins but rather a precipitation of the solutes on the agarose matrix. This view is supported by several reports where the adsorption or precipitation of tRNA or proteins on unsubstituted agarose at high ammonium sulphate concentrations is demonstrated (45,46). The very high capacities of amphiphilic gels in the presence of salting-out ions (22) can also be fit into this explanation. Although agarose does decrease the amount of ammonium sulphate required to precipitate proteins by about 10 % (45), the salt concentrations used in hydrophobic interaction chromatography are normally much lower than the concentration required for precipitation of the proteins. A control experiment with unsubstituted agarose should always be performed, and so far we have not found any protein that binds to the control gel under our experimental conditions. The effect of the alkyl chains is also obvious from experiments with series of gels with alkyl chains of increasing length, where for instance an enzyme under identical conditions binds to hexyl agarose but not to butyl agarose (cf. Ref. 19).

APPLICATIONS

In this section experiments will be described in which neutral alkyl derivatives of agarose synthetized with the glycidyl-ether method have been used for purification of proteins. Many groups using other types of amphiphilic gel matrices are also working with salt-induced hydrophobic interaction chromatography but their applications will not be described here. However, the papers by Låås (47), Rimerman and Hatfield (20), Doellgast and Fishman (21), Raibaud *et al.* (43), Bigelis and Umbarger (48) and Mevarech *et al.* (49) should be kept in mind when the applicability of hydrophobic interaction chromatography is discussed.

Human plasma fractionated on pentyl agarose is shown in Fig. 3. No attempts were made to purify any special plasma component, but analysis of the fractions showed a considerable purification of some of the plasma proteins. Peeters *et al.* (50) have, however, purified lipoproteins from plasma using pentyl agarose. Enzymes from several different sources have been purified in high yield, indicating that the experimental conditions are rather mild. In this respect, the purification of phosphoprotein phosphatase from rat liver on pentyl agarose is interesting (Fig. 4 and Ref. 38).

A low molecular weight preparation obtained after alcohol precipitation has been purified to homogeneity by conventional techniques (51) but the enzyme having the same molecular weight as that found in crude extract has been purified only 60-fold (38). This enzyme is very unstable and purification procedures must therefore be both fast and efficient. Chromatography of crude extract on pentyl agarose gives an 5-10-fold purification of the enzyme in one step, with yields varying from 60-70 % (Fig. 4). The enzyme retains its molecular size and can be further purified by ion-exchange and affinity chromatography to a 400-fold increase in specific activity (38).

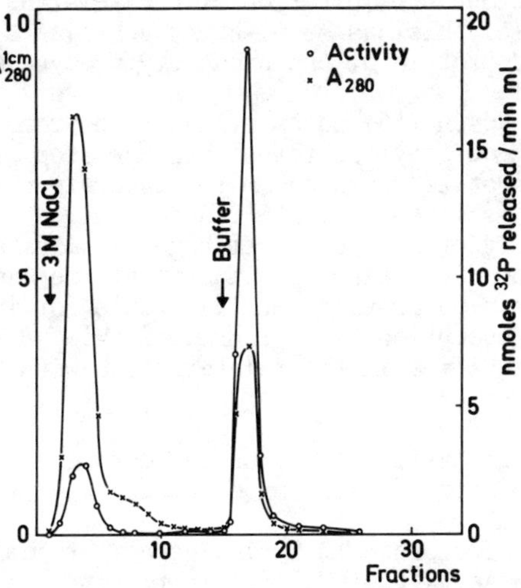

Fig. 4. Phosphoprotein phosphatase from rat liver on pentyl agarose. Sample: 2 ml crude extract in 10 mM imidazole-HCl pH 7.5 + 15 mM β-mercaptoethanol + 2.5 mM $MgCl_2$ + 3 M NaCl. Temperature: 4° and column dimensions: 9 x 1.0 cm I.D. Degree of substitution: 90 mmol/mol galactose. For details concerning enzyme activity measurements and preparation of crude extract, see Ref. 38.

Another labile enzyme, histidine decarboxylase from mouse mastocytoma, has been purified on octyl agarose (19). When crude extract is chromatographed, a 10-fold purification with an 80 % yield is obtained. This enzyme is inactivated at high salt concentrations but is stable in, for instance, 0.6 M potassium phosphate, pH 6.8. A sufficiently strong interaction can be achieved using octyl agarose, thus compensating for the low ionic strength by use of a long alkyl chain. After desorption the enzyme is stable for weeks and can be purified further (52). When histidine decarboxylase from this source is precipitated with ammonium sulphate and the salt is dialysed away, the enzymatic activity is recovered but it drops drastically within a few days after this treatment. This example indicates that the process of adsorption to octyl agarose is not similar to precipitation (see discussion and Ref. 45).

Other enzymes purified by hydrophobic interaction chromatography are cellulases from the fungus *Trichoderma viride* (16), β-amylase from barley meal (39) and tyrosine decarboxylase from *Streptococcus faecalis* (53).

A potential area of application of hydrophobic interaction chromatography is membrane bound proteins. Those parts of the proteins that are embedded in the lipid bilayer presumably contain non-polar amino acid residues. However, when integral membrane proteins are solubilized with detergents, the hydrophobic parts of the proteins are covered with detergent micelles (54). The protein-detergent complex will thus be hydrophilic and have a low affinity for amphiphilic agarose gels. The detergents not bound to proteins will also decrease the capacity of the columns (cf. desorption with detergents). These difficulties can be solved if the detergents can be removed without precipitation of the membrane protein, which then can be adsorbed to alkyl agarose. This has been shown by Liljas *et al.* (55) who have purified the major sialoglycoprotein of the human erythrocyte on pentyl agarose.

Using another approach, Rosén (56) has purified cytochrome *c* oxidase (a membrane protein) from beef-heart on octyl agarose. This one-step procedure gives an enzyme as pure as achieved with other known and more elaborate purification procedures. The purification scheme is given in Table 2 and involves the use of the detergent cholate. This ionic detergent does not form huge micelles and is very efficient for solubilizing membrane bound proteins. The fact that the micelles are small might imply that part of the non-polar sites are available for interaction with the octyl groups. The successful desorption of cytochrome *c* oxidase with Triton X-100 clearly shows that detergents do differ in their interaction with both proteins and amphiphilic agarose derivatives.

TABLE 2 Purification of bovine cytochrome *c* oxidase by hydrophobic interaction chromatography (S. Rosén (56))

1. Addition of cholate (3 % final conc.) to cell homogenate.
2. Precipitation with $(NH_4)_2SO_4$ in two steps. Cholate present.
3. Homogenization of precipitate. Supernatant saved.
4. Application of supernatant to octyl agarose in Tris, 15 % $(NH_4)_2SO_4$ and 1 mM EDTA.
5. Wash with 10 % cholate. Cytochromes desorbed.
6. Addition of 1.5 % Tween 80 to remove cholate. (Cholate disturbes the enzyme assay.)
7. Desorption of cytochrome *c* oxidase with 1 % Triton X-100.

At an early stage in the development of this separation method the possibility of using amphiphilic gels for fractionation of viruses and bacteria was investigated. Using dodecyl or naphtyl agarose it was possible to adsorb satellite tobacco necrosis virus (STNV) and yeast cells at 4 M sodium chloride (16). The particles were desorbed when the salt was removed from the buffer, indicating that STNV and yeast cells possess surface sites of hydrophobic nature. A much more thorough investigation of the interaction between bacteria and alkyl agarose has been undertaken by Edebo and co-workers (57), who have studied two strains of *Salmonella typhimurium* (395 MS and R bacteria) whose surfaces differ in carbohydrate content and charge density. The pathogenic S bacteria have a high carbohydrate content and are uncharged, resist phagocytosis *in vitro*, do not tend to invade HeLa cells *in vitro*, attach poorly to the mouse small intestine *in vitro*, and show no affinity for octyl agarose. The non-pathogenic R bacteria, which have less carbohydrate on the surface, show the opposite affinity for the different mammalian cells and adsorb to octyl agarose at 1 M ammonium sulphate. The bacteria can be desorbed, after lowering the ionic strength, by the addition of 0.01 % Triton X-100. The authors suggest that the effects shown *in vitro* are of importance *in vivo* with respect to, for instance, elimination of invaded bacteria. The experiments also show that hydrophobic interaction chromatography is a simple tool for investigating surface properties of bacteria.

CONCLUSION

For hydrophobic interaction chromatography on neutral alkyl and aryl derivatives of agarose, synthetized with the glycidyl ether method, the following conclusions can be drawn:

1. Hydrophobic interaction chromatography is an efficient separation method which introduces the possibility of separating proteins which differ in "hydrophobicity".

2. Non-charged stable agarose derivatives are obtained with the glycidyl ether method, and the coupling yields are reproducible.

3. The capacity of the amphiphilic gels increases with the salt concentration and the hydrophobicity of the ligand, making it possible to operate with just one or two different types of alkyl derivatives and compensate for low or high intrinsic capacity of the gel with a high or low salt concentration in the adsorption step.

4. Properly chosen adsorption conditions give high recoveries.

5. Hydrophobic interaction chromatography can be used to investigate biopolymers and bacteria with respect to their surface properties.

A grant from "OE och Edla Johanssons vetenskapliga stiftelse" is gratefully acknowledged.

REFERENCES

(1) P. Cuatrecasas, M. Wilchek and C.B. Anfinsen, Selective enzyme purification by affinity chromatography, Proc. Nat. Acad. Sci. USA 61, 636 (1968).
(2) P. Cuatrecasas, Protein purification by affinity chromatography. Derivatizations of agarose and polyacrylamide beads, J. Biol. Chem. 245, 3059 (1970).
(3) R.J. Yon, Chromatography of lipophilic proteins on adsorbents containing mixed hydrophobic and ionic groups, Biochem. J. 126, 765 (1972).
(4) Z. Er-el, Y. Zaidenzaig and S. Shaltiel, Hydrocarbon-coated Sepharoses. Use in the purification of glycogen phosphorylase, Biochem. Biophys. Res. Commun. 49, 383 (1972).
(5) B.H.J. Hofstee, Hydrophobic affinity chromatography of proteins, Anal. Biochem. 52, 430 (1973).
(6) S. Shaltiel and Z. Er-el, Hydrophobic Chromatography: Use for purification of glycogen synthetase, Proc. Nat. Acad. Sci. USA 70, 778 (1973).
(7) R. Axén, J. Porath and S. Ernback, Chemical coupling of peptides and proteins to polysaccharides by means of cyanogen halides, Nature (London) 214, 1302 (1967).
(8) B.H.J. Hofstee, Protein binding by agarose carrying hydrophobic groups in conjunction with charges, Biochem. Biophys. Res. Commun. 50, 751 (1973).
(9) S. Hjertén, Some general aspects of hydrophobic interaction chromatography, J. Chromatogr. 87, 325 (1973).
(10) R. Jost, T. Miron and M. Wilchek, The mode of adsorption of proteins to aliphatic and aromatic amines coupled to cyanogen bromide-activated agarose, Biochim. Biophys. Acta 362, 75 (1974).
(11) K. Broström, S. Ekman, L. Kågedal and S. Åkerström, Covalent binding of proteins to polysaccharides by cyanogen bromide and organic cyanates. III. Structural studies on the linkage region, Acta Chem. Scand. B 28, 102 (1974).
(12) G.I. Tesser, H.-U. Fisch and R. Schwyzer, Limitations of affinity chromatography: solvolytic detachment of ligands from polymeric supports, Helv. Chim. Acta 57, 1718 (1974).
(13) S. Barry and P. O'Carra, Affinity chromatography of nicotinamide-adenine dinucleotide-linked dehydrogenases on immobilized derivatives of the dinucleotide, Biochem. J. 135, 595 (1973).
(14) P. O'Carra, S. Barry and T. Griffin, Spacer arms in affinity chromatography: Use of hydrophilic arms to control or eliminate non-biospecific adsorption effects, FEBS Lett. 43, 169 (1974).
(15) J. Porath, L. Sundberg, N. Fornstedt, I. Olsson, Salting-out in amphiphilic gels as a new approach to hydrophobic adsorption, Nature 245, 465 (1973).

(16) S. Hjertén, J. Rosengren and S. Påhlman, Hydrophobic interaction chromatography. The synthesis and the use of some alkyl and aryl derivatives of agarose, J. Chromatogr. 101, 281 (1974).

(17) A.H. Nishikawa and P. Bailon, Lyotropic salt effects in hydrophobic chromatography, Anal. Biochem. 68, 274 (1975).

(18) B.H.J. Hofstee, Hydrophobic effects in adsorptive protein immobilization, J. Macromol. Sci. -- Chem. A10 (1&2), 111 (1976).

(19) L. Hammar, S. Påhlman and S. Hjertén, Chromatographic purification of a mammalian histidine decarboxylase on charged and non-charged alkyl derivatives of agarose, Biochim. Biophys. Acta 403, 554 (1975).

(20) R.A. Rimerman and G.W. Hatfield, Phosphate-induced protein chromatography, Science 182, 1268 (1973).

(21) G.J. Doellgast and W.H. Fishman, Purification of human placental alkaline phosphatase. Salt effects in affinity chromatography, Biochem. J. 141, 103 (1974).

(22) S. Påhlman, J. Rosengren and S. Hjertén, Hydrophobic interaction chromatography on uncharged Sepharose derivatives. Effects of neutral salts on the adsorption of proteins, J. Chromatogr. 131, 99 (1977).

(23) Tanford, C. (1973) The Hydrophobic Effect. Formation of Micelles and Biological Membranes, Wiley, New York.

(24) H.S. Frank and M.W. Evans, Free volume and entropy in condensed systems. III. Entropy in binary liquid mixtures; partial molal entropy in dilute solutions; structure and thermodynamics in aqueous electrolytes, J. Chem. Phys. 13, 507 (1945).

(25) R. Biltonen, The influence of water on interacting biological systems (1976) in L'eau et les systèmes biologiques (Eds. A. Alfsen and A.J. Berteaud), Centre National de la Recherche Scientifique, Paris, p. 13.

(26) A. Wishnia, On the thermodynamic basis of induced fit. Specific alkane binding to proteins, Biochemistry 8, 5070 (1969).

(27) N. Nichols, R. Sköld, C. Spink, J. Suurkuusk and I. Wadsö, Additivity relations for the heat capacities of non-electrolytes in aqueous solution, J. Chem. Thermodynamics 8, 1081 (1976).

(28) H. Edelhoch and J.C. Osborne, The thermodynamic basis of the stability of proteins, nucleic acids and membranes, Adv. Prot. Chem. 30, 183 (1976).

(29) W. Kauzmann, Some factors in the interpretation of protein denaturation, Adv. Prot. Chem. 14, 1 (1959).

(30) I.M. Klotz, Comparison of molecular structures of proteins: Helix content; distribution of apolar residues, Arch. Biochem. Biophys. 138, 704 (1970).

(31) K. Shibata, Fluorescence dyes bound to hydrophobic regions of proteins (1971) in New techniques in amino acid, peptide and protein analysis (Eds. H. Niederwieser and G. Pataki), p. 343.

(32) R. Yon and R. Simmonds, Protein chromatography on adsorbents with hydrophobic and ionic groups. Some properties of N-(3-carboxypropionyl)-aminodecyl-Sepharose and its interaction with wheat-germ aspartate transcarbamoylase, Biochem. J. 151, 281 (1975).

(33) L. Sundberg and J. Porath, Preparation of adsorbents for biospecific affinity chromatography. I. Attachment of amino group-containing ligands to insoluble polymers by means of bifunctional oxiranes, J. Chromatogr. 90, 87 (1974).

(34) J. Rosengren, S. Påhlman, M. Glad and S. Hjertén, Hydrophobic interaction chromatography on non-charged Sepharose derivatives. Binding of a model protein, related to ionic strength, hydrophobicity of the substituent, and degree of substitution (determined by NMR), Biochim. Biophys. Acta 412, 51 (1975).

(35) M. Wilchek and T. Miron, On the mode of adsorption of proteins to "hydrophobic columns", Biochem. Biophys. Res. Commun. 72, 108 (1976).

(36) H.P. Jennissen and L.M.G. Heilmeyer, Jr., General aspects of hydrophobic chromatography. Adsorption and elution characteristics of some skeletal muscle enzymes, Biochemistry 14, 754 (1975).

(37) B.H. J. Hofstee, Accessible hydrophobic groups of native proteins, Biochem. Biophys. Res. Commun. 63, 618 (1975).

(38) V.P.K. Titanji and S. Påhlman, Non-charged alkyl derivatives of agarose and protamine-agarose in the purification of rat-liver phosphoprotein phosphatases. Submitted.

(39) J.-C. Janson and T. Låås, Hydrophobic interaction chromatography on phenyl- and octyl-Sepharose CL-4B (in press), in Proceedings of the Chemical Society Symposium (Birmingham) on the Chromatographic Fractionation of Macromolecules (Ed. R. Epton).

(40) H.P. Jennissen, Evidence for negative cooperativity in the adsorption of phosphorylase b on hydrophobic agaroses, Biochemistry 15, 5683 (1976).

(41) K. Dahlgren Caldwell, R. Axén, M. Bergwall and J. Porath, Immobilization of enzymes based on hydrophobic interaction. I. Preparation and properties of a β-amylase adsorbate, Biotechnol. Bioeng. XVIII, 1573 (1976).

(42) K. Dahlgren Caldwell, R. Axén and J. Porath, Reversible Immobilization of enzymes to hydrophobic agarose gels, Biotechnol. Bioeng. XVIII, 433 (1976).

(43) O. Raibaud, A. Högberg-Raibaud and M.E. Goldberg, Purification of *E. coli* enzymes by chromatography on amphiphilic gels, FEBS Lett. 50, 130 (1975).

(44) E.E. Schrier and E.B. Schrier, The salting-out behaviour of amides and its relation to the denaturation of proteins by salts, J. Phys. Chem. 71, 1851 (1967).

(45) F. von der Haar, Purification of proteins by fractional interfacial salting out on unsubstituted agarose gels, Biochem. Biophys. Res. Commun. 70, 1009 (1976).

(46) W.M. Holmes, R.E. Hurd, B.R. Reid, R.A Rimerman and G.W. Hatfield, Separation of transfer ribonucleic acid by Sepharose chromatography using reverse salt gradients, Proc. Nat. Acad. Sci. USA 72, 1068 (1975).

(47) T. Låås, Agar derivatives for chromatography, electrophoresis and gel-bound enzyme. IV. Benzylated dibromopropanol cross-linked Sepharose as an amphophilic gel for hydrophobic salting-out chromatography of enzymes wiht special emphasis on denaturation risks, J. Chromatogr. 111, 373 (1975).

(48) R. Bigelis and H.E. Umbarger, Purification of yeast α-isopropylmalate isomerase. High ionic strength hydrophobic chromatography, J. Biol. Chem. 250, 4315 (1975).

(49) M. Mevarech, W. Leicht and M.M. Werber, Hydrophobic chromatography and fractionation of enzymes from extremely halophilic bacteria using decreasing concentration gradients of ammonium sulfate, Biochemistry 15, 2383 (1976).

(50) H. Peeters, V. Blaton and H. Caster, Hydrophobic separation of plasma proteins and lipoproteins (1976), in Prot. Biol. Fluids, 23rd Colloquium (Ed. H. Peeters), Pergamon Press, Oxford, p. 681.

(51) H. Brandt, Z.L. Cafulong and E.Y.C. Lee, Purification and properties of rabbit liver phosphorylase phosphatase, J. Biol. Chem. 250, 8038 (1975).

(52) L. Hammar and S. Hjertén, Mammalian histidine decarboxylase; improved purification procedure. Submitted.

(53) S. Allenmark and B. Servenius, On the purification of bacterial tyrosine decarboxylase (1977) in International Symposium on Affinity Chromatography, Vienna Sept. 20-24, abstract 19.

(54) A Helenius and K. Simons, Solubilization of membranes by detergents, Biochim. Biophys. Acta 415, 29 (1975).

(55) L. Liljas, P. Lundahl and S. Hjertén, The major sialoglycoprotein of the human erythrocyte membrane. Release with a non-ionic detergent and purification, Biochim. Biophys. Acta 426, 526 (1976).

(56) S. Rosén, Purification of beef-heart cytochrome *c* oxidase by hydrophobic interaction chromatography on octyl-Sepharose CL-4B. Submitted.

(57) L. Edebo, J. Hed, E. Kihlström, K.-E. Magnusson, O. Stendahl and C. Tagesson, Bacterial invasion and antigen elimination as a cell surface interaction phenomenon (1977) in Symposium on biology of connective tissue, Uppsala Sept. 4-9, abstract book, p. 128.

THE BINDING OF MYOGLOBIN AND HEMOGLOBIN TO HYDROPHOBICALLY SUBSTITUTED SEPHAROSES

Michael Breitenbach

Institut für Allgemeine Biochemie and Ludwig Boltzmann Forschungsstelle für Biochemie, University of Vienna, Währingerstraße 38, A-1090 Vienna, Austria

INTRODUCTION

In the literature on hydrophobic interaction chromatography there exists a controversy as to the mechanism of binding to the gels introduced by Hjertén (1) and Shaltiel (2), respectively. The former gels, called type 1-gels in this communication (see below for structures of the gels), carry no charge at neutral pH, bind protein more effectively at high salt (1 M NaCl) and high temperature ($25^{\circ}C$) than at low salt (0.01 M NaCl) and low temperature ($4^{\circ}C$). This conforms to predictions based on the theory of purely hydrophobic interactions put forward by Tanford (3), but it must be said that the values to be expected for ΔS, ΔH, and ΔCp of the binding reaction could not be verified up to now and that there are exceptions to the above rules (cf. for instance the binding of sperm whale myoglobin reported below. After binding at low salt this protein can be eluted by 1 M NaCl). The gels introduced by Shaltiel, called type 2-gels here, bind protein more effectively at low salt and at low temperature (for comparison see the main contributions by S.Shaltiel and S. Pahlman, this volume). But in both of the gels, binding of protein clearly depends on the chain length of the hydrophobic alkyl-chain and both, positively and negatively charged proteins can be bound. Both types of gels have proven useful for the purification of proteins.

We have attempted to show that in myoglobin and hemoglobin structural transitions occur after binding to type 2-gels of long (C_{12}) chain length, but not after binding to type 1 gels of the same chain length. The structural transitions are similar to the ones observed after binding of cationic detergents to the proteins.

Experiments with a third type of gels (type 3-gels, see below) show that the binding of myoglobin and hemoglobin to Sepharose-C_6-imidazol and Sepharose-C_6-(1,2,4)-triazol does not occur at the biospecific ligand binding site of the proteins, but is due to a complex hydrophobic and ion exchange interaction.

MATERIALS AND METHODS

Synthesis of the gels: Sepharose 4B (31 g wet weight, corresponding to 1.4 g dry weight) was reacted with epichlorohydrin in 1 N NaOH according to (4), washed thoroughly and reacted with n-dodecanol (5.4 g) in 90 ml of 70 % dioxane/H_2O (0.04 M with respect to NaOH) for 12 h at room temperature. The product was then washed with 500 ml of 70 % dioxane/H_2O and 2 l of H_2O. To get an approximate measure of the ligand density, epoxyactivated Sepharose 4B was reacted under exactly the same conditions with an equivalent amount of ethanol, labelled with 1 x 10^6 dpm of 1-^{14}C ethanol (Amersham). The ligand density thus measured was 1 mM. To get other members of the series of type 1-gels, the corresponding n-alkyl alcohols were used in H_2O or dioxane/H_2O mixtures as appropriate.

For the synthesis of type 2-gels, Sepharose 4B was activated with cyanogen bromide according to (5), washed thoroughly with cold 0.1 M $NaHCO_3$ and then with 80 % pyridine/H_2O, mixed with n-dodecylamine (7 g) in 50 ml 80 % pyridine/H_2O, titrated to pH 9, and stirred slowly at 4^oC for 20 h. After that, the gel was stirred for one hour at room temperature, washed with 80 % pyridine/H_2O, 0.1 M $NaHCO_3$, 0.01 M HCl and 0.5 M NaCl. We have no measurement of ligand density of this gel. For the synthesis of shorter chain-length gels of this series the method of Er-el (2) was used.

Type 3-gels: Aminohexyl Sepharose and carboxyhexyl Sepharose were obtained from Pharmacia. Histamine, N-2-aminoethyl-imidazole and 3-amino-1,2,4-triazole were coupled to carboxyhexyl Sepharose with EDC (1-ethyl-3(3-dimethyl-aminopropyl)-carbodiimide, obtained from Sigma) according to the prescription given by Pharmacia.

All the spectra were measured with a Cary (Model 15) double beam recording spectrophotometer at room temperature between 750 and 350 nm. For the measurement of the spectra of proteins bound to substituted Sepharoses, pure Sepharose 4B was used as a blank. 1 cm quartz cuvettes were used.

Hill plots of the binding of imidazole to the hemoglobin of horse, myoglobin of horse and hemoglobin III of *Chironomus thummi thummi* were measured by determining the amount of hemochrome and hemoglobin by a spectrophotometric method. Dissociation constants were determined at 50 % saturation. All other heme proteins mentioned were obtained from Sigma and used in the oxidized form.

RESULTS AND DISCUSSION

The binding of sperm whale myoglobin to type 1- and type 2-gels will be discussed here. The absence or presence of the above mentioned structural transition can be seen clearly and the monomeric nature of the protein allows an easy interpretation of the spectra. To type 1-gels substituted with dodecyl alcohol (type 1-C_{12}-gel) the protein is bound firmly in 0.01 M sodium phosphate buffer, pH 5.3 at 20° C; It can be eluted with 1 M NaCl and is somewhat retarded again in 4 M NaCl. The spectrum of the bound protein between 400 and 700 nm is identical with the spectrum of sperm whale myoglobin in unsubstituted Sepharose 4B (Fig. 2a) or in buffer (spectrum not shown).

On the other hand, if under exactly the same conditions as above sperm whale myoglobin is bound to the type 2-gel substituted with dodecyl amine, the dark brown solution of the oxidized protein turns red shortly after contact with the gel. It cannot be eluted by changing the ionic strength of the irrigating buffer, but only with

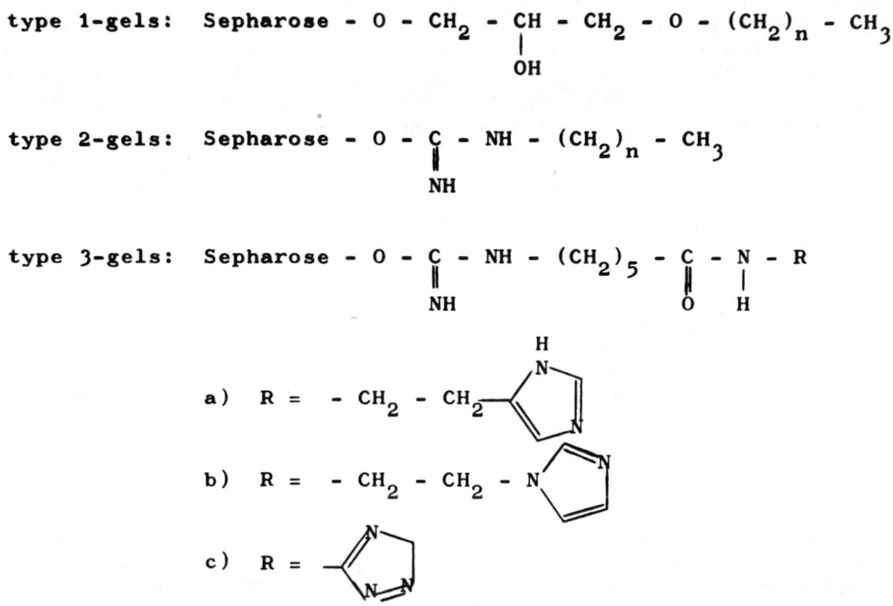

Fig. 1: Structures of the gels used, n = 1, 2, 4, 6, 12.

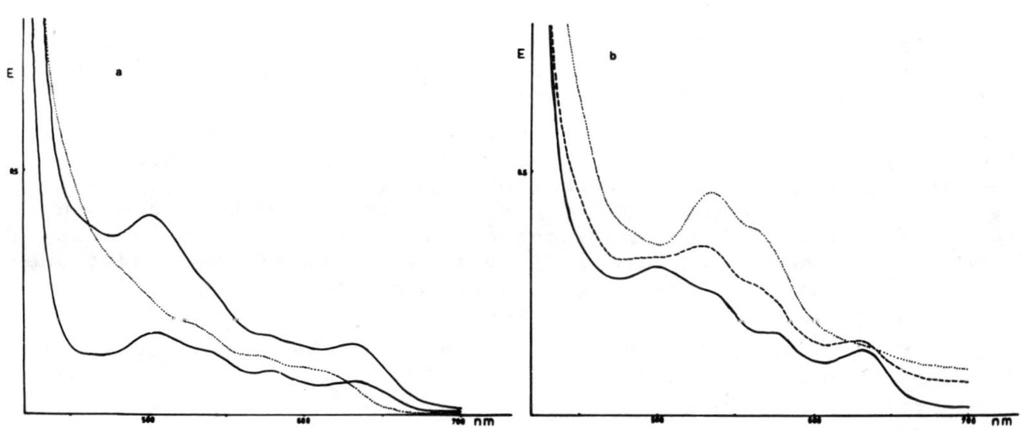

Fig. 2a: Spectra of sperm whale myoglobin in 0.01 N sodium phosphate, pH 5.3 in presence of Sepharose 4B (full line, lower curve), type 1-C_{12}-gel (full line, upper curve) and type 2-C_{12}-gel (dotted).

Fig. 2b: Human hemoglobin in 0.01 N sodium phosphate solution, pH 5.3 (full line), with a 1000 fold molar excess of imidazole, titrated to pH 5.3 (dashed); the same at pH 8.0 (dotted).

sodium dodecyl sulfate (SDS) or 1 % dodecyl pyridinium chloride (lauryl pyridinium chloride). Spectroscopic investigation shows that the peak at 635 nm disappears and that the whole spectrum is similar to but not identical with a hemochrome spectrum (Fig. 2a). Differences between typical hemochromes (peak at 535, shoulder at 565 nm) and a spectrum of the bound myoglobin occur below 500 nm, probably due to the high turbidity of the sample.

Hemochromes of the type described above can be obtained from hemoglobin or myoglobin solutions in either of two ways: a) by adding large amounts (1000 fold molar excess, in a typical case) of heterocyclic bases (e.g. imidazole),which can be bound in the active center of the protein (examples are shown in Fig. 2b) or b) by adding detergents like SDS or lauryl pyridinium chloride. The binding of the detergents is a highly cooperative process (6). After binding of 14 - 20 molecules of lauryl pyridinium chloride a structural transition occurs in the protein. Histidine E7 now occupies the 6th (distal) ligand position on the heme (leading to a hemochrome spectrum very similar to the one obtained with imidazole. If more detergent is added, a second structural transition appears, leading to a spectrum identical with the spectrum of pure hemin in detergent solution.

Our interpretation of the spectra observed after binding to type 2-gels is the following: The ligands on the type 2-gels carrying a positive charge near the point of attachment to the matrix, interact with the hemeprotein in a similar way as does lauryl pyridinium chloride. The cooperativity necessary for the structural transition in free solution is also occurring here as shown by Jenissen (7) and others (cf. the main contribution by S. Shaltiel, this volume). The binding of the protein to the type 1-C_{12}-gel obviously does not involve the formation of a hemochrome. But at present, it can not be decided whether this is due to an intrinsic difference in the binding mechanism or due to different ligand densities on the two gels used. Further studies shall clarify this point.

We performed the same set of experiments with horse hemoglobin and found very similar results, but the spectra cannot be interpreted as easily in this case. In tetrameric hemoglobins binding of only one of the four subunits suffices to bind the whole molecule and mixed-type spectra should be expected. We do not know at present if horse hemoglobin dissociates into monomers under the conditions of binding to the type 2-C_{12}-gel. With the type 1-C_{12}-gel an unchanged hemoglobin spectrum is observed.

A hydrophobic residue 6 C-atoms long is not enough to bring about the spectral change as could be shown with the type 2-C_6-gel (8). But nevertheless strong binding occurred under these conditions.

Chromatographic experiments with human hemoglobin and the type 1-C_{12}-gel show the ionic strength dependence expected for purely hydrophobic interaction. Human hemoglobin interacts less strongly with the gel than does sperm whale myoglobin. It is not bound but only retarded and retardation increases in parallel with NaCl concentration at pH 5.3 and 20°C.

Finally, a few unsuccessful attempts to devise an affinity chromatographic system for the purification of hemoglobins shall be described. The type 3-gels shown in Fig. 1 and aminohexyl Sepharose (control) were used. Preliminary experiments showed that the affinities of horse hemoglobin, horse myoglobin and hemoglobin III of <u>Chironomus thummi thummi</u> for imidazole are rather low, the dissociation constants being 5×10^{-3}, 6.65×10^{-3} and 1.26×10^{-4} M^{-1}, respectively. Furthermore, all substitutions on the imidazole ring with one exception (N-methyl imidazole) destroyed the interaction with the proteins. Therefore gel 3b was synthesized.

The chromatographic behaviour of all the type 3-gels was identical with the behaviour of aminohexyl Sepharose (8). A slight increase in ionic strength from 0.01 to 0.05 M NaCl was enough to elute the bound horse hemoglobin at pH 8.0 and 20°C. There was no difference between hemoglobin and the strong complex cyano methemoglobin. Both could be bound and eluted under the same conditions, indicating a complex type of interaction (the hydrophobic spacer is necessary, cf. 9), but no biospecific interaction.

REFERENCES

1) D.S. Hjertén, J. Rosengren, S. Pahlman, J. Chromatogr. 101, 281 (1974).

2) Z. Er-el, Y. Zaidenzaig, S. Shaltiel, Biochem. Biophys. Res. Commun. 49, 383 (1972).

3) C. Tanford (1973), The hydrophobic effect, formation of micelles and biomembranes. Wiley, New York.

4) L. Sundberg, J. Porath, J. Chromatogr. 90, 87 (1974).

5) R. Axén, J. Porath, S. Ernbäck, Nature (London) 214, 1302 (1967).

6) J. Yonath, G. Blauer, Eur. J. Biochem. 41, 163 (1974).

7) H.P. Jennissen, Hoppe Seylers Z. Physiol. Chem. 357, 1201 (1976).

8) M. Breitenbach, Protides of the Biological Fluids 23, 687 (1974).

ON THE PURIFICATION OF BACTERIAL TYROSINE DECARBOXYLASE

Stig Allenmark and Bo Servenius

Clinical Research Centre, University Hospital, S-58185 Linköping, Sweden

INTRODUCTION

A study of the behaviour of L-(-)-tyrosine apodecarboxylase (TAD),(E.C. 4.1.1.25) on hydrophobic interaction chromatography was undertaken as part of an overall investigation of its purification by electrophoretic and chromatographic methods. TAD is obtained from extracts of cells of *Streptococcus faecalis*. The enzyme is highly dependent upon its cofactor, pyridoxal-5'-phosphate (PLP), and has been used in one of the most reliable enzyme kinetic methods presently available for the determination of PLP in biological fluids (Ref. 1). However, very crude enzyme preparations have been used so far for this purpose and to the best of our knowledge no serious attempts to study the purification and chemical characterization of this enzyme have hitherto been reported.

MATERIALS AND METHODS

Prepurification of Enzyme Material

Freeze-dried cells of specially grown cultures of *Streptococcus faecalis* (Sigma Chem. Co. no. T 4629) were used as the source of TAD investigated. The treatment of the cell material was essentially as described in ref. 1, with the exception that cell rupture was effected by means of a freeze-press technique instead of sonication. The cell debris was removed by centrifugation for 20000 gh, which yielded a clear supernatant. A solution in 0.1 M sodium acetate buffer at pH 5.5 with $A_{280}^{1\ cm} = 2.540$ was used in the experiments; 100-300 µl was taken for each chromatographic run.

Chromatography

All chromatographic runs were performed on uncharged agarose derivatives, *i.e.* materials prepared by the glycidyl ether coupling method (Ref. 2), thus possessing no ion-exchange properties due to electrostatic interactions. The experimental set-up is shown schematically in Fig. 1. In all cases a linear gradient of decreasing ionic strength was started after 1 h of elution with the initial buffer system, which permitted the complete removal of non-adsorbed proteins. All chromatographic operations were carried out at ca. 22°C (room temp.). The carefully washed agarose derivatives were packed in Pharmacia K9/15 columns and preequilibrated with the initial buffer (sodium acetate, pH 6.0).

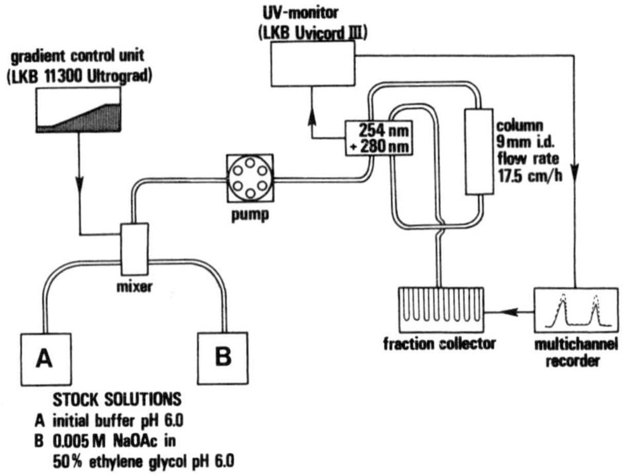

Fig. 1. Block diagram of the experimental set-up used for the chromatographic runs

Enzyme Activity Monitoring

The elution behaviour of TAD on octyl- and phenyl-Sepharose® CL-4B and on butyl-Sepharose 4B[x] (not cross-linked, degree of substitution ca. 84 mmol/mol galactose) with the use of initial buffers of different ionic strengths (0.1 to 1.0 M, pH 6.0) was readily followed by screening the collected fractions for decarboxylase activity in the presence of an excess of PLP. The activity was determined as the initial rate of $^{14}CO_2$ formation from labelled L-(-)-tyrosine, essentially according to the method described in ref. 1. An illustration of the conditions of constant decarboxylation rate used for the activity determinations is shown in Fig. 2.

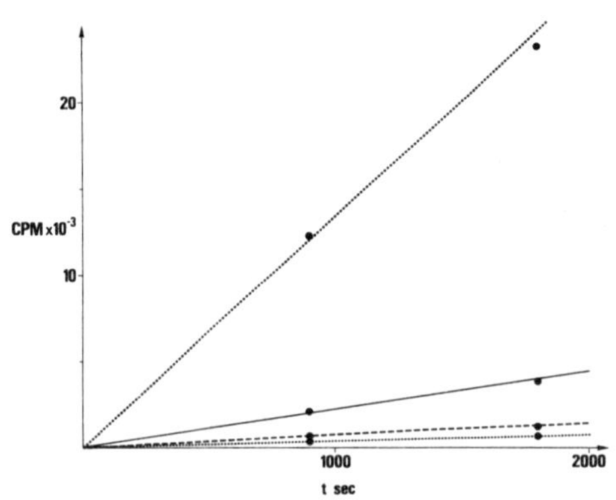

Fig. 2. $^{14}CO_2$ evolution as a function of time corresponding to some consecutive fractions with increasing TAD concentrations obtained on elution from a phenyl-Sepharose CL-4B column

[x]Kindly submitted by Dr. S. Påhlman, Biomedical Centre, Univ. of Uppsala, Sweden

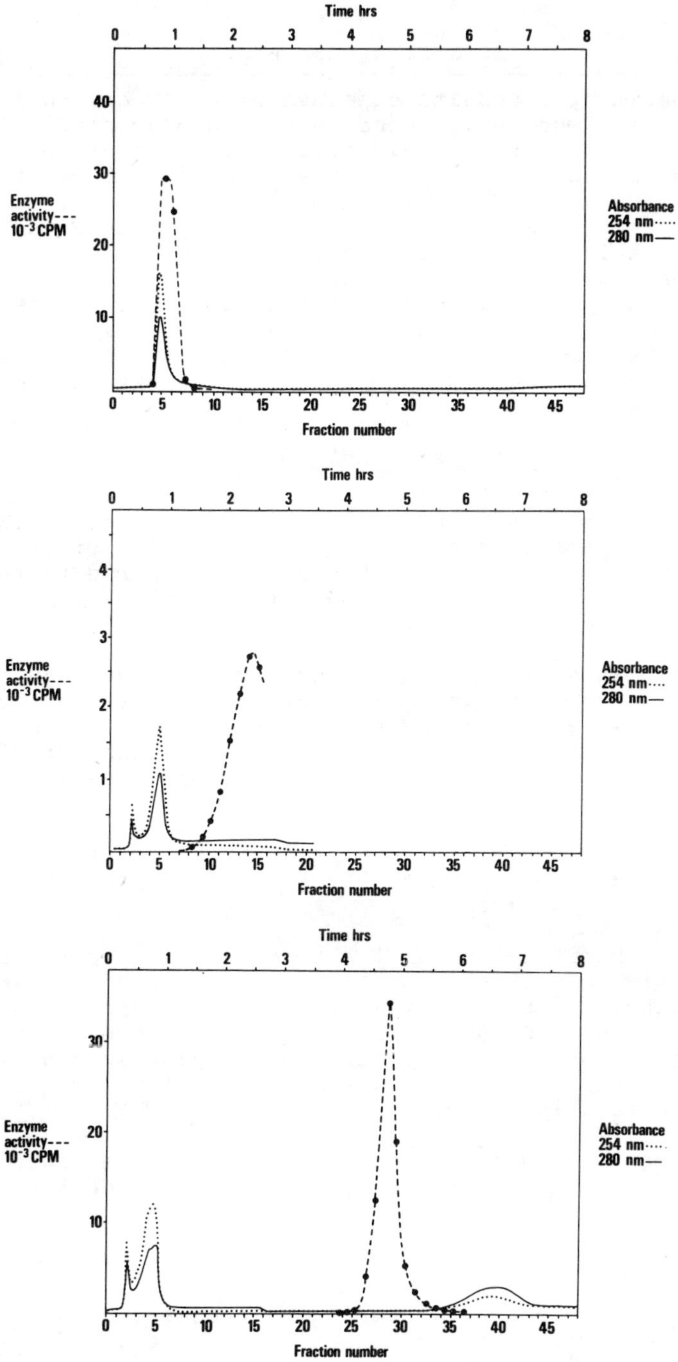

Fig. 3. Elution diagrams from hydrophobic interaction chromatography of TAD on phenyl-Sepharose CL-4B at different conditions for protein adsorption. From the top: a) 0.1 M, b) 0.5 M, c) 1.0 M NaOAc, pH 6.0.

RESULTS

Enzyme Retention Characteristics and Chromatographic Resolution

All the experimental results obtained with TAD are fully consistent with a retention mechanism based solely on hydrophobic interaction between the enzyme and the gel matrix. With all of the conditions used TAD was bound so firmly to octyl-Sepharose CL-4B that no activity was eluted even after 8-9 h. On the other hand, no adsorption occurred under the same conditions when butyl-Sepharose 4B was used. The most suitable gel for purification purposes was found to be phenyl-Sepharose CL-4B, the use of which permitted a change in TAD retention volume by changing the initial buffer concentration in accordance with the general theory of hydrophobic interaction. The results of three different runs with the use of this gel are shown in Fig. 3. The symmetry of the TAD peak and the high degree of resolution are quite evident.

Estimations of Relative Enzyme Activities

The great affinity of TAD for octyl-Sepharose CL-4B was found not to be associated with any significant degree of inactivation of the enzyme. On the contrary it could be shown that it retains almost half of its activity when bound to this gel as compared to that in the free state. When stored in the bound form, however, the activity was found to decrease significantly with time.

The desorbed material in the eluates from the phenyl-Sepharose CL-4B columns showed no significant loss of total enzyme activity. An analysis of the gel after the experiment corresponding to Fig. 3 c showed that a small but significant decarboxylase activity was still present. The reason for this is, however, not clear.

The TAD used in all purification experiments contained very little (if any) PLP. The degree of resolution from cofactor can be seen from the initial rate ratio of 306 found for conditions of added and excluded PLP, respectively.

The degree of purification of the enzyme after the run shown in Fig. 3 c could only be estimated to $\geq 7.5 \times$ because of the very low protein absorbance ($A_{280}^{1\ cm} \leq 0.006$) in the pooled and dialyzed active fractions (25-36). As judged from Fig. 3c, however, the position of the TAD peak in an essentially UV blank region should be interpreted as a considerable purification, the degree of which will be determined more carefully later.

Acknowledgement. This investigation was supported financially by a grant from the *Swedish Natural Science Research Council* which is gratefully acknowledged.

REFERENCES

1. P.R. Sundaresan and D.B. Coursin, Microassay of pyridoxal phosphate using L-tyrosine-1-^{14}C and apodecarboxylase, Methods Enzymol. 18 A, 509 (1970).

2. S. Hjertén, J. Rosengren and S. Påhlman, Hydrophobic interaction chromatography. The synthesis and the use of some alkyl and aryl derivatives of agarose, J. Chromatogr. 101, 281 (1974).

IMPROVED SYNTHESIS OF A FATTY ACID-AGAROSE COMPLEX FOR AFFINITY CHROMATOGRAPHY OF SERUM ALBUMIN

E. Schonne

Laboratoire d'Embryologie, Place Croix du Sud, 5, University of Louvain, 1348-Louvain-la-Neuve, Belgium

INTRODUCTION

In the course of investigations on the isolation of serum albumin, we tried to prepare affinity columns with fatty acids immobilized on agarose according to Peters et al. (I). We intended to synthesize the oleylaminoethylamino-derivative of agarose, but we used I-cyclohexyl-3-(2-morpholinyl-4-ethyl) carbodiimide metho-p-toluene sulfonate (CMC) instead of I-ethyl-3-(3-dimethylaminopropyl) carbodiimide (EDC) as the condensing agent. In these conditions, no coupling was obtained. We then reinvestigated the experimental conditions and devised the following method of agarose activation which integrated experimentally several well known procedures (I-4). The modified method was selected for its ease and simplicity. It requires a minimum of equipment and presents a minimum of risks and of toxicity. Moreover it can be easily extrapolated on a large scale.

METHODS

The materials were obtained from the following sources : agarose was Sepharose 4B, Pharmacia, Uppsala, Sweden; BrCN, ethylene diamine, dioxane, N,N'-dicyclo-hexylcarbodiimide(DCCD) of the highest grade available were obtained from Merck, Darmstadt, W. Germany; N-methyl-2-pyrrolidone, purum from Fluka, Buchs, Switzerland; CMC from Aldrich, Turnhout, Belgium; EDC from Serva, Heidelberg, W. Germany.

Synthesis of Aminoethylamino-agarose

One hundred grams of washed agarose beads were added to an equal volume of

2M sodium carbonate, mixed by stirring and cooled to 4°C. Twelve grams of BrCN were dissolved in 20 ml of a water-miscible organic solvent. N-methyl-2-pyrrolidone was preferred to acrylonitrile for its lower toxicity (3). The solution of cyanogen bromide was added all at once under vigorous stirring to avoid precipitation of cyanogen bromide. The reaction was left to proceed for 7-8 minutes without cooling. The temperature rose to 15 -20°C. The slurry was poured over a Büchner funnel on filter paper. The activated gel was washed with 500 ml of cold (4°C) 0.1 M bicarbonate buffer at pH 10. The gel was suspended in 100 ml of 0.1 M ethylenediamine dissolved in the same buffer and stirred for 20 hours or more at 4°C. It was then filtered and washed with the adequate buffer.

Synthesis of the Oleyl-derivative

Attempts were made to couple oleate to aminoethylamino-agarose by suspension in a 0.1 M solution of the sodium soap of the fatty acid at pH 10 (0.1 M bicarbonate buffer). The suspension was stirred with 50 mg CMC per ml of agarose for 3 days at 37°C. The agarose was washed as described (I). The resulting product contained practically no oleate and did not have any affinity for albumin. Since CMC reacts better at acid pH, attempts were made to work at a lower pH (pH 7 for example) in the presence of 25% ethanol or 20% dimethylformamide and an excess of oleic acid. However no coupling resulted from these modifications. Failure could perhaps be ascribed to a lower reactivity of CMC as compared to the EDC originally used (5, I).

We then took advantage of the fact that agarose gel may be suspended and coupled in dry dioxane without alteration of its properties (2). Dioxane allows the use of DCCD as the coupling agent in the following way : 10 g of aminoethylamino-agarose were washed on a Büchner funnel, first with distilled water, then with dioxane without drying the gel. Then the gel was suspended in 10 ml dioxane. Oleic acid (0.63 ml) was dissolved 10 ml dioxane and 500 mg of DCCD was added under constant stirring. A precipitate formed after 5-10 minutes reaction at room temperature. That solution was added all at once to the agarose suspension and stirred for 4h at room temperature. The derivatized agarose was washed over a Büchner funnel with 6 to 8 volumes of dioxane, followed by the same volume of methanol to remove all the precipitated dicyclohexylurea. The methanol was eliminated by washing the gel again with dioxane and then with the adequate buffer. The unreacted aminogroups were blocked by acetylation with acetic anhydride at pH 7, 0°C, for I hour. After acetylation, the trinitrobenzene test was negative.

RESULTS

The activation of agarose through the described procedure was fast and simple. The handling of the poisonous lachrymator CNBr dissolved in a low toxicity solvent was made much easier and safer. The activation in an homogeneous phase was more reproducible. Fluctuations in pH due to addition of solid CNBr and manual titration were eliminated and a pH meter was not needed. Heat transfer problems during activation were minimized and the reaction temperature offered a good compromise between speed of activation and side reactions (3). The amount of oleic acid bound to the agarose was consistently between 1 and 2 μ moles per ml of agarose. After saturation with a bovine serum albumin solution (1 mg per ml in saline) on a column, and washing with phosphate buffered saline, the oleyl-agarose retained 15 mg of albumin per ml of gel. These properties are similar to those reported previously by T. Peters et al. (I). The gel was used with success for the purification of human, bovine and sheep serum albumin, and also for purification of bovine serum albumin labelled with fluorescamine. A recovery yield of 98% of ^{125}I-labelled human serum albumin was obtained by chromatography on the same material.

REFERENCES

I. T. Peters, Jr., Hiroshi Taniuchi, C.B. Anfinsen, Jr., Affinity chromatography of serum albumin with fatty acids immobilized on agarose, J. Biol. Chem., 248, 2447-2451 (1973).

2. P. Cuatrecasas & I. Parikh, Adsorbents for affinity chromatography. Use of N-hydroxysuccinimide esters of agarose, Biochemistry, II, 2291-2298 (1972).

3. A. Hirotoshi Nishikawa & P. Bailon, Affinity purification methods. Improved procedures for cyanogen bromide reaction on agarose, Anal. Biochem., 64, 268-275 (1975).

4. S. C. March, I. Parikh, P. Cuatrecasas, A simplified method for cyanogen bromide activation of agarose for affinity chromatography, Anal. Biochem., 60, 149-152 (1974).

5. G. E. Means, R. E. Feeney, (1971) Chemical modification of proteins, Holden-Day, San Francisco.

PART IV

MATRIX-BOUND ANTIGENS AND ANTIBODIES

MATRIX-BOUND ANTIGENS AND ANTIBODIES

Tore Kristiansen

Institute of Biochemistry, Biomedical Center, Uppsala, Sweden

INTRODUCTION

Although similar in many respects to other fields of biospecific affinity chromatography, the design and operation of immunosorbents offer some special challenges. One of them is the wide spectrum of affinities among antibody molecules elicited against a single antigen, leading to occasional difficulty in producing an immunoadsorbent that can handle immune complexes of widely different strength at the same time.
Another characteristic is the frequent lack of a specific method to desorb the immune reactants. Often the potential displacers in the form of low molecular weight haptens are unavailable because the immunodeterminants on the antigen have not been identified. In other cases, a macromolecular antigen may be impossible to displace from a particular complex by competing haptens because the complex calls for almost complete molecular integrity of the antigen.
Very often, therefore, only nonspecific desorption methods are feasible.
With strong complexes drastic conditions may be necessary but lead to partial denaturation of the immunoreactants. In recent years immunosorption techniques have developed into extremely useful tools in immunochemistry, yet the field is, and will remain, a tangle of compromises between simplicity of operation, capacity of adsorbents, and recovery of reactants. Recent reviews in the field include refs. 1-6. Since my own experience is mainly with agarose-based immunosorbents and the CNBr method of activation, they will dominate the following discussion.

GENERAL FEATURES OF IMMUNOADSORBENTS

A distinctive feature of antigen-antibody interaction is the fact that the antibody population formed in response to a single antigen may be quite heterogeneous with respect to immunoglobulin classes and subclasses but still show a narrow specificity towards the eliciting antigen. A diversity of antibodies is found not only in antisera pooled from different individuals responding to a multideterminant antigen, but can be observed also in a single animal forming antibody against a single antigenic determinant. For instance, in a series of papers, Haber and collaborators (7-9) describe the isolation of four separate species of antibody molecules specific for the same cellobiose determinant but with different amino acid sequences in the L chains and varying levels of affinity.
The overall functional affinity (avidity) of a discrete population of antibody molecules will vary widely with conditions of antigen administration and the time between immunisation and bleeding. One of the reasons for this is the successive increase in the ratio between IgG and the more avid IgM during the early phase of the immune response, and another is the progressive shift in each animal towards synthesis of more and more avid IgG in the later phase. However, there is evidence that some IgG of low avidity may be formed throughout the immune response (10).

TABLE 1 Variation in antigen-antibody affinity in adult mammals (approximate values compiled from refs. (11÷16))

Days after immunization	Predominant class of immunoglobulin	Range of intrinsic affinity constant K_o in M^{-1}	Amplification factor due to multivalence of IgM	Amplification factor due to divalence of IgG
3-8	IgM	10^4-10^5	5×10^5-10^6	
10-15	IgM + IgG			
15-30	IgG	5×10^2-5×10^4		10^3-10^4
30-40	IgG	5×10^3-10^7		10^3-10^4
after booster at 20-30 days				
300	IgG	5×10^6-5×10^7		10^3-10^4

Table 1 summarises in very approximate terms the distribution of affinities to be expected in adult mammals after various times. A further amplification factor, not listed in the table but somewhere in the range of 10^1-10^3 may have to be considered for interaction on immunoadsorbents with multivalent antigens rather than monovalent haptens. The influence of polyvalency of antigens and antibodies on their interaction has been treated in mathematical terms by Crothers and Metzger (17).
The heterogeneity of an antibody population formed against an antigen carrying several different immunodeterminant groups on the same molecule may lead to functional affinities spanning several orders of magnitude. In practical terms, a wide distribution of affinities can make it difficult to prepare an immunosorbent able to cope with both ends of the affinity spectrum at the same time. An immobilised antigen may be unable to bind the least avid antibody species, while the most avid ones bind so strongly that desorption without irreversible denaturation is almost impossible. On repeated use there may be progressive blocking of antigenic sites by high avidity antibody causing a rapid decline in binding capacity of the adsorbent. Consequently, the inherent heterogeneity of the immune response is of basic importance to the operation of immunoadsorbents. Technical procedures for the immunisation of animals are described e.g. by Livingston (4) and in great detail by Herbert (18).

Immunosorption methods may be divided into three categories at an increasing level of ambition and sophistication:

1. Removal of unwanted antigens or antibodies. Recovery of immunoreactants and regeneration of adsorbent not important.
2. Enrichment or isolation of antigen and antibody populations in relatively abundant supply. Regeneration of adsorbent desirable but not essential.
3. High recoveries of fully active, precious antigens and antibodies on adsorbents of high stability that can be regenerated many times without loss of activity. Adsorbents that recognise subtle differences in molecular parameters. Extensive suppression of secondary adsorption effects.

In cases belonging to the first group preparation of the immunosorbent by direct polymerisation of the antibody or the protein antigen with glutaraldehyde or ethylchloroformate after mixing with an auxiliary protein like bovine serum albumin is a very useful method pioneered by Avrameas and his coworkers (19,20). Although simple and rapid, it is rather wasteful because only a fraction of the binding capacity inherent in the polymerised material is actually put to use. In addition, only batch procedures are practical and regeneration may be difficult. Nevertheless, this approach can sometimes be useful also in group 2, but the potential binding capacity of a given amount of immunoreactant is utilised more efficiently after coupling to an inert carrier or matrix, preferably in bead form providing adequate flow properties for column operation. Also, regeneration of the adsorbent after complete desorption is usually easier with matrix-based immunoadsorbents.

The demands on adsorbents in group 2 are not as tight as in group 3. For instance, the extent to which purified antibody corresponds to distribution of species and avidities present in the starting material may not be of prime importance as long as specificity is retained.

In the third and most exacting group, experience with a few general methods has accumulated to the extent that merits may be balanced against problems and pitfalls. However, strategies for individual systems still have to be largely empirical because parameters of great importance to the precision of immunoadsorbent performance are not amenable to adequate calculation or measurement at the present state of the art. They include for example extent of multipoint binding of ligand under various coupling conditions, true pH value and charge distribution near the matrix backbone and the binding site of a coupled ligand, the degree of cooperative affinity (21) due to hydrophobicity of a spacer, and the relative contributions of hydrophobicity, electrostatic interaction, hydrogen bonding, and other forces to complex formation within a given immunoadsorbent system.

THE MATRIX

The merits and drawbacks of matrices in bioaffinity chromatography have recently been reviewed by Lowe (22).
The four principal matrices for immunoadsorbents so far are cellulose, and beads of porous glass, polyacrylamide, and agarose. Diazo derivatives of cellulose were the first immunoadsorbents of practical value and are still in occasional use. Adsorbents based on cellulose were reviewed ten years ago by Campbell and Weliky (23). The present method of choice involves preparation of bromoacetylcellulose ("BAC") according to procedures described in (2) or (3), which will form covalent bonds with ε-amino groups of lysin in basic solution. Cellulose is cheap and also easy to derivatise but has in our hands shown a tendency towards continuous leakage of soluble carbohydrate-ligand complexes.

Procedures based on derivatives of porous glass are surveyed by Weetall and Filbert (24). Glass is incompressible and completely inert in neutral or acid media but suffers from an inherently high surface charge. This can be reduced but hardly eliminated by suitable chemical treatment. The chemistry involved in preparing derivatives of glass may be considered rather complicated.

Polyacrylamide carries an abundance of carboxamide groups that can be derivatised according to a large number of reactions, reviewed by Inman (25). Avrameas introduced activation of polyacrylamide with glutaraldehyde followed by coupling of ligand in a second step (26). After removal of glutaraldehyde with distilled water the activated beads will bind proteins by way of amino groups. The activated gel can be stored for at least a month at 4° with little decrease in coupling capacity. A maximum of 3 mg protein per ml gel can be coupled, and details of the method are given by Ternynck and Avrameas (26). Sapin et al. (27) and Sapin and Druet (28) used this technique to isolate human IgA, IgM, and IgD and rat and rabbit IgM by adsorption to immobilised anti-immunoglobulins.

A disadvantage of polyacrylamide beads is the deterioration of their mechanical properties when the porosity is increased towards a level allowing sufficient penetration of molecules the size of e.g. IgG. Already on Biogel P 300, where the ratio between elution volume and void volume is only about 1,4 for IgG, the softness of the beads, a minor annoyance in batch procedures, can be a problem in column operation where clogging may occur unless the flow rate is down to the order of 1 $ml/cm^2/h$.

Hydroxyalkyl methacrylate beads under the trade name of Spheron is a very interesting new matrix material developed at the Institute of Macromolecular Chemistry belonging to the Czechoslovak Academy of Sciences. It is apparently chemically and mechanically stable, and can be prepared in a wide range of porosities. The gel is available from Hydron Labs., Inc., New Brunswick, New Jersey, USA. The general structure of hydroxyethyl methacrylate is shown in fig. 1.

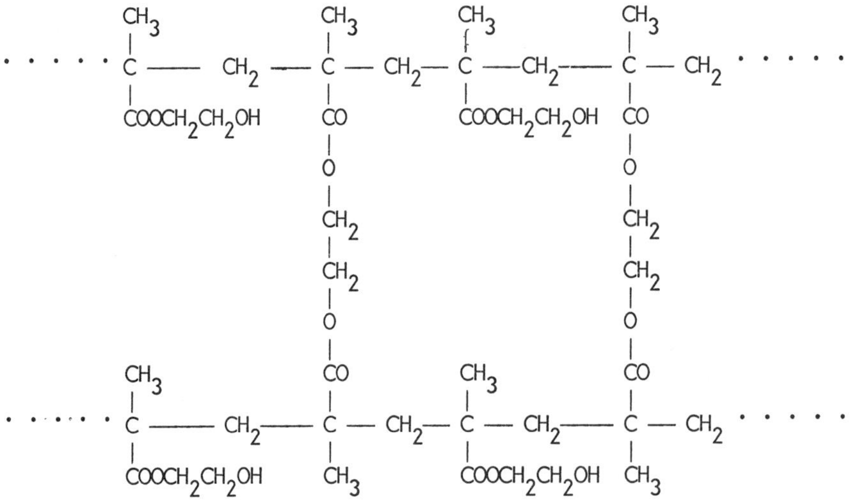

Fig. 1 Structure of hydroxyethyl methacrylate

About 80 per cent of the hydroxyl groups are available for derivatisation. The isolation of antibodies and immunocompetent cells on Spheron-based immunoadsorbents are described by Tlaskalová et al. (29), and its properties and potential in bioaffinity chromatography are reviewed by Turková (30).

Polysaccharide matrices

Cross-linked dextran in the form of Sephadex beads is little used in immunoadsorption due mainly to its low porosity causing exclusion of immunoglobulins and many macromolecular antigens, and there are hardly any advantages to compensate for this drawback.

Beaded agarose does not suffer from this weakness and has in recent years come into prominence as a matrix for immunadsorbents. Covalent coupling of immunoglobulins and protein antigens to agarose can be achieved in several ways, but the most popular procedure is the CNBr method introduced by Axén et al. (31).
In our institute, where this method was first put to use, we realised quite

early that untreated agarose beads were not stable against strong solutions of chaotropic ions like SCN^- or CCl_3COO^- that are sometimes needed to elute an immunoligate. CNBr activation does lead to some stabilisation of the agarose backbone by the formation of linear interchain iminocarbonates (fig. 2), but since weak activation is often preferable to strong activation, the resulting stability may be marginal and results erratic.

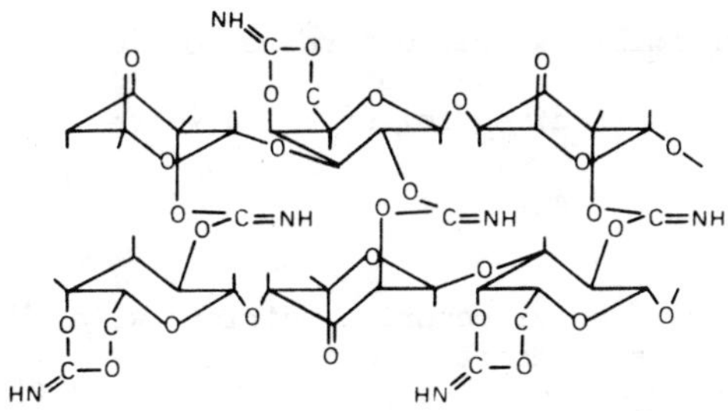

Fig. 2 Hypothetical structure of agarose gel after activation with CNBr. From (32) with permission.

Fortunately, agarose can be readily stabilised by crosslinking with bifunctional reagents like 2,3-dibromopropanol (33) or epichlorohydrin (34) prior to activation with negligible effect on gel porosity.

Agarose-polyacrylamide composite beads recently introduced under the name of Ultrogel are claimed to combine the advantages of polyacrylamide and agarose and to possess better mechanical stability and permit higher flow rates than pure agarose beads. In immunoadsorbent procedures neither property seems to offer problems with conventional, cross-linked agarose, but the added presence of carboxamide groups in Ultrogel, offering a wider scope of coupling methods, may be a genuine advantage. There is little published experience so far with this material in immunoadsorption, although Guesdon and Avrameas (35) obtained good results using Ultrogel activated with glutaraldehyde to isolate rabbit and sheep antibodies to bovine serum albumin, and human and sheep gamma globulin fraction II.

COUPLING OF IMMUNOLIGANDS

Spacers

There is general agreement that a small ligand, e.g. a hapten, should be at a distance of 15-25 Ångström from the matrix backbone in order to be effective, even if there are exceptions to this rule. This requirement can be met by insertion of a molecular bridge, a "spacer", between the ligand and the matrix. It has been gradually realised that the nature of the spacer may have a profound effect on the overall interaction between the ligate and the adsorbent. Soon after the need for a spacer was recognised, many workers in ignorance of evil started using methylene diamines and other spacers of a hydrocarbon nature quite indiscriminately, but with great apparent success. In a series of papers, O'Carra and coworkers (for review see ref. 21) showed that seemingly biospecific adsorption may in fact be due to secondary and nonspecific interaction with the spacer rather than the ligand proper. Presumably, this interaction is mainly of a hydrophobic nature. They have suggested alternative, hydrophilic spacers like the one shown in fig. 3, obtained by bromoacetylation of agarose-linked diaminopropanol followed by reaction with diaminopropanol again.

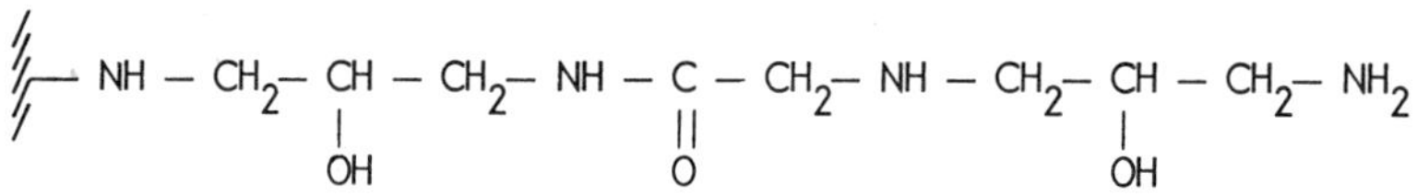

Fig. 3 Hydrophilic spacer according to O'Carra et al. (21)

Details of the preparation are given in ref. (21) where the spacer concept is discussed at length.

The CNBr method

Procedures for activation of polysaccharide matrices with CNBr are widely published, with every laboratory eager to add its own twist. Theoretical and practical aspects of the method are covered in the book "Affinity Chromatography" (36) and a recent chapter in Methods in Enzymology Vol. XLIV (37).
A few points of relevance to immunoadsorbents may be raised.

With immunoglobulins as ligands, a high substitution level can be more of a draw-back than an asset. Comoglio et al. (38) found a progressive and dramatic loss of relative affinity when going from a mass ratio of 0,1 - 0,2 nmol up to 10 nmol IgG per mg Sepharose in a study on the affinity between estradiol and coupled IgG anti-estradiol when the activation level was kept constant and the amount of added antibody varied. There is evidence that the addition of IgG (M.W. 160 000) or IgM (M.W. 900 000) to a strongly activated gel involves a definite risk that the bulky molecule quickly forms a shell of coupled ligand that may impede further penetration into the bead. Using a CNBr:Sepharose 4B dry weight ratio of 6,7:1, David et al. (39) found that bovine lactoperoxidase (M.W. 78 000) and bovine serum albumin (M.W. 70 000) are both attached mainly to the surface of the beads.
In the electron microscope Lasch et al. (40) studied the distribution of ferritin in Sepharose 6B at different levels of CNBr activation. Accumulation of coupled ferritin in the outer part of the bead started at CNBr:Sepharose ratios above 1:1. At the other end of the scale no coupling occurred at ratios below 1:10.

A 1:1 ratio seems a good general recommendation also for immunoadsorbents. Still, for low avidity systems a somewhat higher level of substitution may be desirable. Since the active iminocarbonate will react only with unprotonated amino groups, it might be tried to add the ligand at high concentration to the activated beads at pH 6 or lower where extensive protonation is likely to slow down the reaction and allow the ligand to distribute more evenly before pH is raised to above 8 where coupling will proceed.

Apart from causing a non-uniform distribution of ligands, strong activation will enhance multipoint attachment and locking of ligand-binding sites in positions unfavourable to subsequent interaction with the ligate.
Conversely, an excessive density of functional antigenic sites after coupling can make elution of high avidity antibody adsorbed from solution almost impossible due to the increased probability that both sites on IgG and several sites on IgM will be jointly engaged.

A number of authors have drawn attention to the fact that both the iminocarbonate and the isourea linkages formed in the CNBr reactions are positively charged under physiological conditions and therefore are liable to cause nonspecific adsorption. In a recent study from Saul Roseman's laboratory Schnaar et al. (41) used labelled orthophosphate to assess residual charge and found a complex interdependence of charge, type of ligand, coupling conditions, and amount of coupled ligand. Two mechanisms for the generation of charge were postulated; one being ligand-associated and the other ligand-independent. There was a general increase in ligand-independent charge above pH 10, but no definite pK values could be assigned and the authors stress the need to optimise each system under study.

Another characteristic of the CNBr method observed by many workers is the small but definite leakage of ligand due to rupture of the bonds between ligand or spacer and matrix. Again, the rate of leakage depends on a number of factors like degree of substitution, extent of multipoint binding, pH, and temperature. For example, Schnaar et al. (41) could not detect any leakage of coupled aminothiogalactoside, aminohexanol, or ethanolamine from Sephadex G-25 at 4° in 0,1 N acetic acid, whereas significant leakage did occur at pH 9,5 and room temperature. Parikh et al. (42) discuss the leakage problem at some length. At 4° and pH 7 they found a leakage rate at day 0 for alanine, insulin (bound only by one amino group), and albumin coupled to Sepharose of 0,2, 1, and 0,03 per cent respectively of total ligand bound. At day 25 the figures were 0,03, 0,2, and 0,006 per cent. The lower figure for albumin may in part be accounted for by multipoint binding, and indicates that conditions for coupling may be optimal when there is simultaneous attachment to more than one site, but not to so many that the active sites become distorted or inaccessible. The risk of shell formation by the ligand as discussed above must also be considered.
In case of unacceptable leakage, the problem may be mitigated by combination of the CNBr method and the utilisation of a multifunctional spacer of the type introduced by Wilchek and Miron (43). This consists of polyacrylic hydrazide forming a veritable network unaffected by occasional ruptures of individual bonds anchoring it to the matrix (fig. 4)

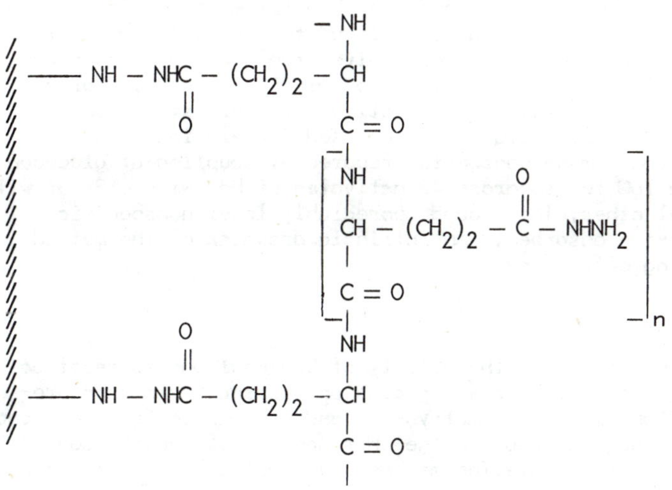

Fig. 4 Multifunctional polyhydrazide spacer according to Wilchek and Miron (43)

The hydrazide can be further derivatised by several reactions to form stable conjugates. (See ref. (43) for further references.)

The bisepoxide (bisoxirane) method (44,45)

Bisepoxides react with amino groups and hydroxyl groups, and hydroxylic matrices can be activated to give oxirane ethers:

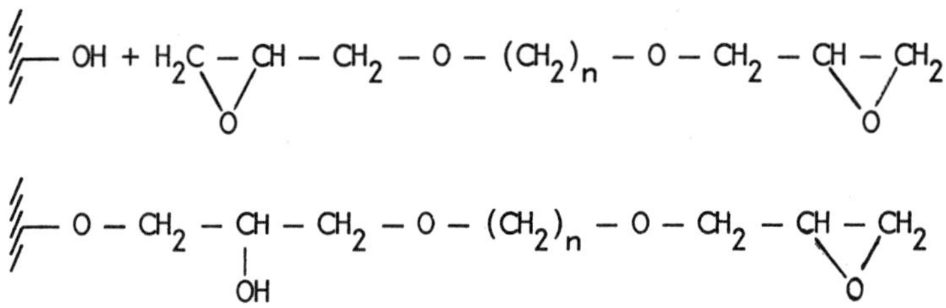

In the coupling stage, amino compounds ($R\text{-}NH_2$) give alkylamines:

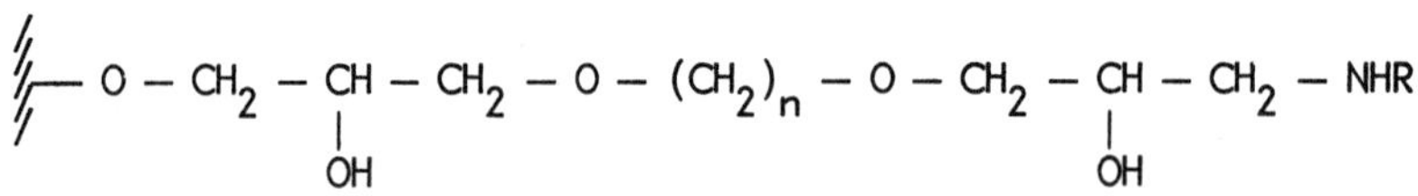

and hydroxyl-containing ligands (R-OH) are bound by another ether bond:

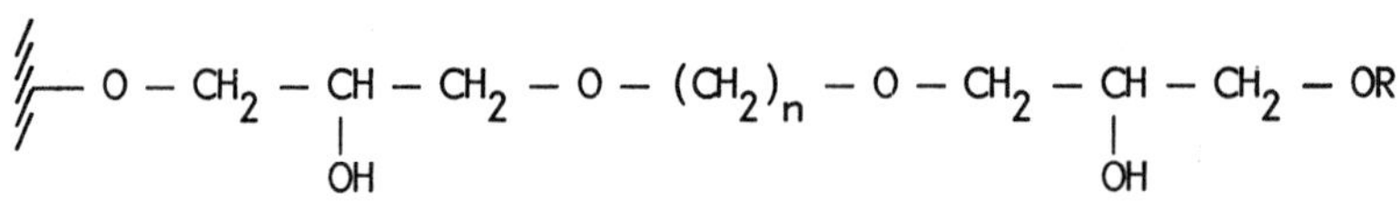

Amino groups react faster than hydroxyls and the pH optimum is between 11 and 12. However, proteins can be coupled by treatment for up to 4 days at pH 8,5-9 and room temperature. Residual oxirane groups are quite stable and must be blocked by treatment with 2 M ethanolamine, hydroxylamine, or glycine at pH 8,5 for a minimum of 24 h at room temperature. The derivatives are very stable under neutral and basic conditions, but acid cannot be used for elution.
Murphy et al. (46) compared immunoadsorbents prepared by coupling of glucagon and rabbit anti-glucagon IgG to Sepharose 4B activated either with CNBr or with 1,4-butanediol diglycidyl ether. They found appreciably lower nonspecific adsorption on the bisepoxide adsorbent. An intrinsic drawback of the method is the rather low coupling efficiency.

Benzoquinone

Brandt et al. (47) took advantage of the ability of benzoquinone to react both with amino groups in proteins and hydroxyl groups on polysaccharides in preparing derivatives of ribonuclease, chymotrypsin, and albumin on Sepharose and Sephadex. Coupling efficiency was good in the pH range 7-11,5 and the conjugates were apparently highly stable. The method was adopted by Ternynck and Avrameas (48) for the conjugation of antigens and antibodies to erythrocytes and various soluble markers, but there is no published experience with immunoadsorbents so far.

METHODS TO ATTENUATE ANTIGEN AND ANTIBODY AFFINITY

Interactions between a multivalent antigen and multivalent IgM or bivalent IgG collected late during the immune response may be characterised by intrinsic affinity constants in the range 10^{10} - 10^{11} M^{-1}, and reasonably efficient desorption from immunoadsorbents may require conditions severe enough to cause extensive destruction both of ligand and ligate.

With antigen as the ligand, the interaction may sometimes be attenuated by moderate chemical modification of the antigen molecule. For example, Hoffmann and Kuo (49) were faced with the problem of isolating rabbit antibody to dinitrophenyl-lysine with the high average K_o of 1,2 x 10^9 M^{-1}. When DNP-lysine was coupled directly to Sepharose only 23 per cent of adsorbed antibody could be eluted with dinitrophenol. A remarkable feature of this system is the adsorption efficiency of the ligand in the absence of a spacer. Recognising the considerable contribution by the lysyl part of the antigen to the free energy of binding to antibody, the authors tried several ways to modify or eliminate the lysyl structure. By reacting fluorodinitrobenzene with aminoethylamino-Sepharose a DNP-aminoethylamino-Sepharose was prepared that gave a 57 per cent recovery of anti-DNP.

Murphy et al. (50) in working with isolation of rabbit anti-glucagon antibodies found it worthwhile to attempt reducing the affinity level of the system in order to achieve a better recovery. Modification of glucagon after coupling to Sepharose was carried out by either of three methods: oxidation of Met_{27} to methionine sulphoxide with H_2O_2, alkylation of Trp_{25} with hydroxybenzyl bromide, and nitration of Tyr_{10} and Tyr_{13} with tetranitromethane. Detailed changes in the affinity of the modified ligands were complex, but in general elution of antibody originally requiring 4 M urea could be achieved in 0,15 M NaCl at pH 11 after the modifications. Similar modifications are often feasible, but fortunately the planning of specific strategies must be up to the ingenuity of the individual investigator.

An obvious but inelegant approach to the attenuation of antibody affinity might be simply to take advantage of the generally observed reduction in overall antibody affinity with increasingly strong activation of the matrix, keeping the amount of added antibody constant, or the addition of increasing amounts of antibody, keeping the level of activation constant as in the study by Comoglio et al. (38) already referred to.

A more attractive proposition to reduce affinity would be to split the IgG molecule and use separate F(ab) fragments as ligands rather than intact IgG, thereby reducing average affinity by 2-3 orders of magnitude but still utilising most of the antibody binding sites instead of quenching them out. After conventional papain digestion F(ab) fragments can be isolated by immunoadsorption as suggested by e.g. Lopatin and Voss (15) or by ion exchange chromatography on DEAE-cellulose as described by Fey (51).

DESORPTION OF IMMUNOADSORBENTS

Specific desorption

When adsorbent systems involve haptenic ligands, antibody can often be released more or less efficiently with hapten or a hapten analogue even if it may be hard to remove from the antibody afterwards.

Also, when immunodeterminant groups on macromolecular antigens are known or suspected, specific desorption can be successful. For instance, Kaplan and Kabat (52) desorbed anti-A antibody from polymerised blood group substance A with the immunodeterminant sugar N-acetyl-D-galactosamine. Another example, previously referred to, is the work of Cheng et al. (7) who could isolate a series of slightly different antibodies to pneumococcal polysaccharides by gradient elution from an agarose-based adsorbent with cellobiose.

Nonspecific desorption

The forces that make immune complexes stick together are the same as those that operate in other types of specific biomolecular interactions; a medley of electrostatic and apolar (hydrophobic) interactions, van der Waals and London dispersion forces, and hydrogen bonding (53). The relative contributions of these factors vary within and between Ig classes and subclasses and from one animal to the next (54). The different desorbing agents available influence these forces according to specific patterns and if the contribution of the binding forces is not known, an adequate procedure must be found by trial and error.
No matter which procedure is decided upon, there is an important practical expedient that is sometimes overlooked: antibody adsorbed to columns should always be eluted backwards. Otherwise high avidity antibody tending to accumulate on the inlet side will have to interact with antigen throughout the column before it is eluted, if it is, indeed, eluted at all. A strong desorber, able to propel it all along to the outlet, may cause severe damage both to the antibody and the ligand.

Adjustment of pH. Antigen-antibody complexes tend to weaken or dissolve at high or low pH. Desorption at or below pH 2,6 is a useful method with systems of weak or moderately strong interaction. Glycine-HCl buffer or 1 M propionic acid are often used. The mechanism of desorption is not entirely clear. As discussed by Crook et al. (55) several factors are involved, and the rate of protonation may be especially important. With lysozyme as an antigen they found an increase in desorption efficiency with protonation rate in the pH interval 4-2,2.
In unpublished experiments in our own laboratory with desorption of anti-A antibody to blood group substance A, a system characterised by an extremely strong affinity, we applied a continuous sinusoidal gradient between pH 1,5 and 4,5. Whenever the gradient passed pH 2,5-2,8 downwards or upwards some antibody was released, but neither above nor below. Nowhere between pH 4,5 and 1,5 was antibody released when pH was kept constant.
If desorption at low pH is found to be marginally efficient, a succession of steep gradients may be tried, rather than a constant level.

Basic desorption at pH up to 11 may sometimes be useful. Chidlow et al. (56) obtained essentially complete recovery of collagen peptides desorbed from Sepharose-coupled antibody by 1 M ammonia. Desorption of α- mannosidase from Sepharose-linked antibody was described by Paus, who recovered 70 per cent of applied enzyme activity with desorption at pH 10,6 in 1 M NaCl (57).
With systems in need of stronger agents for desorption, pre-elution at basic pH may be useful for removal of nonspecifically adsorbed material. In experiments with desorption of alphafetoprotein from Sepharose-antibody, Zoller and Matzku (58) compared the efficiency of pre-elution at acid, neutral, and basic pH, with markedly superior results at pH 10 in NaCl before desorption in 3 M SCN^-.

Chaotropic ions. The term "chaotropic ion" was coined by Hamaguchi and Geiduschek in a study on the denaturation of DNA by various salts (59).
In physical chemistry the Hofmeister series or lyotropic series defines a series of ions according to their tendency to precipitate lyophilic substances from colloidal solution. Their major effect is on the structure of water and thereby on the ability of water to accomodate solutes. At one end of the series are precipitating, anti-chaotropic ions like SO_4^{--} promoting association, folding, and coiling of biomolecules in watery solution. At the other end are the chaotropic ions like SCN^-, counteracting the same processes and increasing solubility.

The chaos-inducing properties of these ions refer to their ability to break up the structure of water primarily by disrupting hydrogen bonds. Since antigen-antibody complexes are largely dependent on the integrity of water structure, they tend to dissolve in strong solutions of chaotropic ions even at neutral pH. However, the tertiary structure of the immunoreactants may also be irreversibly affected, resulting in partial or complete denaturation and inactivation.
The physical chemistry of chaotropic ions is treated by Dandliker and de Saussure (60) and the aspects that apply particularly to antigen-antibody interaction by Levison et al. (61). The chaotropic properties of haloacetates are discussed by Hanstein et al. (62).

Chaotropic ions of interest in immunoadsorption can be arranged from left to right according to decreasing chaotropicity:

$$CCl_3COO^- \geqq SCN^- > CF_3COO^- > ClO_4^- > I^- > Cl^-$$

with chloride being only weakly chaotropic. Phosphate, ubiquitous as a buffering ion in immunochemistry, is an anti-chaotropic ion but usually not present in concentrations that will noticeably oppose the effect of strong solutions (2-3 M) of chaotropes like SCN^- or CCl_3COO^-.
Chaotropic ions used in neutral solution in concentrations up to 3 M are valuable desorbers of immunoadsorbents. It should be realised, however, that the binding sites tend to be shielded from rapid access of even strongly chaotropic ions, as evidenced by the fact that chaotropes are much more efficient in preventing the formation of immune complexes than in breaking those already formed. Especially when working with high-affinity systems it may be better to use a moderate concentration of chaotrope and perform the desorption in consecutive stages, leaving the adsorbent to equilibrate between elutions.
This strategy was used in our laboratory in work involving desorption of late-phase human anti-A antibody to blood group substance A, using 3 M CF_3COO^-. Despite the extremely high affinity, an overall recovery of 40-55 per cent was obtained (33).

Urea and guanidine hydrochloride. The ability of these substances to break hydrogen bonds makes them sometimes effective as desorbing agents, although usually with considerable loss of antibody activity. Their net effect on antigen-antibody complexes is similar to that of chaotropic ions, although it is not clear whether the effect is primarily on the solute or mainly an indirect one mediated by a change in water structure (60).

REFERENCES

1. H.H. Weetall, Insolubilized antigens and antibodies, p. 597 in M.L. Hair (ed.) (1972), The Chemistry of Biosurfaces, Vol. II, Marcel Dekker, New York

2. S. Fuchs and M.Sela, Immunoadsorbents, section 11 in D.M. Weir (ed.) (1973), Handbook of Experimental Immunology, Blackwell Scientific Publications, Oxford

3. J.B. Robbins and R. Schneerson, Immunoadsorbents, p. 703 in W.B. Jakoby and M. Wilchek (eds.) (1974), Methods in Enzymology, Vol. XXXIV, Academic Press, New York

4. D.M. Livingston, Immunoaffinity chromatography of proteins, Ibid., p. 723

5. J. Porath and T. Kristiansen, Biospecific affinity chromatography and related methods, p. 95 in H. Neurath and R.L. Hill (eds.) (1975), The Proteins, 3rd.ed. Vol.I, Academic Press, New York

6. T. Kristiansen, Matrix-bound antigens and antibodies, p. 19 in E. Ruoslahti (ed.) (1976), Immunoadsorbents in protein purification, Scand.J. Immunol. Suppl. 3, Universitetsforlaget, Oslo

7. W.C. Cheng, K.J. Fraser and E. Haber, Fractionation of antibodies to the pneumococcal polysaccharides by affinity chromatography, J. Immunol. 111, 1677 (1973)

8. F.W. Chen, L.E. Cannon, M.N. Margolies, A.D. Strosberg and E. Haber, Purification, specificity, and hypervariable region sequence of anti-pneumococcal polysaccharide antibodies elicited in a single rabbit, Ibid. 117, 807 (1976)

9. E. Haber, M.N. Margolies and L.E. Cannon, Origins of antibody diversity: Insights gained from amino acid sequence studies of elicited antibodies, Cold Spring Harbour Symp. Quant. Biol. 41, 647 (1976)

10. A.J.L. Macario and E.C. de Macario, Synthesis of low affinity antibody molecules during the decline of the immune response, Immunochemistry 11, 619 (1974)

11. P. Abramoff, M.F. La Via and N.K. Brien, Antibody induction and production, p. 136 in P. Abramoff and M.F. La Via (1970), Biology of the Immune Response, McGraw-Hill, New York

12. H.N. Eisen and G. Siskind, Variation in affinities of antibodies during the immune response, Biochemistry 3, 996 (1964)

13. C.L. Hornick and F. Karush, Antibody affinity - III. The role of multivalence, Immunochemistry 9, 325 (1972)

14. E.W. Voss, Jr. and M.M. Sigel, Valence and temporal change in affinity of purified 7S and 18S nurse shark anti-2,4-dinitrophenyl antibodies, J. Immunol. 109, 665 (1972)

15. D.E. Lopatin and E.W. Voss Jr., Avidity in immunosorption of IgG antibodies, Immunochemistry 11, 333 (1974)

16. Y.T. Kim, T.P. Werblin and G.W. Siskind, Distribution of antibody affinities - II. Fractionation of antibody with respect to its hapten binding affinity, Ibid. 11, 685 (1974)

17. D.M. Crothers and H. Metzger, The influence of polyvalency on the binding properties of antibodies, Ibid. 9, 341 (1972)

18. W.J. Herbert, Laboratory animal techniques for immunology, section A.3 in D.M. Weir (ed.) (1973) Handbook of Experimental Immunology, Blackwell Scientific Publications, Oxford

19. T. Ternynck and S. Avrameas, Polymerization and immobilization of proteins using ethylchloroformate and glutaraldehyde, p. 29 in E. Ruoslahti (ed.) (1976), Immunoadsorbents in Protein Purification, Scand. J. Immunol. Suppl.3, Universitetsforlaget, Oslo

20. Reference deleted

21. P. O'Carra, S. Barry and T. Griffin, Interfering and complicating adsorption effects in bioaffinity chromatography, p. 108 in W.B. Jakoby and M. Wilchek (eds.) (1974), Methods in Enzymology Vol.XXXIV Academic Press, New York

22. C.R. Lowe, Affinity chromatography: The current status, Int. J. Biochem. 8, 177 (1977)

23. D.H. Campbell and N. Weliky, Immunosorbents: preparation and use of cellulose derivatives, p. 365 in C.A. Williams and M.W. Chase (eds.) (1967), Methods in Immunology and Immunochemistry Vol. I, Academic Press, New York

24. H.H. Weetall and A.M. Filbert, Porous glass for affinity chromatography applications, p. 59 in W.B. Jakoby and M. Wilchek (eds.) (1974), Methods in Enzymology Vol. XXXIV, Academic Press, New York

25. J.K. Inman, Covalent linkage of functional groups, ligands and proteins to polyacrylamide beads, Ibid., p. 30

26. T. Ternynck and S. Avrameas, Polyacrylamide - protein immunoadsorbents prepared with glutaraldehyde, FEBS Letters 23, 24 (1972)

27. C. Sapin, A. Massez, A. Coutet and P. Druet, Isolation of normal human IgA, IgM and IgG fragments by polyacrylamide beads immunoadsorbents, J. Immunol. Meth. 9, 27 (1975)

28. C. Sapin and P. Druet, Isolation of rat and rabbit IgM from normal serum using anti-human μ antibody-polyacrylamide beads immunoadsorbents, Ibid. 12, 355 (1976)

29. H. Tlaskalová, L. Tučková, M. Křivaková, J. Rejnek and J. Čoupek, Receptor-specific fractionation of immunocompetent cells and purification of antibodies on hydroxyalkyl methacrylate immunosorbents, Immunochemistry 12, 801 (1975)

30. J. Turková, Immobilization of enzymes on hydroxyalkyl methacrylate gels, p. 66 in K. Mosbach (ed.) (1976), Methods in Enzymology Vol. XLIV, Academic Press, New York

31. R. Axén, J. Porath and S. Ernback, Chemical coupling of peptides and proteins to polysaccharides by means of cyanogen halides, Nature 214, 1302 (1967)

32. M. Joustra and R. Axén, Stability of the binding groups generated by CNBr activation of agarose, p. 525 in H. Peeters (ed.) (1976), Protides of the Biological Fluids, Vol. 23, Pergamon Press, Oxford

33. T. Kristiansen, Studies on blood group susbstances. V. Blood group substance A coupled to Sepharose as an immunosorbent, Biochim. Biophys. Acta 362, 567 (1974)

34. J. Porath, J.-C. Janson and T. Laas, Agar derivatives for chromatography, electrophoresis and gel-bound enzymes, J. Chromatog. 60, 167 (1971)

35. J.L. Guesdon and S. Avrameas, Polyacrylamide-agarose beads for the preparation of effective immunosorbents, J. Immunol. Meth. 11, 129 (1976)

36. C. Lowe and P. Dean (1974), Affinity Chromatography, Wiley, London

37. J. Porath and R. Axén, Immobilization of enzymes to agar, agarose, and Sephadex supports, p. 19 in K. Mosbach (ed.) (1976), Methods in Enzymology Vol. XLIV, Acad. Press, New York

38. S. Comoglio, A. Massaglia, E. Rolleri and U. Rosa, Factors affecting the properties of insolibilized antibodies, Biochim. Biophys. Acta 420, 246 (1976)

39. G.S. David, T.H. Chino and R.A. Reisfeld, Binding of proteins to CNBr-activated Sepharose 4B, FEBS Letters 43, 264 (1974)

40. J. Lasch, M. Iwig and R. Koelsch, Studies of the distribution of proteins bound to CNBr-activated Sepharose 6B at the electron-microscopic level, Eur. J. Biochem. 60, 163 (1975)

41. R.L. Schnaar, T.F. Sparks and S. Roseman, Cyanogen bromide activation of polysaccharides. Effects of reaction conditions on cationic charge and ligand content, Anal. Biochem. 79, 513 (1977)

42. I. Parikh, S. March and P. Cuatrecasas, Topics in the methodology of substitution reactions with agarose, p. 77 in W.B. Jakoby and M. Wilchek (eds.) (1974), Methods in Enzymology Vol. XXXIV, Acad. Press, New York

43. M. Wilchek and T. Miron (1974), Polymers coupled to agarose as stable and high capacity spacers, Ibid. p. 72

44. J. Porath (1974), General methods and coupling procedures, Ibid. p. 13

45. L. Sundberg and J. Porath, Preparation of adsorbents for biospecific affinity chromatography, J. Chromatog. 90, 87 (1974)

46. R.F. Murphy, J.M. Conlon, A. Imam and G.J.C. Kelly, Comparison of non-biospecific effects in immunoaffinity chromatography using cyanogen bromide and bifunctional oxirane as immobilising agents, Ibid. 135, 427 (1977)

47. J. Brandt, L.-O. Andersson and J. Porath, Covalent attachment of proteins to polysaccharide carriers by means of benzoquinone, Biochim. Biophys. Acta 386, 196 (1975)

48. T. Ternynck and S. Avrameas, A new method using p-benzoquinone for coupling antigens and antibodies to marker substances, Ann. Immunol. (Inst.Pasteur) 127 C, 197 (1976)

49. L.G. Hoffmann and C.-Y. Kuo, Immunoadsorbents for the isolation of high-affinity anti-hapten antibodies in high yield, J. Immunol. Meth. 15, 101 (1977)

50. R.F. Murphy, A Imam, A.E. Hughes, M.J. McGucken, K.D. Buchanan, J.M. Conlon and D.T. Elmore, Avoidance of strongly chaotropic eluents for immuno-affinity chromatography by chemical modification of immobilized ligand, Biochim. Biophys. Acta 420, 87 (1976)

51. H. Fey, A simple procedure for the production of F(ab) from bovine IgG as an absorbent in the preparation of class-specific anti-immunoglobulin, Immunochemistry 12, 235 (1975)

52. M.E. Kaplan and E.A. Kabat, Studies on human antibodies. IV. Purification and properties of anti-A and anti-B obtained by absorption and elution from insoluble blood group substances, J. Exp. Med. 123, 1061 (1966)

53. F. Karush, Immunologic specificity and molecular structure, Adv. Immunol. 2, 1 (1963)

54. E. Benjamini, M. Shimizu, J.D. Young and C.Y. Leung, Immunochemical studies on tobacco mosaic virus protein. IX. Investigation on binding and antigenic specificity of antibodies to an antigenic area of tobacco mosaic virus protein, Biochemistry 8, 2242 (1969)

55. N.E. Crook, J. Stephen and H. Smith, Recovery of bound homologous protein antigens from disulphide-linked immunosorbents at neutral pH, Immunochemistry 9, 945 (1972)

56. J.W. Chidlow, A.J. Bourne and A.J. Bailey, Production of hyperimmune serum against collagen and its use for the isolation of specific collagen peptides on immunosorbent columns, FEBS Letters 41, 248 (1974)

57. E. Paus, Immunoadsorbent affinity purification of the two enzyme forms of α-mannosidase from *Phaseolus vulgaris*, FEBS Letters 72, 39 (1976)

58. M. Zoller and S. Matzku, Antigen and antibody purification by immuno-adsorption: elimination of non-biospecifically bound proteins, J. Immunol. Meth. 11, 287 (1976)

59. K. Hamaguchi and E.P. Geiduschek, The effect of electrolytes on the stability of the deoxyribonucleate helix, J. Amer. Chem. Soc. 84, 1329 (1962)

60. W.B. Dandliker and V.A. deSaussure, Stabilization of macromolecules by hydrophobic bonding: Role of water structure and of chaotropic ions, p. 1 in M. Hair (ed.) (1971), The Chemistry of Biosurfaces, Vol. I, Marcel Dekker, New York

61. S.A. Levison, F. Kierszenbaum and W.B. Dandliker, Salt effects on antigen-antibody kinetics, Biochemistry 9, 322 (1970)

62. W.G. Hanstein, K.A. Davis and Y. Hatefi, Water structure and the chaotropic properties of haloacetates, Arch. Biochem. Biophys. 147, 534 (1971)

COMPARISON OF SEPHAROSE AND CELLULOSE AS A SUPPORT FOR ANTIBODY AND ANTIGEN

P. D. Weston and R. Scorer

Wellcome Reagents Ltd., Langley Court, Beckenham, Kent, England

ABSTRACT

IgG was offered at an increasing level to CNBr activated Sepharose and cellulose. The Sepharose bound almost all the protein offered whereas the binding to the cellulose quickly reached a plateau. The antibody activity measured by radioimmunoassay reached a maximum at a relatively low level of Sepharose substitution whereas the cellulose behaved differently. The two supports were also compared with respect to the activity of IgG acting as antigen.

INTRODUCTION

Our interest in the attachment of antibody to solid phase for use as second antibody in radioimmunoassay has led us to examine Sepharose and cellulose for their suitability. Earlier workers have coupled first antibody to cellulose and Sephadex (Wide, 1969), and also to Sepharose (Gardner et al, 1974; Bolton & Hunter, 1973; Comoglio et al 1976). This paper presents our observations of IgG attached to solid phase acting as antibody and also as antigen.

MATERIALS & METHODS

Coupling Procedure

Sepharose CL - 4B (Pharmacia), cellulose (Merck Microcrystalline Art 2330) and fibrous cellulose (Whatman CF11) were activated with CNBr (Aldrich) by the method of Porath et al. 1973. Activation was for 10 mins with stirring at 4° - 10°C using 13 g of the solid phase, (sucked dry weight) CNBr (90 mg/ml, 17 ml) and buffer (39 ml.) Purified IgG from donkey anti-rabbit IgG serum for the antibody experiments and human IgG (KABI) for the antigen work were dialysed against 0.2M trisodium citrate adjusted to pH 6.5 with HCl. Coupling was performed in duplicate for several levels of protein and was by rotation for 2 hours at room temperature then overnight at 4°C. The extent of attachment was calculated from the OD_{280} of the supernatant after coupling, compared with controls for each level using non activated support.

Antibody Activity

After coupling to antibody IgG the supports were thoroughly washed with saline (0.15M) on a Buchner funnel, then sucked dry and a weighed amount suspended in pH 7.4 phosphate buffer 0.04M containing NA_2EDTA 0.01M; NaCl 0.15M, NaN_3 0.02M, BSA 5g/l, polyvinylpyrrolidone, M.W. 7000,000, 30g/l and Tween 20, 1.4 ml/l. The antibody activity was assessed by using the suspensions as second antibody in radioimmunoassay. A number of glass test tubes were filled with the first incubation mixture:-

100 µl I-125 Digoxin 2ng/ml, 100,000 cpm.
100 µl rabbit anti-digoxin serum diluted 1:20,000 in buffer.
100 µl phosphate buffer, as above.

The tubes were kept at room temperature 30 mins then 1 ml of the Sepharose 1/100 or cellulose 1/500 suspension added; after vortex mixing the tubes were stood a further 30 mins then centrifuged for 15 mins at 2000g, the supernatant was aspirated to waste and the pellets counted in an NE 1600 gamma counter. Results are expressed as a % of the number of counts in the precipitate/total counts available.

Antigenic Activity

As an auxiliary reagent, specific antibody was prepared by passing purified immunoglobulins (25 ml, 33 mg/ml) from a sheep anti-human IgG serum down a column of human IgG (1g) coupled to Sepharose CL - 4B (100g) poured on top of a column of Sephadex G25 (300 ml). After washing, the absorbed antibodies were eluted with 3M Na CNS at pH 6.8 and yielded 29 mgs. The eluted protein was unstable and prior to use was centrifuged to remove turbidity (12,000 rpm, 30 mins in MSE High Speed 18).

Sepharose and cellulose with varying levels of human IgG attached were washed with saline then 5g suspended in saline (15 ml). Aliquots of the suspensions (2 ml) were mixed with the purified specific antibody (2 ml) and rotated together with controls at 37°/30 mins in 5 ml glass bottles. The bottles were centrifuged, the supernatants aspirated and respun to clarify, then the OD_{280} measured.

RESULTS

Using the relatively high concentrations of CNBr for activation it was found that Sepharose bound nearly 100% of the IgG offered to it, whereas the binding by the cellulose was limited (Fig.1). When the antibody activity was measured the two supports behaved differently (Figs. 2 and 3), the Sepharose passing through a maximum but the cellulose reaching a plateau. The cellulose made much better use of antibody since suspensions 5 times more dilute gave similar binding of antigen.

Fibrous cellulose bound less protein (maximum 1.5 mgs IgG/gram) than the microcrystalline material and was also a less effective support for IgG acting as antigen than Sepharose substituted to the same level (Fig. 4).

DISCUSSION

It was surprising to find that above the level of 3-4 mgs IgG/gram Sepharose additional bound IgG was ineffective as antibody. As higher levels of IgG were coupled to the Sepharose antibody activity actually diminished, suggesting that crowding of IgG on the beads was preventing the action of antibody. For the cellulose the activity reached a maximum at a higher level of substitution 7mg/gram cellulose and in contrast to the Sepharose-antibody steric hindrance of antibody action was not observed.

When IgG was tested for its activity as antigen a similar situation was discovered. Increased levels of bound IgG failed to give increased activity beyond about 3 mg/gram. Thus for this particular protein (IgG) acting either as antibody or antigen

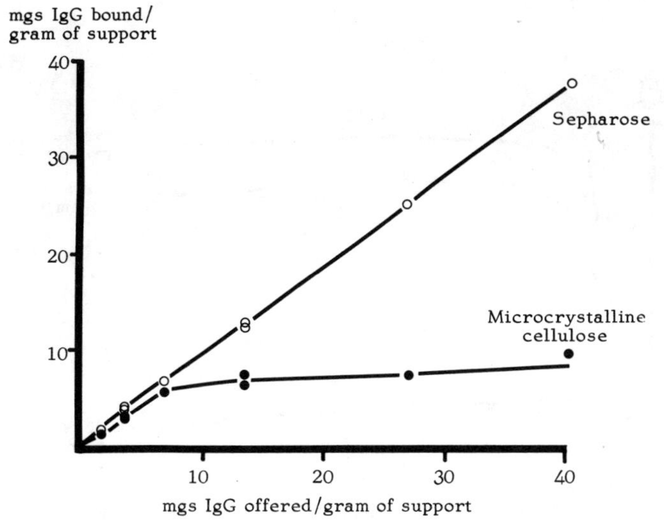

Fig. 1. Binding of IgG to CN.Br activated supports.

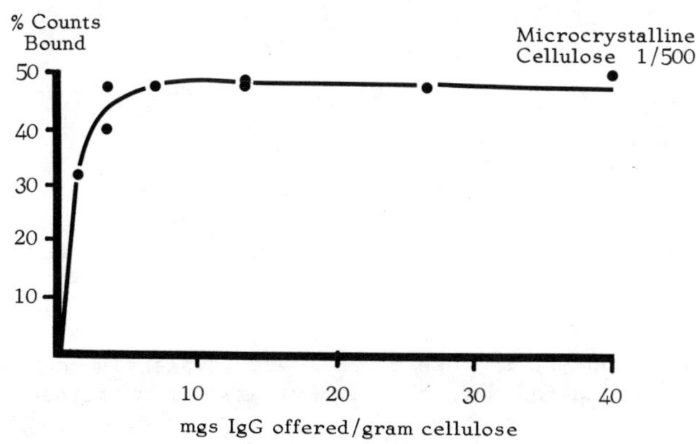

Fig. 2. Antibody activity of IgG bound to cellulose.

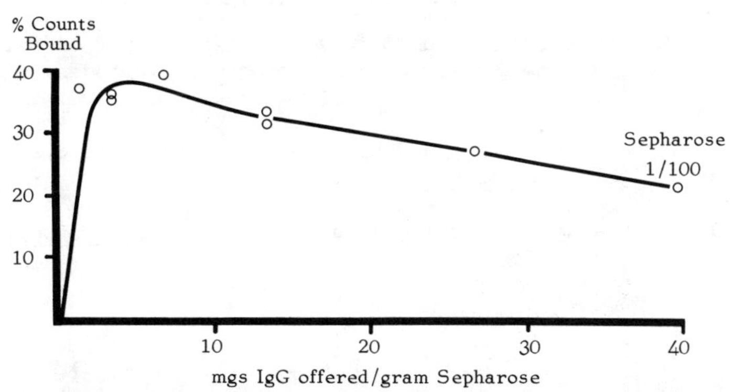

Fig. 3. Antibody activity of IgG bound to Sepharose.

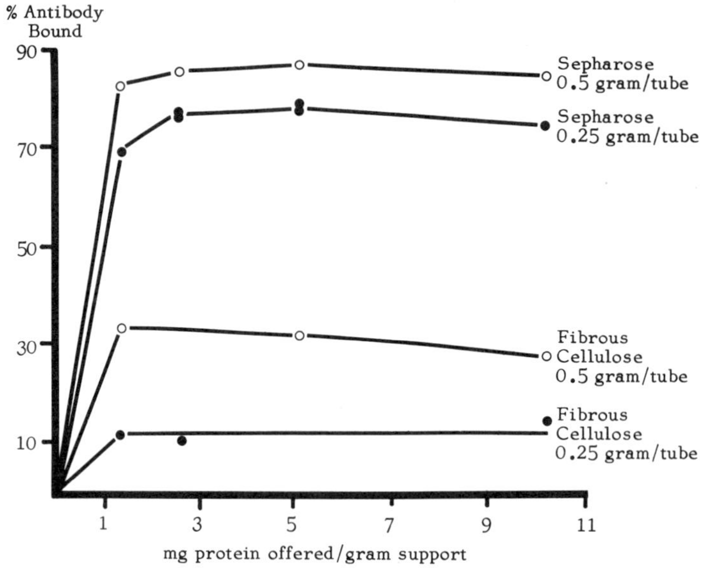

Fig. 4. Antigenic activity of IgG bound to different supports.

there seems no point in coupling the protein at the high levels that CNBr activation allows; relatively low levels of substitution (3 mg/gram) giving immunoadsorbents working at their optimum activity whether the support is Sepharose or cellulose.

REFERENCES

A.E. Bolton and W.H. Hunter, The use of antisera covalently coupled to agarose, cellulose and Sephadex in radioimmunoassay systems for proteins and haptens, Biochem. Biophys. Acta 329, 318 (1973)

S. Comoglio, A. Massaglia, E. Rolleri and U. Rosa, Factors affecting the properties of insolubilised antibodies, Biochem. Biophys. Acta 420, 246 (1976).

J. Gardner, G. Bailey and T. Chard, Observations on the use of solid-phase-coupled antibodies in the radioimmunoassay of human placental lactogen, Biochem. J. 173, 469 (1974).

J. Porath, K. Aspberg, H. Drevin and R. Axen, Preparation of cyanogen bromide-activated agarose gels, J. Chromatog. 86, 53 (1973).

L. Wide, Radioimmunoassays employing immunosorbents. Acta Endocrinol. 63 Suppl. 142, 207 (1969).

APPROACH TO THE SEQUENCE ANALYSIS OF PROTEINS WITH HLA ACTIVITY BY AFFINITY CHROMATOGRAPHY

Claude Bieva* and Maurice Wikler**

*Department of Immunology, Hospital St Pierre, 322 rue Haute, 1000 Brussels, Belgium
**Department of Immunology, Free University of Brussels, 67 rue-aux-Chevaux, 1640 Rhodes-Ste-Genèse

ABSTRACT

Elevated amounts of beta$_2$-microglobulin and HLA antigens are found in patients with tubular proteinuria. A method was developped to remove by affinity chromatography all urinary proteins except HLA antigens. The immunoadsorbent was obtained by coupling horse antibodies against human serum proteins together with monospecific anti-glycoproteins antibodies to Sepharose. Several passages of urinary proteins on the immunoadsorbent column allowed to recover a highly pure fraction of HLA antigens which were not adsorbed on the column. Both beta$_2$-microglobulin and HLA activity measured by radioimmunoassay and inhibition of lymphocytotoxicity, were preserved during the isolation process. Radiolabeled preparations of the isolated material showed two peaks on SDS-gel electrophoresis ; approximate molecular weights of 33,000 and 12,000 were identified for the two non-covalently linked subunits of HLA antigens. The large protein had HLA specificity and the small protein was assessed to be beta$_2$-microglobulin.

The major serologically defined histocompatibility antigens, called respectively HLA and H2 in man and in the mouse, are cell-surface products of genes in a chromosome region governing several immunological functions.

Previous reports have shown that HLA antigens are composed of two polypeptide subunits of 33,000 and 12,000 daltons. The smaller subunit is beta$_2$-microglobulin. It has amino-acid sequence homology with the CH3 domain of immunoglobulin IgG.

Proteins with HLA activity cannot be isolated from cell cultures, serum or urine in pure form by any single step procedure. The most widely adopted separation technique have employed ion exchange chromatography, gel filtration, preparative gel electrophoresis and other biochemical methods (Table 1). Those isolations procedures are time consuming and the compromise between purity and yield, common to almost all protein purification problems, result in a poor overall yield of proteins with HLA activity.

Our aim was to prepare sufficient amount of HLA proteins from various origin in order to compare their structural relationship. The starting material most suitable for this purpose seems to be urine of patients with tubular proteinuria. As normal human serum contains beta$_2$-microglobulin and HLA antigens in trace amounts, an anti-NHS serum will include a negligible amount of anti-beta$_2$-

microglobulin and anti-HLA antibodies.
An immunoadsorbent prepared with such immunoglobulins coupled to Sepharose will not retain the HLA and beta$_2$-microglobulin molecule present in high amounts in the urine of patients with tubular proteinuria. All other proteins of the preparation will be adsorbed on the column. Patients with tubular proteinuria excrete urinary well defined proteins which are recognized by antisera against serum proteins.

The present work uses a 6-amino-hexyl Sepharose coupled to polyspecific normal serum antibodies by means of glutaraldehyde as cross-linking agent. In comparison with the previous system using BrCN-Sepharose directly coupled to the immunoglobulins, we observed an increased capacity of the AH-Sepharose preparation. In a typical experiment, three hundred milligrams of proteins were added to 15 gr of glutaraldehyde AH-Sepharose. Forty-eight milligrams of urinary proteins were passed through the column and five successive passages are needed to adsorb all the proteins except beta$_2$-microglobulin and proteins with HLA activity.
The adsorbed urinary proteins can only be completely removed from the immunoadsorbent column by use of 5M Guanidine hydrochloride (acetic acid and urea 3M were insufficient to elute all the proteins). The progress of the purification was tested by Ouchterlony analysis and immunoelectrophoresis. We don't observe any lines of precipitation with an antiserum against whole normal human serum. One line only appears with an anti-beta$_2$-microglobulin antiserum. SDS-polyacrylamide gel electrophoresis reveals two bands after staining with Coomassie Blue, the first with approximate molecular weight 33,000 also stained with Schiff's reagent, the second with 12,000 molecular weight only stained with Coomassie Blue.
SDS-polyacrylamide gel of the radiolabeled purified proteins showed two major peaks. This radiolabeled preparation could be fractionated at low pH into two components on a G-100 Sephadex column. The radioactivity of the beta$_2$-microglobulin peak represent 10 to 20 % of the total radioactivity.
The lymphocytotoxicity show that proteins of 33,000 molecular weight were able to inhibit the lymphocytotoxicity of all corresponding antisera. The isolation procedure achieved a 250 fold increase in antigenic specificity. No cytotoxic activity was detected in peak II of the Sephadex column. A radioimmunoassay inhibition test of the purified products show that peak I was devoid of inhibitory capacity, fraction II inhibit only anti-beta$_2$-microglobulin antiserum.

In conclusion this method of purification offers several attractive features :

- the minimisation of handling and concentrating procedures is important since proteins with HLA activity are notoriously subject to denaturation during long and difficult isolation methodology
- the high yield of HLA obtained is attractive in allowing the purification of amounts sufficient to approach the amino acid sequence.

TABLE I

METHODS OF PURIFICATION

- HLA antigens purified from cell, serum, urine.

Shimada, Nathenson, 1967	- papaïn solubilization
Kahan, Reisfeld, 1971	- sonication
Reisfeld et al, 1974	- extraction with 3M KCl
Snary et al, 1974	- use of ionic detergents
Springer, Strominger, 1973	- use of non-ionic detergents
	- ion exchange chromatography
Miyakawa et al, 1971	- gel filtration
	- preparative electrophoresis (acrylamide, agarose ...)
Bernier, 1973	- isoelectric focusing
Reisfeld, Peterson, 1975	- affinity chromatography.

REFERENCES

- C. BIEVA and M. WIKLER, A simplified method of isolation of pure HLA antigens and beta$_2$-microglobulin, Acta Clinica Belgica, 31, Suppl. 8 (1976).
- B.A. CUNNINGHAM, J.L. WANG, I. BERGGARD and P.A. PETERSON, The complete amino acid sequence of beta$_2$-microglobulin, Biochem. 12, 4811 (1973).
- P.E. EVRAIN, P.A. PETERSON, L. WIDE and I. BERGGARD, Radioimmunoassay of beta$_2$-microglobulin in human biological fluids. Scand. J. Clin. Lab. Invest. 28, 439 (1971).
- W.M. HUNTER and F.C. GREENWOOD, Preparation of Iodine-131 labeled human growth hormone of high specific activity. Nature 194, 495 (1962).
- L. RASK, L. OSTBERG, J.B. LINDBLOM, Y. FERNSTEDT and P.A. PETERSON, The subunit structure of transplantation antigens. Transplant. Rev. 21, 85 (1974).
- M. ROBERT, C. VINCENT and J.P. REVILLARD, Presence of HLA antigens and beta$_2$-microglobulin in tubular proteinuria. Transplantation 18, 89 (1974)
- K. WEBER, M. OSBORN, The reliability of molecular weight determination by dodecyl sulfate-polyacrylamide gel electrophoresis. J. Biol. Chem. 244, 4406 (1969).

BIO- AND IMMUNOSPECIFIC AFFINITY CHROMATOGRAPHY OF PLASMA MEMBRANE ANTIGENS FROM HUMAN KIDNEY AND RENAL CELL CARCINOMA

Jürgen E. Scherberich, Cornelia Gauhl, Werner Mondorf

Dept. Nephrology, Z.I.M. Dept. Microbiol. Virology, Z.H., University D 6000 Frankfurt am Main, West Germany

INTRODUCTION

Concanavalin A (Con A) and wheat germ agglutinin (WGA) have been introduced as promising tools for investigating structure and function of biomembrane receptor sites. Suitable lectins, that have been made insoluble through fixation on highmolecular matrices, allow rapid isolation of essential membrane surface glycoproteins by biospecific affinity chromatography (1,2). In similar manner, plasma membrane antigens can be isolated by immunosorption (3). Specific antigens fixed by antibody-gel can be highly purified from a heterogeneous suspension after desorption by acidic or neutral buffers containing "chaotropic ions" (4).

In the present work studies were carried out to analyse the glycoprotein profile of plasma-membranes from human kidney cortex as well as from renal cell carcinoma. Constituents solubilized from the surface of the membranes were fractioned by affinity chromatography and characterized by immunochemical and ultrastructural investigation.

METHODS

Plasma-membranes (PM) were isolated from renal adenocarcinoma (PMCa) and human kidney cortex (PMK) by differential- and density gradient centrifugation techniques, that have been previously described (5,6). PM were incubated with trypsin, bromelain (INC-K-K), and papain (Serva) respectively at 37°C for 20 min (ratio enzyme :PM protein=1:20). The suspensions were then centrifuged at 50 000xg (30 min, 2°C). If not mentioned otherwise the supernatants were used for further studies.

Biospecific affinity chromatography: Columns (1.5 x10 cm) were packed with WGA- and Con A Sepharose 6MB respectively and equilibrated with phosphate buffer (0.1 M, pH7.6). After having applied the samples, the proteins not retained by the columns were eluted and fractioned. Biospecifically bound constituents were then desorbed (T 280 nm) with buffer containing 10% (w/v) a-methyl-D-glucopyranoside (Con A-) or N-acetylglucosamine (WGA Sepharose).

Immunospecific affinity chromatography: Sepharose 4B was activated with CNBr according to the method described by Axen et al. (7) and Cuatrecasas (8). Medium activated (Sepharose: CNBr 10:1, w/w) as well as highly activated beads (Sepharose:CNBr 1:1, 2:1 w/w) were prepared. Antisera directed against PMK (5) and their membrane surface proteins were covalently linked to the activated gel during an incubation period of 18 hrs at 4°C. Coupling of antisera to medium activated beads was performed at pH 7.8, to highly activated Sepharose at pH 6.8. Soluble PMconstituents were applied to the columns and eluted with PBS. Immunospecifically adsorbed antigens were eluted either with 0.1 M glycine-HCl buffer, pH 2.3 or with 1.5 M CCl_3COONa buffer, pH 6.8 (4).

The activity of alkaline phosphatase (E.C 3.1.3.1), gammaglutamyltranspeptidase (E.C 2.3 2-)

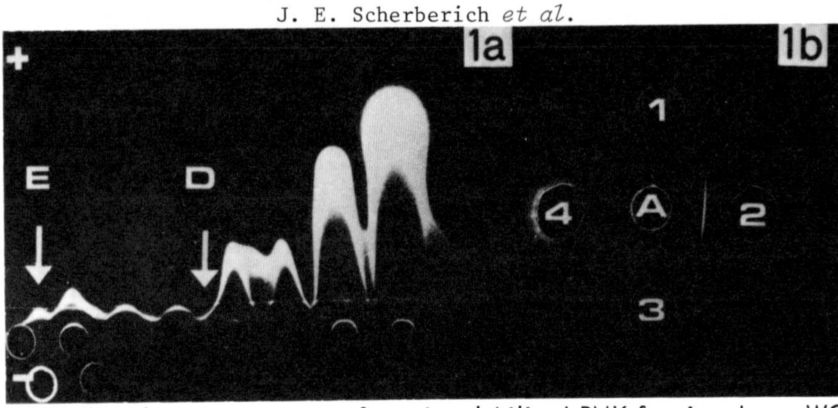

Fig. 1A: Fused rocket electroimmunoassay of papain solubilized PMK fractioned on a WGA-Sepharose column. Aliquots of eluate and desorbate fractions were pipetted into wells and electrophoresed (14 hrs, 10 Vcm^{-1}) against 5% anti-PMK serum from the rabbit; 0.8% agarose; E=elution-, D=desorption fractions; a linear gradient (2-10%) N-acetylglucosamine in phosphate buffer, pH 7.6 was used to desorb biospecifically bound PM surface glycoproteins (staining with ala-ß-naphthylamide and Fast Blue B). 1 B: Double immunodiffusion of papain solubilized PMK and PMCa separated on WGA-Sepharose; A: anti-PMK serum; 1=eluate PMK, 2=desorbate PMK; 3=eluate PMCa, 4=desorbate PMCa. The major PMK surface glycoprotein (2) is lacking in comparable fractions of PMCa (staining for ala-aminopeptidase).

and alanineaminopeptidase (E.C 3.4.11-), substrate: ala-p-nitranilide) was determined with modifications according to Bergmeyer (9). The amount of protein was measured according to Lowry et al (10). Electroimmunoassay was carried out as described by Axelsen et al (11). PAA disc electrophoresis was performed in 7.5% PAA gels at pH 8.6 and processed according to Maurer (12). Isoelectric focusing was done using PAA Ampholine plates, pH range 3.5-10. Staining for ala-aminopeptidase was carried out as has been previously described (5).
Ultrastructural studies: Formvar coated copper grids (300 mesh) covered with a fine film of carbon were placed on top of a droplet of the sample to investigate. After an adequate adsorption time the grids were handled in same manner for fixation (PBS+1% Formalin), washing (PBS) and staining (phosphotungstic acid), pH 6.8. Thoroughly dried grids were then examined in an Elmiskop 102 (Siemens) at a magnification of x 100 000.

RESULTS

The increasing amount of plasma membranes prepared from human kidney (PMK) was monitored by the rising activity of PM marker enzymes γ-glutamyltranspeptidase (GGTP), alkaline phosphatase (AP) and alanineaminopeptidase (AAP) and by ultrastructural criteria (5). In the case of PM from renal adenocarcinoma (PMCa) the appearance of the fine-structure and the rise of specific GGTP activity was used as indicators.
Biospecific affinity chromatography: Supernatants of proteolytic digested PMK and PMCa were applied respectively to columns packed with ConA as well as with WGA-Sepharose. The gels were washed with buffer until the absorbance at 280 nm was less than 0.1. Two peaks were eluted (PMK, PMCa). Changing of the buffer system gave rise to a desorption peak of approximately 15% square area from that of the elution peak (PMK). However, in the case of PM treated with trypsin a peak over a smaller area (desorption peak 3-6%) was integrated. In the eluate and desorbate fractions obtained activities of GGTP, AAP, AP were measured. The results are summarized in table 1. Since supernatants of papain and bromelain digested PMK and PMCa were characterized by low AP activity the elution profile of this enzyme could not be satisfiably reproduced. No significant differences were observed in the enzyme patterns of PMK whether bromelain or papain were used for solubilization. However, treatment with trypsin was not capable neither to release AAP nor GGTP activity from PMK, while papain and bromelain caused cleavage of more than 90% of total AAP and approxima-

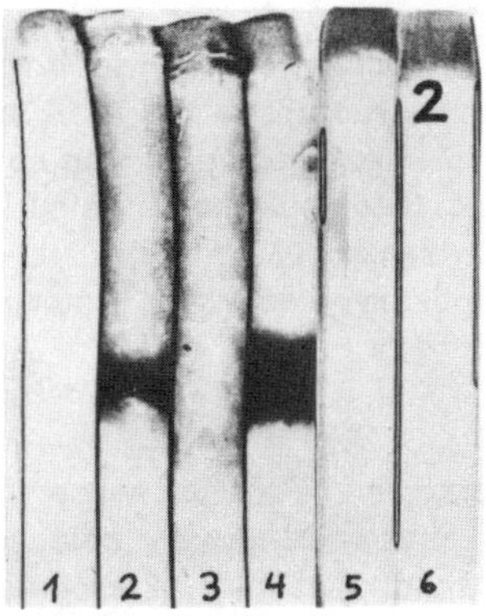

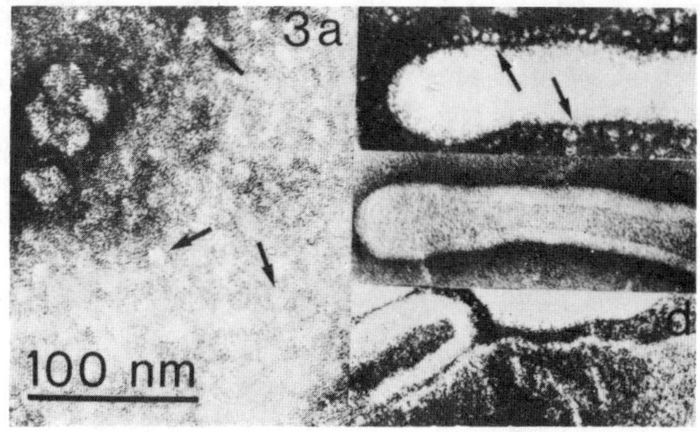

Fig. 2: PAA-disc electrophoresis; PMK and PMCa were solubilized by bromelain and separated on Con A as well as WGA-Sepharose; 1=PMK eluate(ConA); 2=PMK desorbate(ConA); 3=PMK eluate (WGA); 4=PMK desorbate(WGA); 5=PMCa desorbate(ConA); 6=PMCa desorbate(WGA). Fig. 3A: negative staining of papain solubilized PMK desorbed from an immunosorbent column(anti-AAP-Sepharose). Desorption buffer: 1.5M CCl_3COONa, pH6.9. The fraction was composed of 5 nm particles; a similar finding was made with PMK fractions desorbed from WGA-, and Con A Sepharose. Soluble WGA and Con A caused the particles to aggregate(arrow). Particles were found to be randomly distributed on the surface of untreated PMK (3b); proteolytic digestion(papain, bromelain) selectively released the particles from PMK, remaining a smooth PMK surface(3c). No comparable particles were found on the surface of untreated PMCa, thus indicating depletion of structural glycoproteins (3 d).

Fraction	Con-A Sepharose 6 MB		WGA-Sepharose 6 MB	
	eluate	desorbate	eluate	desorbate
AAP(PMK)	25 %	75 %	5 %	95 %
GGTP(PMK)	97 %	3%	24 %	76 %
AAP(PMCa)	not present in significant amounts			
GGTP(PMCa)	81 %	19 %	48 %	52 %

Table 1: Distribution of PM marker enzymes in eluates and desorbates of PM treated with bromelain. Enzyme activities are expressed in % of total enzyme activity recovered.

tely 30% of total GGTP activity. Fused rocket electroimmunoassay (fig. 1A) provides evidence that AAP is a major glycoprotein of the PMK surface. The molecular weight of AAP (PMK-desorbate) was determined to 2.4×10^5 daltons (PAA-SDS disc electrophoresis, analytical gel-filtration; 5). The isoelectric point of AAP was 8. 2. No comparable immunoprecipitation pattern shown in fig 1 was obtained with PMCa components eluted from Con A- and WGA-Sepharose. Glycoproteins (PMCa) desorbed from the columns gave no AAP positive staining of immunoprecipitates as compared to PMK when tested against anti-PMK sera (fig. 1B). The observation that PMCa might be characterized by a depletion of AAP was also supported by PAA-disc electrophoresis (fig. 2). The total amount of glycoproteins (pooled and concentrated desorbate fractions) was less in the case of PMCa as compared to PMK (Coomassie staining).

Immunospecific affinity chromatography: Most of the soluble constituents of PMK were retained on the anti-PMK immunosorbent column. Moreover, in the case of anti-AAP-serum beads, AAP was selectively adsorbed on the gel matrix and could be desorbed without loss of enzyme activity at high purity: only one strong band was observed in PAA disc electrophoresis as well as in electroimmunoassay against polyspecific PMK-antibody (enzyme and protein staining). No major changes in these profiles were obtained when instead of glycine-HCl buffer 1.5 M CCl_3COONa buffer, pH 6.9 was used for desorption. AAP active material was not retained on anti-AAP immunosorbent columns when PMCa (n=4) solubilized by papain or bromelain were applied to the gel. However, two faint immunoprecipitates were noticed when PMCa components desorbed from an anti-PMK immunosorbent column were run against PMK antibody, thus indicating a cross reaction of antigens .

Ultrastructural and histological studies: As shown in fig. 3a the main part of PMK glycoproteins isolated by affinity chromatography (desorbates) consisted of 5 nm particles. Through negative staining of the untreated PMK these particles ("knobs ")were found to be randomly distributed on the membrane surface (fig. 3b), sticking to thin, linear constituents, which were themselves attached to the peripheral membrane matrix. The fine structure of the PMK surface appeared smooth after limited proteolysis (papain, bromelain), fig. 3c. Trypsin was not effective removing the knobs. However, none of the particles described could be detected in desorbate fractions of PMCa (WGA-, Con A-, anti-AAP-Seph.). Untreated PMCa had a smooth surface (fig. 3d) without any of the discrete globular constituents observed on PMK. Histochemistry showed loss of AAP activity in 5 of the 6 tumors studied.

DISCUSSION

The results present evidence that the surface of PMK is characterized by the presence of discrete globular proteins, which are sensitive to proteolytic digestion. Selectively released from the membrane surface, the constituents exhibit high lectin binding affinity (2) especially to WGA, indicating presence of glucose, mannose, galactose (ConA) as well as N-acetylglucosamine residues (WGA). The main part of Con A or WGA receptors of PMK is equivalent to a leu-(ala-)-splitting enzyme as shown by biochemical and immunological methods. Since GGTP of PMK shows no significant affinity to Con A and the fine structure of the desorbed surface particles does not correlate with GGTP activity, it can be assumed that the enzyme is not as close to the PM surface as the AAP. Surface constituents break off from PMK under pathological conditions in vivo (13). Such PMK antigens, eliminated at an increased rate, can be isolated from urine by immunosorption (3). The smooth surface area of PMCa and the fact that the major glycoprotein of PMK can not be demonstrated after lectin-and immunospecific chromatography indicate a depletion of this antigen in the tumor PM. However, further studies suggest that a small portion of AAP exists in PMCa as a cryptoantigen (6). Neoplastic conversion of cells parallels a variety of changes in plasma membrane properties (14)that especially involve the Con A agglutinating system. Modifications in the carbohydrate composition and loss or partial depletion of membrane surface antigens may be in part responsible for altered cell recognition, which allows evation of the tumor cell from immune surveillance mechanisms.

REFERENCES

1) CUATRECASAS, P., PARIKH, I: Methods in Enzymology 34, 653 (1974)
2) SCHERBERICH, J.E., MONDORF, W.: Prot. Biol. Fluids 23, 575 (1975)
3) SCHERBERICH, J.E., FALKENBERG, F., STEFANESCU, T., MONDORF, W: In: Chromat. of synthetic and biopolymers (Ed. R. Epton), Ellis Horwood Ltd 1977 (in press)
4) PORATH, J., KRISTIANSEN, T: The Proteins (Neurath, Hill) Academic Press NY p 95 (1975)
5) SCHERBERICH, J.E., FALKENBERG, F., MONDORF, W, MÜLLER, H, PFLEIDERER, G: Clin. Chim. Acta 55, 179 (1974)
6) SCHERBERICH, J.E, MONDORF, W, KÜSSNER, B, JACOB, R, GRÜNWALD; R: Prot. Biol. Fluids, 24, 497 (1976)
7) AXEN, R., PORATH, I., ERBACK, S.: Nature, 214, 1302 (1967)
8) CUATRECASAS, P.: J. Biol Chem. 245, 3059 (1970)
9) BERGMEYER, H.: Methoden d. enzymatischen Analyse, Verlag Chemie, Weinheim 1970
10) LOWRY, O.H., ROSEBROUGHT, H.J, FARR, A.L., RANDALL, R.J: J. Biol. Chem. 193, 265 (1951)
11) AXELSEN, N.H. et al.: Manual of quantitative immunoelectrophoresis, Univ. forl. Oslo 1973
12) MAURER, H.R.: Polyacrylamide disc electrophoresis, de Gryter Verlag Berlin 1971
13) SCHERBERICH, J.E., MONDORF, W., FASSBINDER, W., KOCH, K: Proc. Europ. Dialysis and Transplant. Ass. 13, Pitman Medical, p 159 (1976)
14) NICOLSON, G.L., POSTE G.: N. Engl. J. Med. 295, 253 (1976)

THE PURIFICATION OF MELANOMA ANTIGEN FROM HUMAN URINE

K. B. Cooke and Claire Bennett

Department of Chemical Pathology, Westminster Medical School, London SW1P 2AR
Great Britain

SUMMARY

1) Patients suffering from malignant melanoma or the associated benign condition of halo naevus excrete a specific melanoma associated protein in their urine.
2) Purification of this protein by gel chromatography, high voltage electrophoresis and salt fractionation gave low yields because of contaminating proteases in the urine.
3) Affinity chromatography of native urine on a column of rabbit anti-melanoma antiserum bound to AH-Sepharose 4B gave a single electrophoretic component in good yield.
4) 3M KCNS and 1 M propionic acid pH 2.5 were equally effective as desorbents.
5) Rechromatography on a dummy column resulted in significant loss of specific protein.

INTRODUCTION

The melanoma specific protein (M.S.P.) found in the urine of patients with malignant melanoma or the benign condition of halo naevus is of interest for two main reasons. Firstly the assay of such a protein could be of value in the diagnosis of the disease and of prognostic value in assessing response to therapy. Secondly the protein is related to components of altered naevus cells, as distinct from normal melanocytes, so that its characterisation will help to elucidate the changes accompanying malignant transformations in cells. Two groups of workers (Jehn et al, 1970; Carrel and Theilkaes, 1973) have described the preparation of a specific protein from the urine of melanoma patients. This was characterised as a β globulin, mol. wt. 40-60,000. Bennett, Cooke and Geck (1975) showed that a protein could be readily detected in the urine of advanced melanoma patients but not in patients with the closely related neuroblastoma and we have recently shown that it also occurs in patients with halo naevus (Staughton et al, 1977). In order to set up a quantitative assay for M.S.P. we have examined the possibility of purifying the protein by affinity chromatography and our results are presented here.

METHODS AND MATERIALS

1. Heterologous anti-human melanoma antiserum (R.A.M.A.) was raised in rabbits using neuraminidase treated stored fresh-frozen tumour cells (Ray et al, 1975) and the harvested antiserum absorbed with pooled human red blood cells,lymphocytes (23 patients), acetone powders of normal skin and normal organs (heart, kidney

and lung), insolubilised serum proteins (Avrameas and Ternynck, 1969), insolubilised urine solids (Bennett, Cooke and Geck, 1975) and normal mole. The absorbed antiserum titred at 1:25 in the indirect fluorescence test.

2. Preparation of Immune Adsorbent 9 ml RAMA was added to 25 g pre-swollen AH-Sepharose 4B (Pharmacia Ltd) in 100 ml distilled water, 800 mgs Ethyl-3 dimethyl-amino propyl carbodiimide HCl was added and adjusted to pH 5.2. The reaction mixture was stirred overnight at 4°C and then washed succesively on a buchner funnel with 1 l. 0.1 M $NaCO_3$/HCl pH 10.7 containing 1 M NaCl, 1 l. di Na H Citrate/HCl, pH 3.0 containing 1M NaCl, 5 l. distilled water twice and 1 l. Miller -Golder (M.G.) buffer pH 7.4, I=0.2. The material was packed in a 4 cm dia. column at 7 psi and washed with M.G. buffer until the eluate pH was 7.4. This gave a column 6 cm high, 75 ml volume, with a capacity for 5-7 l. melanotic urine.

3. Urine Protein Enrichment All concentrations and dialyses were performed in a DC2 ultrafiltration vessel using a H*l* DP 10 hollow fibre cartridge (Amicon Ltd).
A. 25 l. pooled melanotic urine was concentrated to 2 l., dialysed against 0.2 M NH_4OH and further concentrated to 100 ml. This concentrate was applied to a 74 x 2.5 cm dia. column of Sephadex G10 (Pharmacia Ltd) and eluted with 0.2 M NH_4OH. The unretarded peak was dialysed against tris-citrate buffer, pH 8.6, then reconcentrated to 200 ml, applied to a 80 x 10 cm dia. column of Sephadex G150 and eluted with the same buffer. The peak at 2,200-2,800 ml elution volume was reconcentrated to 50 ml and separated on a Hannig VAP I electrophoresis cell in the same buffer, sample application 2 ml/hr, over 25 hours. The MSP positive fractions were pooled and reconcentrated to 100 ml.
B. 25 l. melanotic urine was concentrated to 0.5 l., the concentrate made 3M with respect to KCl, and stirred overnight at 4°C. The preparation was centrifuged at 136,000 r.c.f. for 20 mins and the supernatant dialysed against M-G buffer and reconcentrated to 100 ml.

4. Affinity Chromatography 5 l. urine, or urine concentrate in M-G buffer, was applied to the 75 ml column at 7 p.s.i. and washed through until the eluate fell below 0.05 OD at 280 nm. Specific elution then followed with either 3M KCNS or 1 M propionic acid pH 2.5 until the eluate absorbance at 280 nm fell to that of the eluting buffer. This usually required 1-2 l. buffer. The column was regenerated with M-G buffer.

RESULTS

Figure I shows the electrophoretic pattern on cellulose acetate, of urine concentrates prepared according to the procedures A (5) and B (1) together with the unretained protein pattern (6) and the propionic acid eluate (7) of preparation A from the affinity column. Both urine concentrates contained significant amounts of lysozyme (slow γ mobility) and in both preparations M.S.P. was only detectable by immunological methods. After the application of 5 l. untreated urine the unretained peak (6) resembled the material applied, whilst specific desorption resulted in a single band with α_1 mobility (2) and yielded 3 mg protein. Rechromatography on a dummy column of insolubilised non-immune rabbit serum (3) left no detectable protein in the eluate. Whilst the results shown here depict elution with 1 M propionic acid essentially the same results were obtained with 3 M KCNS without any improvement in yield.

DISCUSSION

The purification of melanoma specific protein presents a particular problem since its excretion accounts for less than 0.5% of total urine protein whilst it has a similar electrophoretic mobility and molecular size to the major protein constituents of urine. The low concentration (ca. 500 ng/ml) means

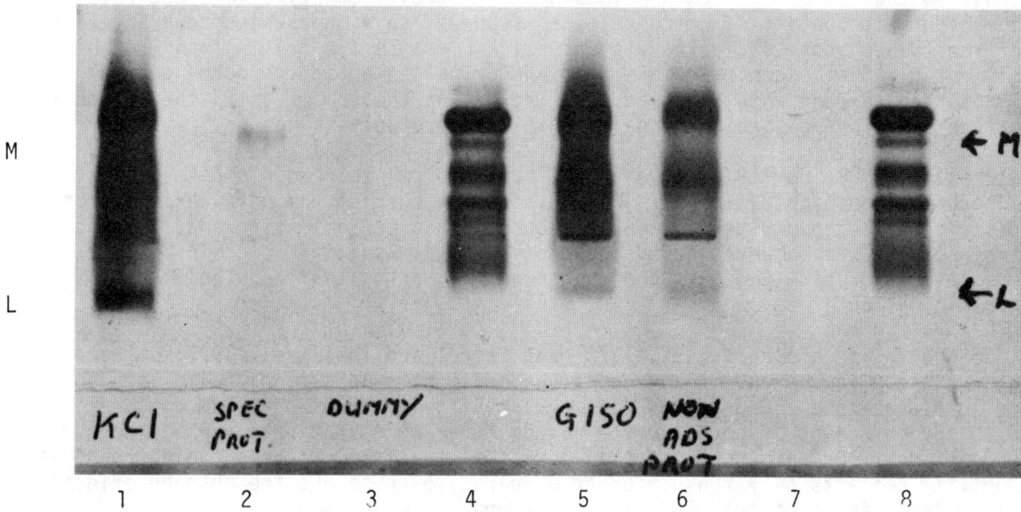

Figure I Electrophoretic pattern of M.S.P. and various urine concentrates

1) Urine concentrate by KCl fractionation. 2) M.S.P. eluted from affinity column. 3) M.S.P. rechromatographed on inert column. 4 & 8) Human serum marker. 5) Urine concentrate from gel filtration. 6) Unretained peak from affinity column. 7) Specific eluate of 5 from affinity column. "M" indicates mobility of M.S.P. (α_1 globulin) and "L" indicates mobility of Lysozyme (Plasminogen, Mol. wt. 80,000 has the same mobility).

(DIAGRAMMATIC ILLUSTRATION BELOW OF PHOTOGRAPH ABOVE)

that large volumes of urine have to be processed. Urinary proteolytic enzymes concentrate with M.S.P. so that Carrel and Theilkaes (1973) comment on the extreme lability of their preparations. Our own 25 l. pools have always yielded MSP detectable only by immunolgical methods, presumably due to proteolysis losses. Direct application of urine to the affinity column offers many advantages provided that pH 6.0 - 8.0 is maintained. Proteolytic enzymes are not retained, the high salt content of urine helps to suppress any ion-exchange characteristics of the AH-Sepharose 4B, urine pigment sticks less to the column than when highly pigmented urine concentrates are used and urine can be added to the column and washed through when received, storing MSP on the column until the column is saturated before desorbing; again minimising proteolysis losses.

Preston and Price (1977) have used a dummy column to remove non-specifically adsorbed protein contaminants. We had anticipated that the unfavourable protein distribution of our material would produce marked contamination and followed their procedure. However dummy column also removes specific protein by non-specific adsorption and this phenomenon was so marked that no significant protein was obtained.

Both Jehn and his coworkers (1970) and Carrel and Theilkaes (1973) obtained a β_1 globulin whereas in our hands the protein is an α_1 globulin both on cellulose acetate and gradient polyacrylamide gels. Our antiserum does not react with any component of normal urine nor do M.S.P. preparations have any colour suggestive of melanin contamination. We feel that the discrepant electrophoretic mobility reflects only the different preparative route and the changed effects of proteolysis using this modified procedure.

ACKNOWLEDGEMENTS

The authors wish to thank the Cancer Research Campaign and the Research Committee of the North West Thames Regional Health Authority for their generous support of this project and Professor J. R. Hobbs for his encouragement and stimulating discussions.

REFERENCES

Avrameas, S and Ternynck, Therese, Immunochem. 6, 53-66 (1969)

Bennett, Claire, Cooke, K.B. and Geck, P., Prot. Biol. Fluids, 24, 667-70 (1976)

Carrel, S and Theilkaes, L., Nature, 242, 609-10 (1973)

Jehn, U.W., Nathanson, L., Schwartz, R.S. and Skinner, Mary, New Eng. J. Med. 283, 329-33 (1970)

Preston, Valerie and Price, M.R. Biochem. Soc. Transact. 5, 123-5 (1977)

Ray, P. K., Thakur, V.S. and Sundaram, K., Eur. J. Cancer, 11, 1-8 (1975)

Staughton, R., Copeman, P.W.M., Cooke, K.B. and Bennett, Claire. Proc. Roy. Soc. Med. 1977 (In Press).

BENZYLPENICILLOYL (BPO)-SPECIFIC AFFINITY AND SUBSEQUENT ION EXCHANGE CHROMATOGRAPHY: CHARACTERIZATION OF ANTIBODY SUBFRACTIONS

Otto Scheiner*, Heinrich Stemberger**, Dietrich Kraft* and Gerhard Wiedermann**

**Institute of General and Experimental Pathology, Univ. Vienna, Austria*
***Institute for Specific Prophylaxis and Tropical Medicine, Univ. Vienna, Austria*

INTRODUCTION

Affinity chromatography is a well known tool for isolation of monospecific antibodies. In the present study it should be investigated whether antibody fractions with different immunological and physicochemical properties could be obtained by different desorption procedures and chromatographic measures.

MATERIAL AND METHODS

Immunization procedures: 20 rabbits were immunized by a course of intramuscular injections every ten days. The first injection consisted of 10 mg of antigen in 2 ml of complete Freund's adjuvant and the subsequent four of 10 mg of antigen in 2 ml of incomplete Freund's adjuvant. The animals were bled ten days after the last injection and their sera were pooled. The antigen was prepared by incubation of Na-Penicillin G in 1 M K_2CO_3 overnight at 4°C as described by Locher (1), leading to a conjugate BPO_{40}-BGG.
Affinity chromatography: BPO-coated Sepharose was synthesized by incubation of Aminohexylsepharose 4B with Na-Penicillin G at pH 10,0 for one hour at 37°C (2). 100 ml of the affinity gel was equilibrated with 50 mM Tris-HCl buffer, pH 8,0, in a Büchner funnel and then incubated with 100 ml anti-BPO-serum for one hour at 37°C with gentle shaking. The gel was transferred to a chromatographic column and washed with increasing saline concentrations (Fig. 1). The hapten (BPO-ε-aminocaproic acid, BPO-EACA) eluted antibodies obtained were subjected twice to chromatography on Sephadex G-25 at pH 2,7 causing the removal of the hapten from the antibody. Subsequent treatment of the affinity gel with 0.1 N acetic acid (pH 2.7) led to a second protein peak (acid eluted antibodies, Fig. 1).
Ion Exchange Chromatography: Hapten and acid eluted antibodies were dialyzed against ethylenediamine/acetic acid buffer (50 mM, pH 7.0, starting buffer). QAE-Sephadex A-50 was equilibrated with starting buffer and a pH-gradient was run in 16 hours with 0.33 M Na-acetate, pH 4.0, as end buffer. This procedure resulted in five antibody subfractions (Fig. 2), three of hapten eluted antibodies (H1, H2, H3) and two of acid eluted antibodies (A1, A2). Subsequently, the antibody subfractions were dialyzed against isotonic phosphate buffered saline (PBS), pH 7.2.

Molecular sieving chromatography on porous glass beads: This method was described by Frisch-Niggemeyer (3). Porous glass with a pore diameter of 177 Å ± 5% was used. This chromatographic procedure separated the IgM from the IgG. Albumin contamination was removed by ammoniumsulphate precipitation (Fraction A2).
Immunoelectrophoresis: A method according to Scheidegger (4) was used. Subfractions were tested with anti-rabbit-IgG, -IgM and -IgA.
Red cell linked antigen-antiglobulin reaction (RCLAAR): This test was performed according to Coombs (5). All subfractions were tested with anti-rabbit-IgG and anti-rabbit-IgM.
Passive haemagglutination (pHA): pHA was performed using BPO-coated human erythrocytes according to Levine's technique (6).
Antibody-dependent cellular cytotoxicity (ADCC): Lymphocytes were prepared from normal human donors (7). BPO-coated ^{51}Cr-labelled HeLa-cells were used as targets (8). The effector to target cell ratio was 40:1. ADCC-activity was calculated by plotting the cytotoxic index (c_i) as a function of log c of antibody (8) and defined as the concentration of antibody resulting in a c_i of 0,25.
Passive immunhaemolysis (PIH): Densely hapten coated sheep erythrocytes, i.e. more than 450.000 antigenic determinants per cell (9), were mixed with various concentrations of anti-BPO-IgG and guinea pig serum as a source of complement. The mixture was incubated for one hour at 37°C and subsequently centrifuged. The OD_{541} of the supernatant was measured and PIH-activity was defined as the concentration of antibody causing 50% haemolysis.
Influence of anti-BPO-IgG on IgM-induced complement dependent haemolysis: Minimally BPO-coated sheep red blood cells, i.e. appr. 190.000 determinants per cell (9), were mixed with anti-BPO-IgM obtained from subfraction A2 (sufficient to induce 80% haemolysis), and with increasing amounts of IgG-subfractions and with guinea pig serum (Fig. 3).
Dissociation constant: Equilibrium dialysis was done according to Kabat and Mayer with some modifications (10). Dissociation constant was calculated with respect to the reaction $HAb \rightleftharpoons H + Ab$, where Ab, H and HAb represent free antibody, free hapten and hapten bound antibody, respectively.

RESULTS AND DISCUSSION

Affinity chromatography was used for the preparation of monospecific anti-BPO antibodies. An appropriate gel could be easily prepared (2). The concentration of hapten was 14 μMol per ml of packed gel as could be calculated from coupling experiments with labelled hapten. After incubation of the anti-BPO serum with the affinity gel nearly all unspecifically bound protein could be removed by washing with increasing ionic strength up to 2 M (Fig. 1). Detachment of hapten elutable antibodies was observed at a concentration of 7×10^{-5} M BPO-EACA in a single clear cut peak. Subsequent acid elution (pH 2,7) caused further detachment of anti-BPO Ig. Ion exchange chromatography led to splitting in the five subfractions, H1, H2, H3 (hapten elution) and A1, A2 (acid elution) (Fig. 2).

The composition of the subfractions was examined using the RCLAAR and immunoelectrophoresis. According to the tests performed, H1, H2, H3 and A1 proved to be pure IgG, whereas A2 contained some IgM and slight amounts of albumin due to it's hydrophobic nature. A2 was purified by glass bead chromatography (resulting in pure anti-BPO-IgM) and ammoniumsulphate precipitation. Re-examination with the RCLAAR and immunoelectrophoresis showed the recovery of pure IgG.

Previous studies of Wiedermann and co-workers (9) showed that the ability of IgG for complement dependent haemolysis of hapten coated sheep erythrocytes depends on the hapten density on the cell surface. Minimally hapten coated sheep erythrocytes (~ 190.000 determinants per cell) can only be lysed by IgM but not by IgG. IgG-molecules need more than appr. 450.000 determinants per cell for optimal complement dependent lysis. Obviously IgG is able to block complement dependent IgM-induced lysis of minimal hapten coated sheep erythrocytes.

Thus, subfractions H1 - A2 were tested in this system using anti-BPO IgM obtained from A2 (sufficient to induce 80% haemolysis) (Fig. 3). H1, H2 and A1 showed - as expected - a blocking effect augmented by increasing amounts of IgG. No effect could be seen with normal rabbit-γ-globulin (NRG). In contrast and unexpected, however, H3 and A2 showed an additional lytic effect (up to 100% lysis) even with the lowest IgG concentration.

To elucidate other immunological properties of these IgG subfractions, activities were tested with regard to pHA, ADCC, PIH and dissociation constant K (Table 1).

The results presented clearly indicate:

1) Each antibody fraction exhibits it's own spectrum of activity. The assumption that hapten eluted antibodies represent a unique fraction, independent from acid eluted antibodies and vice versa, is supported by the fact that "early antisera" (blood taken ten days after the first antigen injection) revealed no hapten elutable antibodies (11). Acid eluted antibodies, however, showed a pattern in affinity and ion exchange chromatography very similar to the corresponding hyperimmune antibodies (11).

2) Affinity and subsequent ion exchange chromatography is a useful tool for separating distinct IgG fractions with different lytic and blocking properties, at least in our system.

3) Blocking effects in our complement dependent system are due to the concentration of antigenic determinants on the target cell surface. Each antibody fraction seems to require a characteristic minimal ("critical") hapten density to act as a lytic antibody. The best illustration of this fact seems to be fraction H1 (with it's superior agglutinating property, it's maximal activity in ADCC and it's low dissociation constant), which shows the best lytic activity in PIH only if the density of BPO-groups on the cell surface is sufficient. In case of minimally BPO-coated targets the same antibody fraction exhibits the strongest blocking effect on IgM induced complement dependent haemolysis.

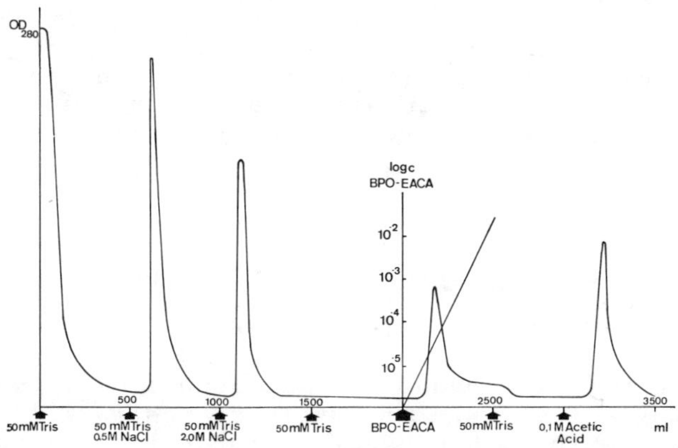

Fig. 1: BPO-specific affinity chromatography: elution pattern

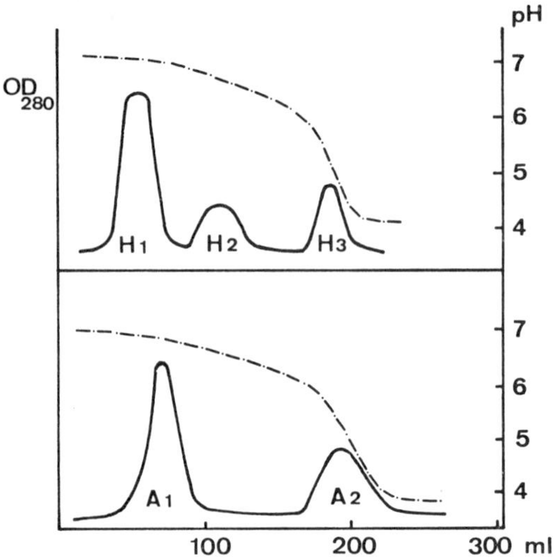

Fig. 2: Ion exchange chromatography (QAE-Sephadex A-50)

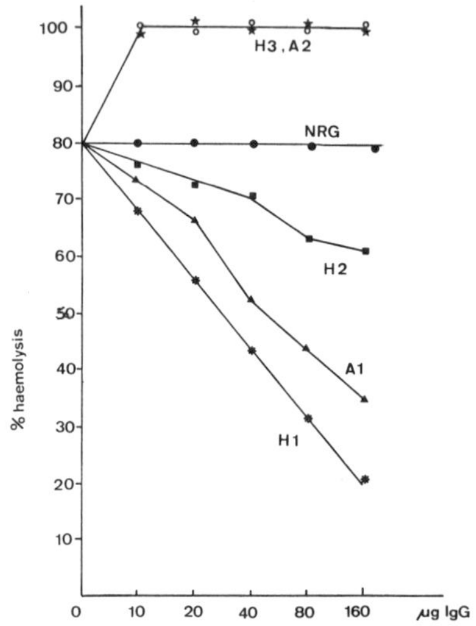

Fig. 3: Influence of IgG (H1, H2, H3, A1, A2) and normal rabbit gammaglobulin (NRG) on complement dependent IgM-induced haemolysis

ANTIBODY FRACTION	PHA pg/ml	PIH ng/ml	ADCC ng/ml	K [Mol/l]
H1	40	493	7,6	$1,2 \times 10^{-8}$
H2	80	560	13,2	$0,7 \times 10^{-8}$
H3	240	533	36,3	$4,4 \times 10^{-8}$
A1	500	1866	27,8	$3,4 \times 10^{-8}$
A2	98 000	560	87,9	$6,7 \times 10^{-8}$

Table 1:
pHA: concentration of IgG, just leading to a visible agglutination
PIH: concentration of IgG causing a 50%-haemolysis
ADCC: concentration of IgG leading to a cytotoxic index of 0,25
K: dissociation constant

SUMMARY

BPO-specific antibodies were obtained by means of affinity chromatography using BPO-coated aminohexylsepharose 4B as adsorbent. Antibodies were detached by specific and subsequent acid elution. Ion exchange chromatography of these antibody fractions on QAE Sephadex A-50 resulted in 3 subfractions of hapten eluted and 2 subfractions of acid eluted antibodies. Each of these antibody fractions exhibited its own spectrum of activity with regard to passive hemagglutination, antibody-dependent cellular cytotoxicity, passive immune hemolysis and dissociation constant. Furthermore the various antibody fractions exhibited lytic or blocking properties as tested in an IgM/complement (C) system on minimally coated SRBC. Blocking properties of antibody fractions in our C-dependent system were influenced by the density of antigenic determinants on the target cell surface and the avidity of the respective antibody. It thus could be shown that antibodies with low dissociation constant and excellent lytic activity under optimal conditions, exhibited a pronounced blocking effect when minimally coated target cells were used. Antibodies with high dissociation constant exhibited lytic rather than blocking properties under the same conditions.

ACKNOWLEDGMENT

This study was supported by grants (No 2023, 2024 and 3271) from the "Fonds zur Förderung der wissenschaftlichen Forschung" (Austria).

REFERENCES

(1) G.W.Locher, C.H.Schneider, A.L.de Weck, Z.Immun.Forsch. 141, 256 (1971)
(2) H.Stemberger, G.Wiedermann, W.Frisch-Niggemeyer, M.M.Müller, Mh.Chem. 108, 1 (1977)
(3) W.Frisch-Niggemeyer, F.Heinz, H.Stemberger, Immunität und Infektion 2, 231 (1974)
(4) J.J.Scheidegger, Int.Archs.Allergy appl.Immun. 7, 103 (1955)
(5) R.R.A.Coombs, A.W.Howard, L.S.Mynors, Br.J.exp.Path. 34, 525 (1953)
(6) B.B.Levine, M.Z.Fellner, V.Levitska, J.Immunol. 96, 707 (1966)
(7) Report of WHO/IARC - Sponsered Workshop on Human B and T Cells, London 15th-17th July 1974, Scand.J.Immunol. 3, 521
(8) H.Stemberger, O.Scheiner, G.Wiedermann, D.Kraft, O.Förster, J.Immunol.Method (accepted for publication)
(9) G.Wiedermann, M.M.Müller, O.Förster, H.Stemberger, Int.Archs. Allergy appl.Immun. 49, 483 (1975)
(10) O.Scheiner, H.Stemberger, D.Kraft, G.Wiedermann, J.Immunol. Method (accepted for publication)
(11) G.Wiedermann, H.Stemberger, O.Scheiner, D.Kraft, Cell Immunol. (submitted)

IMMUNOSORBENT SEPARATION OF IgG AND IgM FOR THE RADIOIMMUNOASSAY OF SPECIFIC ANTIBODIES

John A. Diment and J. Pepys

Cardiothoracic Institute, Fulham Road, London, S.W.3.

INTRODUCTION

The development of the radioallergosorbent test (RAST) for specific IgE (1) has given stimulus to the search for convenient general methods for measuring specific antibody titres in individual antibody classes. Prior to the development of the RAST and its recent modifications for measurement of specific IgG and IgM (2), most methods depended on prior physical separation of each antibody class and this made application to large numbers of sera impractical. A method has been recently described (3) which uses a solid phase antiglobulin to achieve separation of one class of antibody and the binding capacity of this separated antibody, still bound to the immunosorbent, can be measured by counting its uptake from an excess of a labelled antigen. This method has been applied to measurements of the antigen binding capacity of IgG and the present work was undertaken to extend this to IgM.

MATERIALS AND METHODS

Sepharose anti-IgG, anti-IgM and anti-IgA

These were prepared by cyanogen bromide activation of Sepharose 4B and reaction with heavy chain specific globulins (Dakopatts, Copenhagen) as previously described for Sepharose anti-IgG (3). Sepharose was also coupled to normal rabbit globulins for use as a control in the same manner.

Labelled pigeon γ-globulin and IgG, IgM and IgA

Pigeon γ-globulin was prepared as previously described (4) and labelled with I-125 to a known specific activity between 5 - 6 μCi/mg using the Chloramine T method. The myeloma immunoglobulins were a gift from Dr. Drew of the University of Birmingham and were labelled in the same manner but to a higher specific activity. Each of the labelled immunoglobulins were adsorbed with immunosorbents of the other two classes of antibody and used in the tests with 0.2 μl of a pooled normal human serum.

The Assays

Sepharose anti-IgM which had been blocked by the addition of a IgG/IgA mixture was diluted to give 15 μl of packed gel per assay tube in 0.5 ml of PBS. Cord serum (200 μl) was added to each tube. After agitation, 1 μl of test serum was added and the tubes capped and rotated overnight. Subsequently the gel was washed 4 times with PBS containing Tween 20 with centrifugation at 1500 g for 5 minutes to settle the gel between washes. ^{125}I-pigeon γ-globulin was added to the tubes (10 - 18 ng in a suitable volume, ca 150,000 cpm) together with 1 ml of 15% rabbit serum PBS. The tubes were rotated overnight then washed 5 times. The IgG assay was performed using the same labelled antigen by the method used for subtilisin (3). To ensure antigen was present in excess, if the counts bound exceeded 80% of those added, the assay was repeated using less test serum in the first step. The specificity assays were performed by rotating the labelled immunoglobulin with the immunosorbent in the presence of 10% rabbit serum followed by washing with PBS-Tween. Sepharose normal rabbit immunoglobulins were used as control with each labelled immunoglobulin.

RESULTS AND DISCUSSION

Table 1 shows that Sepharose anti-IgG, anti-IgA and anti-IgM varied in the cleanness of their binding specificity. Anti-IgG bound little or no IgA or IgM but the anti-IgM tested bound significant amounts of IgG and IgA.

Table 1 - Binding of labelled immunoglobulins to anti-immunoglobulin gels

Sepharose	IgG	IgA	IgM
anti IgG	100%	0.01%	0%
anti IgA	0.45%	100%	n.a.
anti IgM	10.3%	4%	100%
anti IgM IgA/IgG blocked	0.02%	0.08%	77%

As IgM occurs in serum at about only one tenth the level of IgG and a serum may contain high levels of specific IgG and IgM, this commercial anti-IgM, even though raised against μ-chain could not be used in the assay without further adsorption. Similar results were obtained with anti-IgM from another source when coupled to Sepharose. The specificity of the anti-IgM was improved by adsorption with insolubilised IgG before coupling but it was found more convenient to block the unwanted binding by addition of an IgG/IgA mixture that was available, and to perform the assay in 15% cord serum which gave no precipitins in double diffusion against anti-IgM. These procedures reduced the IgM binding of the anti-IgM gel by 23% compared with the unblocked material but the improvement in specificity achieved was crucial to the success of the assay.

Table 2 - Binding of pigeon globulin by the serum of pigeon breeders

IgG	Mean level of 18 symptomatic pigeon fanciers 50 μg/ml (range 0-200 μg/ml)
IgM	Mean of 7 symptomatic pigeon fanciers whose serum showed significant titres 0.11 μg/ml (range 0.02 - 0.2 μg/ml)

Detection limit 0.02 μg/ml.

Table 2 shows the results obtained with some pigeon fanciers when their serum IgG and IgM binding of pigeon γ-globulin was measured. Some of the sera showed 1000 times more specific IgG binding than specific IgM. The residual binding of IgG and IgA may still determine the detection limit of the assay in such sera, since if low counts are found it is not possible to determine whether this represents a true low titre of specific IgM, or if the gel is picking up traces of IgG or IgA with a high binding capacity for the antigen.

REFERENCES

1. L. Wide, H. Bennich and S. Johansson, Diagnosis of allergy by an in-vitro test for allergen antibody, Lancet, ii, 1105 (1967).

2. M. Viljanen, K. Granfors and P. Toivanen, Radioimmunoassay of class-specific antibodies, Immunochem, 12, 699 (1975).

3. B. Jacoby and J. Pepys, The antigen-binding capacity of serum IgG, J.Immunol. Methods, 11, 37 (1976).

4. J. Diment and J. Pepys, Avian erythrocyte agglutination tests with the sera of bird fanciers, J. Clin. Path., 30, 29 (1977).

ELUTION BY GLUCOSE OF BOVINE IMMUNOGLOBULINS ADSORBED ON CROSS-LINKED DEXTRAN

J. J. Pahud, K. Schwarz and H. Hilpert

Research Department, Nestlé Products Technical Assistance Co. Ltd., 1000 Lausanne, Switzerland

In our study of bovine immunoglobulins applied to local passive immunization of the gastro-intestinal tract, the specific antibody activity of cow's milk against pathogenic *E.coli* was stimulated by active immunization of pregnant cows with a conventional polyvalent vaccine. The specificity and antibody activity of the immune colostral whey proteins, was demonstrated *in vitro* and *in vivo* (Demierre et al., 1975, Hilpert et al., 1977). One of our main objectives being to obtain maximal yields of specific antibodies, we have been investigating ways to improve the vaccination procedure. Our attention was drawn to a recent analysis of murine antibodies with specificity for α-1,3 dextran (Hansburg et al., 1976). Mice primed with Dextran B 1355 in complete Freund's adjuvant, and boosted with *E.coli* strain B, produced an average of 15mg/ml of anti-dextran antibodies with some animals exceeding 40mg/ml. Conversely these antibodies were reactive against *E.coli* B. It was thus demonstrated that the α-1,3 epitope was cross-reactive with antigenic determinants on *E.coli* B. We assumed that these findings could be applied to our problem, hoping that the α-1,3 epitope might be shared by pathogenic *E.coli*. First it was necessary to determine if Hanburg's dextran - *E.coli* protocol would result in the production of antibodies reacting with pathogenic *E.coli* and interfering with their pathomechanism. This point is still under study and will not be presented here. The next step was to examine the blood serum and milk whey from immunized and non-immunized cows, for the possible presence of anti-dextran activity caused by *E.coli* antigens. To approach this second aspect of the question we decided to use the commercial cross-linked dextran B 512 gels, even though most of the glucosidic bonds were α-1,6 with less than 5% of α-1,3 linkages. At this symposium we wish to discuss the selective interactions observed between immunoglobulins and cross-linked dextran. Our work was largely inspired by the pioneer studies of Kabat (1953, 1961), Schlossman (1962) and their coworkers concerning the antibody combining site with dextran-antidextran systems.

Materials and Methods

Serum and colostral milk samples were collected from normal or immunized cows. Vaccination was performed during the gestation period with a formolized polyvalent *E.coli* vaccine (Demierre et al., 1975). Colostral whey was obtained after skimming and precipitation of casein at pH 4.6 with citric acid.

Cross-linked dextran B 512 gels are supplied by Pharmacia Fine Chemical as the known Sephadex G series. Affinity chromatography on the gel types G-25 to G-200 was performed in PBS (0.01M phosphate pH 7.2-0.15M NaCl) and standard columns 1.6 x 20cm were used. Automation of affinity chromatography was monitored with LKB Ultrograd^R set up as a process programmer (Pahud et al., 1976).

Total protein adsorption on Sephadex was determined with suspensions in buffer corresponding to 1ml of packed gel for each test. After incubation with the serum sample and thorough washing with buffer, 1ml of 0.2M NaOH was added to the sedimented gel, and the protein concentration was measured in the supernatant by the Folin-Ciocalteu method.

The sugars tested in this study were obtained from Fluka AG, Buchs, Switzerland, with exception of α-D-isomaltose and α-D-kojibiose which were available from Koch-Light., Colnbrook, England. The comparative inhibition tests were performed with serum and 10mM sugar solutions in PBS.

Electrophoresis was carried out on cellulose acetate with the Beckman Microzone system. Immunoelectrophoresis (in 1% agar and 0.025M barbital, pH 8.3) was applied to the protein fractions collected; specific and polyvalent antisera were produced in rabbits.

Results and Discussion

Protein adsorption on Sephadex suspensions was determined on types G-25 Medium, G-25 Superfine and G-75 Superfine. After saturation of the gels with bovine serum, the amounts of adsorbed proteins were estimated roughly as 200μg, 300μg and 400μg per ml of the respective sedimented gels. The high degree of cross-linking in Sephadex G-25 seemed to decrease the number of sites necessary for interaction. A greater total external surface of the gel beads favoured the adsorption as in the superfine grades, indicating a large molecular weight for the proteins involved. When dialyzed serum was filtered through a Sephadex column, it was initially observed that part of the adsorbed proteins (about 30-40%) could be eluted by 10 mM glucose. On Microzone-electrophoresis they were shown to be immunoglobulins of restricted heterogeneity, migrating in the β_2 region (Fig. 1). The immunoglobulins recovered belonged to the IgG_1 subclass as demonstrated by immunoelectrophoresis (Fig. 2). After glucose application, the proteins remaining on the column could be eluted by acidification at pH 2.5-3.0. In Kabat's experiment, inhibition reached 40% with 0.1M glucose and did not exceed 60% with higher concentrations. Other determinants were involved which were not cross-reactive with glucose. There seemed to be no significant difference between normal and hyperimmune anti-*E.coli* serum, thus the vaccination was probably not responsible for the induction of anti-dextran activity. The occurence of these antibodies could be explained by the natural challenge of cattle with streptococci, pneumococci and *E.coli* which possess antigenic determinants cross-reacting with dextran. We were unable to obtain a precipitin reaction between the eluted immunoglobulins and soluble dextran, even when applying reversed single radial immunodiffusion (Richter et al., 1972). It should be mentioned that the recovered fractions were poorly soluble after dialysis and lyophilization. Binding of immunoglobulins to Sephadex could be prevented by previous adsorption of the samples, by addition of glucose to the solutions, or by the use of borate buffer (Svenson et al., 1970).

Elution or inhibition of the binding was tested with various sugars and compared to glucose. With minor exceptions, our observations were generally in agreement with Kabat's findings in the human anti-dextran system. Specificity for the α-1,6 linkage was obvious since isomaltose was better, and the other diglucoses, maltose (α-1,4) or kojibiose (α-1,2), were less efficient than glucose (nigerose α-1,3 could not be obtained for testing). Melibiose and raffinose had effects practically equal to glucose whereas α-methyl-glucoside, sucrose, mannose and galactose were less active. Lactose and α-methyl-mannoside did not cause any detectable inhibition. Antibodies with specificity for the α-1,3 linkage, if present, appeared to be impossible to demonstrate in a system where the α-1,6 epitope so largely predominated. The use of Sephadex as an adsorbent determined the selection of antibodies with higher affinity for sugars of the isomaltose series. These antibodies were then able to mask any activity against other determinants on dextran. For this purpose, highly branched dextrans would have been more appropriate.

It is difficult to make a direct comparison between specific desorption in affinity chromatography and hapten inhibition, as carried out in the customary procedure. The latter is measured in a state of equilibrium, whereas in affinity chromatography a steady state is never reached. Thus, inhibition activity is detectable at lower hapten concentrations on the affinity gel than in the test tube.

According to our examination of literature concerned with the sugar transport systems in various tissues (Silverman, 1976; Kasahara et al., 1977) affinity chromatography with cross-linked dextrans, or other types of insolubilized polysaccharides, would seem to offer a practical approach to the isolation of the plasma membrane proteins specialized in this transport. Similar gels have already been extensively used for the isolation of lectins and glycosidases (reviews by Lowe and Dean, 1974; Wilchek and Jakoby, 1974).

As a practical conclusion to this study, we would like to emphasize that the presence of natural anti-dextran antibodies should be considered whenever gel filtration of biological fluids is performed on polymerized dextran.

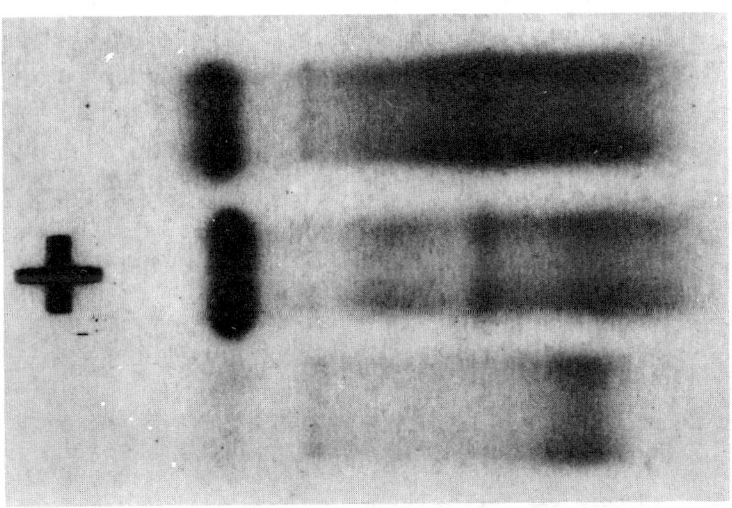

Fig. 1 Electrophoresis on cellulose acetate. Top : native bovine serum; middle : unadsorbed serum proteins; bottom : proteins eluted by 10mM glucose.

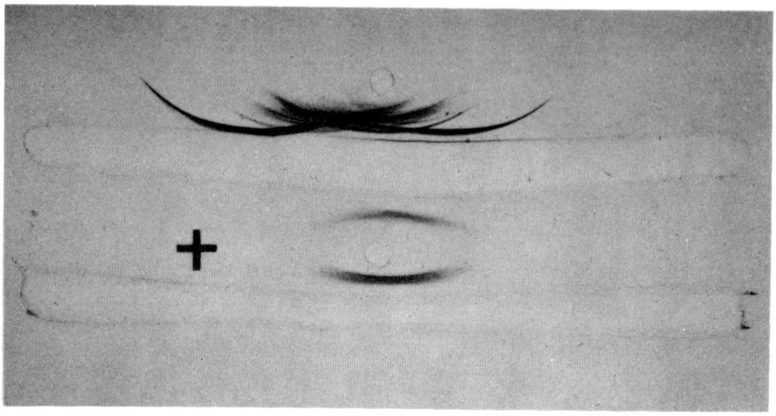

Fig. 2 Immunoelectrophoresis
Upper well : total bovine serum; lower well : serum IgG_1 eluted by 10mM glucose. Middle trough : antiserum against total serum; lower trough : antiserum against bovine IgG_1.

References

Demierre, G., Rivier, D., Hilpert, H., Gerber, H., and Zinkernagel, R., Pathol. Microbiol. 42, 137 (1975).

Hansburg, D., Briles, D.E., and Davie, J.M., J. Immunol. 117, 569 (1976).

Hilpert, H., Gerber, H., Amster, H., Pahud, J.J., Ballabriga, A., Arcalis, L., Farriaux, F., de Peyer, E., and Nussle, D., XIII Symp.Swed.Nutr.Found. Almqvist and Wiksell International, Stockholm, 182 (1977).

Kabat, E.A., and Berg, D., J. Immunol. 70, 514, (1953).

Kabat, E.A., and Mayer, M.M., Experimental Immunochemistry, 2nd ed., Charles C Thomas, Springfield, Ill., 1961.

Kasahara, M., and Hinkle, P.C., Biochemistry of membrane transport, Springer Verlag. FEBS Symp. 42, 346 (1977).

Lowe, C.R., and Dean, P.D.G., Affinity Chromatography. Wiley, New York, 1974.

Pahud, J.J., and Schwarz, K., Science Tools 23, 40 (1976).

Richter, W., and Kagedal, L., Int. Arch. Allergy 42, 885 (1972).

Schlossman, S.F., and Kabat, E.A., J. Exp. Med. 116, 535 (1962).

Silverman, M., Biochim. Biophys. Acta 457, 303 (1976).

Svensson, S., Hammarström, S.G., and Kabat, E.A., Immunochemistry 7, 413 (1970).

Wilchek, M., and Jakoby, W.B., Methods Enzymol. 34, 3 (1974).

PURIFICATION OF ENZYME-LABELLED CONJUGATE BY AFFINITY CHROMATOGRAPHY

Margareta Lannér*, Robert Bergquist*, Jan Carlsson**, Gunnel Huldt*

**Department of Parasitology, National Bacteriological Laboratory, Stockholm, Sweden*
***Department of Biochemistry, University of Uppsala, Sweden*

ABSTRACT

The present paper concerns the application of concanavalin A for the purification of enzyme labelled anti-human immunoglobulin conjugates intended for use in an enzyme-based immunoassay. The quality of an enzyme conjugate is related to its relative content of complexed enzyme-immunoglobulin while the occurrence of unbound enzyme and free immunoglobulin contribute to serious side effects. The carbohydrate moiety of horse-radish peroxidase is shown to adhere to concanavalin A in contrast to unlabelled immunoglobulin. Based on this fact a chromatographic two-step method yielding concentrated enzyme-immunoglobulin complexes of high purity have been devised. After the removal of unbound enzyme by conventional gel filtration, labelled immunoglobulin is further purified by affinity chromatography.

INTRODUCTION

Enzyme-based immunoassays (EIA) are presently very popular for a wide variety of applications in medical diagnostic microbiology. The enzyme-linked immunosorbent assay (ELISA) was published by Engwall and Perlmann in 1971 (1). Independently in the same year a similar method was reported by van Weemen and Schuurs (2). The reagent in ELISA consists of enzyme-labelled antibodies (conjugate), which are allowed to react with the antigen, immobilized by adsorption to a plastic surface, e.g. the wall of a test tube. The enzyme is often horse-radish peroxidase (HRP), in which case end products of the reaction are visualized by the addition of a dye. The reaction mixture takes on a colour, the intensity of which is in the early phase of enzyme-substrate reaction directly proportional to the amount of conjugate bound to antigen. The result is read spectrophotometrically. When used for serologic diagnosis of infections in man the conjugate is prepared of anti-human antibodies and used in an indirect application. In this case patient serum is added to a test tube with purified antigen adsorbed to its inner surface. If the sample contains the specific antibodies sought for, they will react with the antigen and subsequently be detected when conjugate and dye is added.

It has been increasingly clear that no EIA can easily give reproducible quantitative results. The quality of the conjugate is a crucial factor in this respect. It is quite evident that a conjugate contaminated both by unlabelled antibodies and free enzyme molecules, which is often the case, will not behave in a standardized way. Moreover, the size of enzyme-antibody complexes often range from 200.000 daltons to several million daltons. Efforts to minimize these effects and thereby create a better quality of conjugate will be presented here.

METHODS

Conventional Chromatography

Two different HRP labelled conjugates were used for the studies. One conjugate was obtained from a commercial source (A) and the other one was prepared at our laboratory (B). Both conjugates were directed against human immunoglobulin. When preparing conjugate B the enzyme was first activated with glutaraldehyde to obtain a reactive link to be coupled with immunoglobulin in a second step as described by Avrameas et al. (3). The enzyme was HRP, type VI (SIGMA, Missouri, USA) with a 2.9 RZ-value, which is an expression of the ratio of hemin to protein content. Pure crystalline peroxidase is reported to have an RZ of 3.0. The anti-human antibodies were purified on an immunosorbent according to Wofsy et al. (4). Primary analyses of the conjugates were obtained by chromatographical fractionation. A 3.9 mg sample of undiluted conjugate A was gel filtrated using a Sephacryl S-200 gel (Pharmacia, Uppsala, Sweden), which has an exclusion limit around 250.000 daltons and a fractionation range stretching down to 5.000 daltons. The column measured 80 cm with an inner diameter of 2 cm. Conjugate B was investigated on a 80 x 1.6 cm Sepharose 6 B (Pharmacia) column. A sample containing 43 mg of undiluted conjugate was applied. In both cases fractions of 2.5 ml were collected at a flow rate of 6 ml/ cm^2/h. The elution buffer was a 50 mM sodium phosphate buffer, pH 7.0, supplied with 100 mM sodium chloride and 1 per cent butanol. Fractions were assayed for protein at 280 nm and for enzyme at 403 nm.

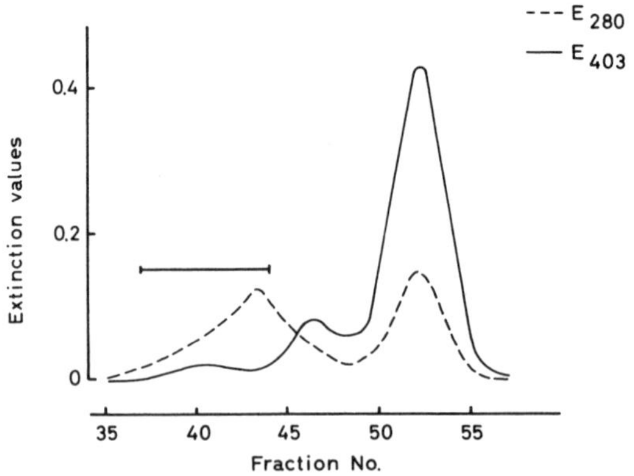

Fig. 1. Gel filtration on Sephacryl S-200

As can be seen in fig. 1, Sephacryl S-200 distinguishes relatively well between monomers (molecular weight = 40.000 daltons) and dimers (80.000 daltons) of free enzyme but unlabelled immunoglobulin is not separated from the labelled fraction. When calculated from the graph it is evident that as much as 95 per cent of enzyme is uncoupled (15 per cent in the form of dimers and 80 per cent as monomers). We were not interested in separation of dimeric enzyme from monomeric enzyme but it was found that the gel also distinguishes well between immunoglobulin and free enzyme. To avoid free enzyme as much as possible only fractions comprising the first eluated part of the first peak were collected for further purification.

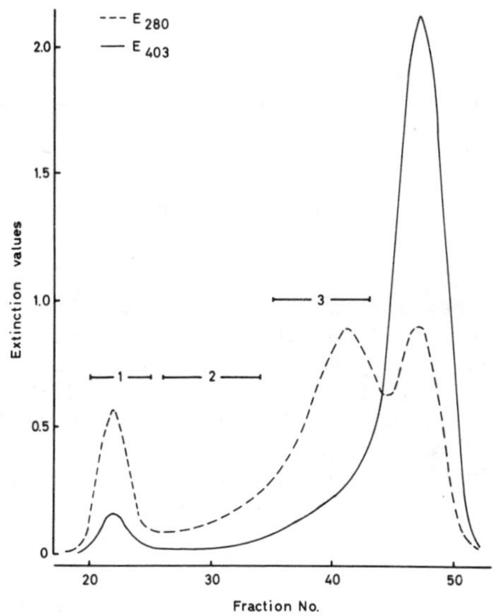

Fig. 2. Gel filtration on Sepharose 6 B

In contrast to fig. 1 only one enzyme peak is seen in fig. 2 since Sepharose 6 B does not separate the dimers of free enzyme from the monomers. Recently, Sepharose 6 B has been substituted by Sephacryl S-200 at our laboratory to obtain a better separation between free enzyme and immunoglobulin. Calculations based on the graph of fig. 2 indicate that about 80 per cent of the total amount of enzyme applied does not bind to immunoglobulin. Peak 3 contains about 15 per cent of the total amount of enzyme and 50 per cent of the total immunoglobulin. The fractions of peak 3 were collected, concentrated and saved for affinity chromatography.

Affinity Chromatography

When unlabelled sheep anti-human-antibodies were subjected to chromatography on Con A-sepharose (Pharmacia) 90 per cent of the material did not adsorb to the column, using the buffer described above for adsorption. The remaining 10 per cent could be eluted with the addition of 10 mM α-Methyl-D-mannoside to the buffer. This material should theoretically contain mainly antibodies of the IgA and IgM classes (9) but the fraction was not investigated further. Since it has been shown that HRP can be purified using Con A (5.6) it was thought feasible to separate HRP labelled antibodies from unlabelled antibodies by affinity chromatography on Con A-Sepharose.

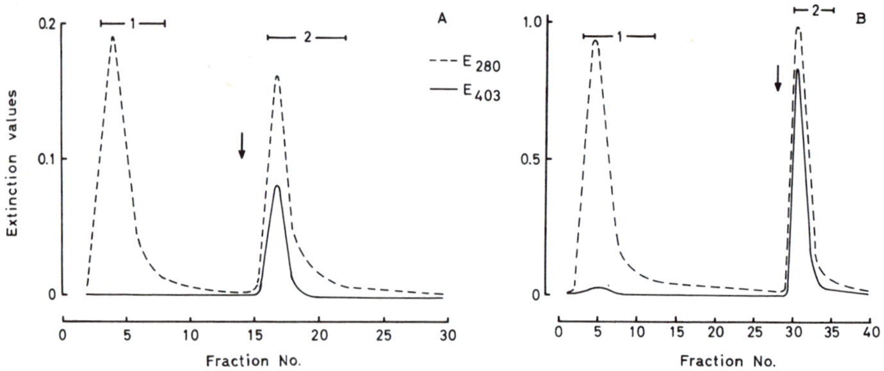

Fig. 3. Affinity chromatography on Con A-Sepharose

Samples of the two conjugates were applied to a 3 x 0.8 cm Con A-Sepharose column in separate experiments using 0.96 mg of conjugate A and 4.3 mg of conjugate B. Adsorbed material was subsequently eluted by addition of 10 mM of α-Methyl-D-mannoside to the buffer as indicated by the arrows in fig. 3. A generous amount of adsorption buffer was allowed to pass the column before desorption. Fractions of 1 ml were collected at a flow rate of 0.25 ml per minute. As can be seen in the figure, both conjugates contained much material (about 50 per cent), which did not adsorb to the column.

RESULTS OF AFFINITY CHROMATOGRAPHY

CONJUGATE	CONJUGATE FRACTION	RECOVERY % OF TOTAL PROTEIN	RECOVERY % OF TOTAL ENZYME	ENZYME/ANTIBODY MOLAR RATIO	ELISA ACTIVITY*
A	UNTREATED	100	100	0,6	0,13
	PEAK 1	51	21	0,2	0
	PEAK 2	49	79	0,9	0,17
B	UNTREATED	100	100	1,3	0,13
	PEAK 1	53	3	0,1	0
	PEAK 2	44	68	2,1	0,23

* ACTIVITY DEFINED AS INCREASE OF EXTINCTION AT 449 NM/MIN/µG

Peaks 1 and 2 from both runs were tested for enzyme-immunological activity using direct ELISA (7). Tubes were coated with human gammaglobulin (Kabi, Stockholm, Sweden) by incubation with 1 ml of a solution containing 1 µg per ml. Incubation with conjugate at a concentration of 1.65 µg per ml was performed at room temperature and the amount of enzyme bound to the adsorbed antigen was measured after 4 hours (8). Hydrogen peroxide with 5-amino-2-hydroxybenzoic acid was added at pH 6 and at room temperature. The enzyme-substrate reaction was stopped after 2 minutes by the addition of 1 M NaOH and the samples were measured at 449 nm in a Hitachi model 100-60 spectrophotometer.

Results of the affinity chromatography are summarized in the table. The yields of the chromatographic runs were near 100 per cent. As can be expected the molar ratios of enzyme to antibody increased after removal of the unlabelled antibodies. When conjugate A was eluted no enzyme was detected in the fractions of peak 1, but upon concentration this pooled material was shown to contain 21 per cent of the total amount of eluted enzyme.

Conjugate A increased its activity only slightly. This could be explained if the material withdrawn from the original sample was not immunoglobulin, but, according to the manufacturer, no stabilising protein is added. If partially inactivated by aggregation the withdrawn material could still be immunoglobulin. In that case the Con A HRP link would possibly not be strong enough to hold the complexes, which would account also for the finding of enzyme in peak I. Conjugate B, in contrast, increased its activity by 73 per cent. A negligible loss of labelled antibody (< 1 per cent of total eluted activity) was recorded when the unadsorbed material from the two runs was investigated. Overloading of the column (5) was ruled out since further labelled antibody could not be eluted by desorption after re-run of this unadsorbed material on the column. Inaccessibility to Con A because of steric hindrance of this particular material may be a possible explanation.

The present results indicate that it would be feasible to concentrate the active ingredients of HRP conjugates by specific adsorption of labelled immunoglobulin to Con A. The removal of free immunoglobulin does not only enhance the activity of a conjugate by concentration since unlabelled antibodies due to steric advantage compete favourably with labelled antibodies when binding to antigen.

REFERENCES

(1) E. Engvall and P. Perlmann, Enzyme-linked immunosorbent assay (ELISA). Quantitative assay of immunoglobulin G, Immunochemistry 8, 871 (1971).

(2) B.K. van Weemen and A.H.W.M. Schuurs, Immunoassay using antigen-enzyme conjugates, FEBS letters 15, 232 (1971).

(3) S. Avrameas and T. Ternynk, Peraxidase labelled antibody and Fab conjugates with enhanced intracellular penetration, Immunochemistry 8, 1175 (1971).

(4) L. Wofsy and B. Burr, The use of affinity chromatography for the specific purification of antibodies and antigens, Journal of Immunology 103, 380 (1969).

(5) S. Avrameas and B. Guilbert, Biologically active waterinsoluble protein polymers. Their use for the isolation of specifically interaction proteins, Biochimie 53, 603 (1971).

(6) M. Wagner, Reiningung von Meerrettich-Peroxydase durch Affinitäts chromatographie an Concanavalin A-Agarose, Acta biol.med.germ. 34, 1429 (1975).

(7) E. Engvall and I. Ljungström, Detection of human antibodies to Trichinella spiralis by enzyme-linked immunosorbent assay, ELISA, Acta path. microbiol. scand. Sect. C, 83, 231 (1975).

(8) E. Ruitenberg, P.A. Steerenberg, B.J.M. Brosi and J. Byus, Reliability of the enzyme-linked immunosorbent assay (ELISA) for the serodiagnosis of Trichinella spiralis infection in conventionally raised pigs, Journal of Immunological Methods 10, 67 (1976).

(9) K. Aspberg and J. Porath, Group-specific adsorption of glycoproteins, Acta Chem. Scand. 24, 1839 (1970).

KILLED *STAPHYLOCOCCAL AUREUS* CELLS AS SUPPORT FOR ANTIBODY MEDIATED AFFINITY CHROMATOGRAPHY

Rudolf Varró, Ildikó Barna-Vetró and Andor Jagicza

Human Institute for Serobacteriological Research and Production, Budapest, Hungary

ABSTRACT

Antibodies for different antigens were attached to protein A prepared from killed Staphylococcus aureus (PASA) and the resulting adsorbents were used for specific binding of the corresponding antigen. PASA was succesfully used for preparing effective solid-phase second antibody reagents and for negative immunoadsorption to remove impurities.

INTRODUCTION

Protein A containing stabilized killed Staphylococcus aureus cells are uniform particles with the capacity of binding most of the mammalian IgG via their Fc region. Therefore the killed bacteria are suitable as natural preformed support for antibody mediated affinity chromatography; specific adsorption is caused by the attached antibody. Thus PASA has a very wide area of use. As ligand binding does not require covalent coupling laboratory work is rather unsophisticated.

MATERIALS AND METHODS

Stabilized killed Staphylococcus aureus suspensions were prepared as described by Kronvall (1). A strain isolated from a patient was used. This strain contained 30 % more protein A than strain Cowan 1 (2). Anti-guinea pig-IgG, anti-rat-IgG, anti-goat-IgG, and anti-horse-radish peroxidase (HRPO) were raised in rabbits according to standard techniques. All other antisera were commercial products of our Institute.

The enzyme HRPO was purchased from Reanal, Hungary, and was further purified by gel electrophoresis. Horse antibodies to tetanus toxoid were purified by immunoadsorption using our own method (3). C-reactive protein (CRP) was prepared as described by Hokama and Riley (4). Antisera were adsorbed to PASA, using 0.1 ml rabbit anti-serum for 1 ml of 10 % bacterial suspension (1). HRPO activity was measured by diaminobenzidine (DAB) conversion (5). Immunoelectrophoresis and single radial immunodiffusion were performed according to standard methods.

RESULTS

Immobilization of HRPO

Anti-HRPO rabbit serum was adsorbed to PASA as described above. A suspension consisting of normal rabbit serum and bacteria served as control. The centrifuged pellet of 1 ml of both suspensions was incubated with 1 ml HRPO, RZ = 2.2. The coated bacteria were washed repeatedly with phosphate buffered saline (PBS), pH 7.2, until no enzyme activity was detected in the supernatants. The washed cells were reconstituted to 1 ml with PBS and tested for peroxidase activity. As demonstrated in fig. 1 the specific antibody coated PASA bound HRPO since different amounts of the reconstituted suspension showed different DAB converting rate. The control suspension displayed no enzyme activity proving that non-specific binding was not involved.

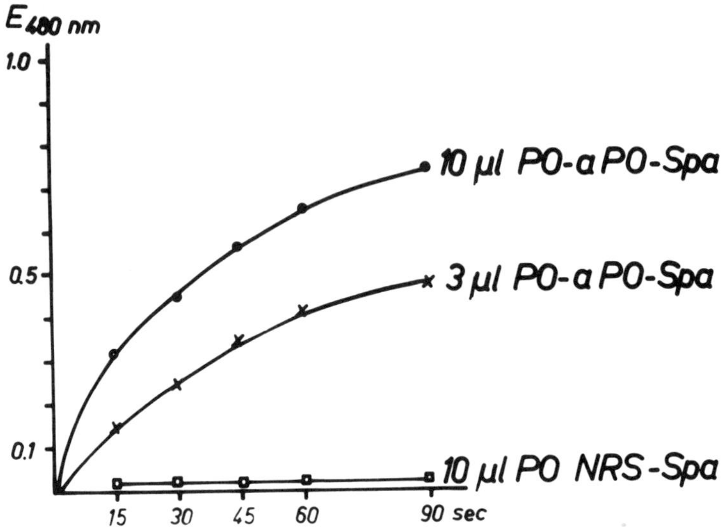

Fig. 1. Demonstration of HRPO immobilization. Different amounts of enzyme/anti-enzyme/staphylococcal suspension (PO-aPO-SPA) showed different DAB conversion. The control suspension enzyme/normal rabbit serum/ staphylococcal suspension (PO NRS-SPA) had no HRPO activity.

Preparation of solid-phase second antibodies

PASA itself can be used as solid-phase anti-IgG reagens for radioimmunoassay (RIA) (1). We investigated bacterial suspensions coated with anti-IgG antibodies. Rabbit anti-goat-IgG, -anti-rat-IgG, and -anti-guinea pig-IgG were adsorbed to PASA. 1 ml of tenfold diluted goat, rat and guinea pig serum was mixed with the pellet of 50, 100, 200 and 400 µl amounts of the corresponding suspensions. After incubation the supernatants were controlled for remaining IgG by single radial immunodiffusion. The amount of suspension needed to bind 90 % of total IgG is compiled in table 1.

TABLE 1: The binding of IgG by anti-IgG PASA suspensions

Antibody bound to PASA	Amount of anti IgG-PASA to adsorb 90 % of IgG from 1 ml tenfold diluted corresponding serum
Anti-rat IgG	200 µl
Anti-goat IgG	400 µl
Anti-guinea pig IgG	50 µl

All three suspensions bound the corresponding IgG fractions in relation to the activity of PASA attached antibody. This type of bacterial suspensions can be succesfully applied in radioimmuno-assays using antibodies unreactive to protein A. Killed staphylococci can be rapidly separated from the fluid phase and the anti-body-PASA suspensions are easy to handle.

Negative immunoadsorption

In two experiments the ability of coated PASA to remove unwanted impurities can be demonstrated.

CRP is purified from human sera by conventional methods. The final product contains some human IgG contaminant. To remove the IgG the following approach was used: Increasing amounts of the suspension prepared from anti-human IgG and PASA were centrifuged. Each pellet was mixed with 1 ml of the CRP solution. The IgG content of the supernatant was determined by single radial immunodiffusion. 200 µl of the anti-IgG-PASA suspension was required to remove all contaminating IgG. The adsorption is shown on fig. 2.

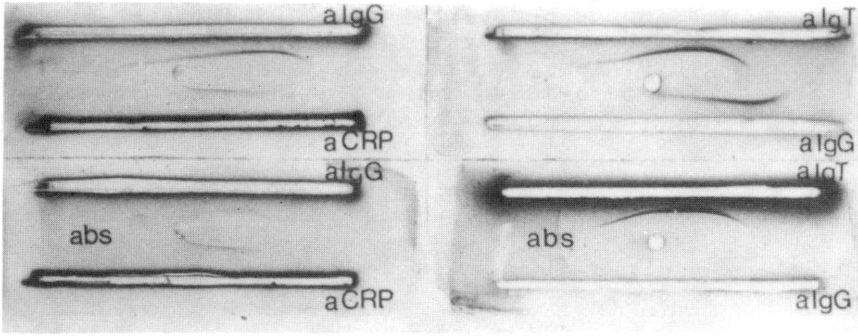

Fig. 2. Negative immunoadsorption by coated PASA
Left: Immunoelectrophoresis (IEP) of contaminated and purified CRP. Right: IEP of contaminated and purified horse IgT (Anode to the right for each IEP slide).

The second example concerns the purification of horse IgT. Antibodies purified by immunoadsorption from hyperimmune antitoxic horse sera usually belong to both the IgG and the IgT class. This is the case with anti-tetanus antibodies (3). Horse IgG can be separated easily from the eluate by conventional separation method. It is difficult, on the other hand, to prepare IgT devoid of IgG. Anti-horse IgG rabbit serum was adsorbed to PASA and series of adsorption mixtures were set up analogously to the CRP purification scheme. The right side of fig. 2 demonstrates that all contaminating IgG could be removed by this method. It is of interest to mention that neither horse IgG nor IgT do react with protein A.

DISCUSSION

Although bound ligands can not be recovered from the PASA type adsorbents, a variety of methodologies can profit by the use of this inexpensive support. Enzyme immobilization can be performed via the anti-enzyme antibody, provided it does not reduce the enzymatic activity. In our investigations no decrease of HRPO activity could be demonstrated in the active suspension one month after preparation. Additionally, no enzyme activity could be measured in the supernatant. Protein A unreactive antibodies can be bound to PASA coated with the corresponding antiserum, which is prepared in a PASA-reactive species. Negative immunoadsorption has wide application in difficult purification steps. Another advantage of the PASA method are the low costs combined with minimal labwork. It is not too complicated to prepare big amounts of bacterial suspensions, so pilot plant or industrial scale use seems to be suitable as well.

REFERENCES

(1) G. Kronvall, A rapid slide-agglutination test for typing pneumococci by means of specific antibody adsorbed to protein A-containing staphylococci. J. Med. Microbiol. 6, 187 (1973).

(2) R. Varró and A. Jagicza, Connection between antibiotic resistance and protein A content in Staphylococcus aureus. Communication at the annual meeting of the Hungarian Microbiological Society, Sopron, 1975.

(3) R. Varró, A study of immunospecifically isolated tetanus antibodies. Ann. Immunol. Hung. 17, 93 (1973).

(4) Y. Hokama and R.F. Riley, Purification of C-reactive protein, an acute phase protein of human serum. Biochim. Biophys. Acta 74, 305 (1963).

(5) V. Herzog and H.D. Fahimi, A new sensitive colorimetric assay for peroxidase using 3,3'-diamino benzidine as hydrogen donor. Anal. Biochem. 55, 554 (1973).

THE USE OF AFFINITY CHROMATOGRAPHY FOR PURIFICATION OF ENZYME - ANTIBODY CONJUGATES

Rudolf Varró and Ildikó Barna-Vetró

Human Institute for Serobacteriological Research and Production, Budapest, Hungary

ABSTRACT

Horseradish peroxidase (HRPO) labelled antibodies were purified by affinity chromatography on Concanavalin A, anti-HRPO and Staphylococcal protein A adsorbents. The ability of these adsorbents to react with various products of the labelling reaction was determined. Based on these results we suggested a sequential purification method for the removal of free antibody and unreacted HRPO.

INTRODUCTION

Recently enzyme labelled antibodies were shown to be applicable as reagents for non-isotopic immunoassay and as immunohistochemicals. A variety of coupling reactions has been described, yielding the conjugate and different monomeric and polymeric side products. These contaminants might interfere with the proper use of the enzyme-labelled antibodies causing false results (1). Purification of the conjugates can be performed by conventional separation methods, e.g. salt precipitation, gel chromatography, electrophoresis, etc. The resulting preparation is not pure. Direct affinity chromatography on the corresponding antigen adsorbent might result in the loss of high-avidity labelled antibodies and by this method no distinction can be made between free and bound antibody. These considerations led us to test group specific affinity methods and immunoadsorption directed towards the enzyme moeity for effective purification.

MATERIALS AND METHODS

HRPO was purchased from Reanal, Hungary, and purified to RZ = 2.2 by preparative electrophoresis. CNBr-activated Sepharose and Con A-Sepharose were products of Pharmacia AB, Sweden. Sheep-anti-rabbit-IgG serum and rabbit-anti-human-Fab serum were raised according to standard methods.

Stable killed Staphylococcus aureus suspensions were prepared according to Kronvall (2) using a strain isolated from a patient (3). Antibodies were purified by salt precipitation and subsequent ion-exchange chromatography. Conjugation was performed by the periodate oxidation method of Nakane and Kawoi (4). An anti-HRPO coated staphylococcal reagent was made by reacting 0.1 ml of rabbit-anti-HRPO serum with 1 ml of a stable 10 % bacterial suspension. The conjugates were controlled by agarose gel electrophoresis using protein and enzyme staining. Binding activity of conjugates was tested by their ability to adsorb on staphylococcal cells coated with the corresponding antigen, i.e. normal rabbit serum and normal human serum. Peroxidase activity was determined by the diamino-benzidine (DAB) conversion method (5).

RESULTS

In the first set of experiments the rate of binding of the reaction mixtures was investigated. The sheep globulin-HRPO conjugate with anti-rabbit-IgG activity was applied to Con A Sepharose and to anti-HRPO Staphylococcus complex, the rabbit globulin-HRPO conjugate to Staphylococcal cells according to the following schedule:
50, 100, 200 and 400 μl of adsorbent suspensions (10 % in case of bacterial suspensions; 1 g wet weight/2 ml buffer in case of Con A) were centrifuged, then mixed with 1 ml of the 1:10 diluted labelling mixtures. After appropriate incubation the supernatants were checked for peroxidase activity. The time course of the DAB conversion of the reaction mixture and of the supernatants of the 50 and 400 μl adsorbent samples are plotted on Fig. 1. for all three adsorbents.

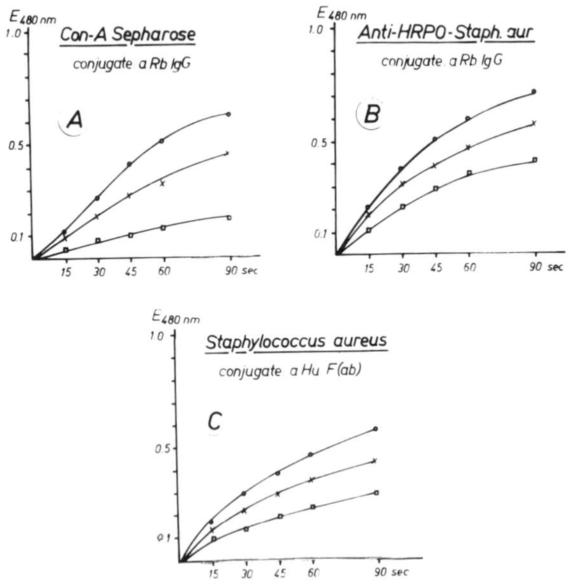

Fig. 1. Binding of conjugates to different adsorbents. DAB conversion of the starting material (o—o), of the supernatants of 50 μl (x—x) and 400 μl (□—□) adsorbent samples is plotted for each adsorbent

In each case the binding capacity was proportional to the adsorbent volume, suggesting that:

- -the reaction mixture can be separated into glycoproteins (HRPO and HRPO-antibody conjugate) and Con A non-reactive material.
- -anti-HRPO is able to bind free and antibody bound enzyme, whereas free antibody remains unbound.
- -Staphylococcal bacteria bind IgG in both the labelled and unlabelled form.

These suggestions were confirmed by the patterns of the supernatants in agarose electrophoresis.

Subsequently, elution of the adsorbed materials was carried out. In these studies an anti-HRPO-Sepharose adsorbent prepared by the CNBr method was used instead of the antibody-staphylococc complex. The RZ values of the eluates are compiled in Table 1.

TABLE 1: RZ values (E_{403}/E_{280}) of HRPO-antibody-conjugates eluted from different adsorbents

Conjugate	Adsorbent	Eluant	RZ value of Starting material	RZ value of Eluate
anti-rabbit IgG (sheep)	Con A	MGP*	0.348	0.420
anti-rabbit IgG (sheep)	anti-HRPO	glycine**	0.348	0.470
anti-human Fab (rabit)	Protein A	glycine**	0.320	0.285

* MGP = 10 % methyl-α-D-glucopyranoside
** Glycine = 0.2 m glycine HCl - 0.5 m NaCl, pH = 2.4

The RZ-values of the eluates from Con A and anti-HRPO adsorbents are increased in comparison with the starting reaction mixture. This could be explained by the lack of binding capacity to free antibody. The higher value obtained with the anti-enzyme adsorbent reflects the fact that non-glycoprotein isoenzymes are present in the preparation. The eluates from the protein A adsorbent possess decreased RZ-values due to the fact that both free and labelled IgG react with protein A, whereas free enzyme remains unbound. Electrophoretic analysis using enzyme staining shows indeed presence of both cathodic (free HRPO) and fast moving components (conjugate) in the Con A and anti-HRPO eluates. On the other hand in the protein A eluate only the fast moving fraction was detectable.

DISCUSSION

These data show that group specific affinity chromatography is another way to improve the quality of enzyme-labelled antibodies. We suggest that sequential use of two of the adsorbents, one for the removal of unlabelled antibodies (Con A or anti-HRPO), another for the elimination of free enzyme (protein A), yield pure und functionally active conjugates. Since quantitative aspects have not yet been worked out it is difficult to state whether affinity methods could effectively compete with conventional separation prodecures with regard to yield. Combined affinity chromatography is a valuable purification method in cases where enzyme labelled antibodies without contaminants are requested.

REFERENCES

(1) L.A. Sternberger, Immunocytochemistry, Prentice-Hall, Englewood Cliffs, N.J. 1974.

(2) G. Kronvall, A rapid slide-agglutination test for typing pneumococci by means of specific antibody adsorbed to protein A-containing staphylococci, J. Med. Microbiol. 6, 187 (1973).

(3) R. Varró and A. Jagicza, Connection between antibiotics resistance and protein A content in Staphylococcus aureus. Communication at the annual meeting of the Hungarian Microbiological Society, Sopron, 1975.

(4) P. Nakane and A. Kawoi, Peroxidase labelled antibody. A new method of conjugation, J. Histochem. Cytochem. 22, 1048 (1974).

(5) V. Herzog and H.D. Fahimi, A new sensitive colorimetric assay for peroxidase using 3,3'-diamino benzidine as hydrogen donor, Annal. Biochem. 55, 554 (1973).

PURIFICATION OF DD-CARBOXYPEPTIDASES FROM *STREPTOMYCES* STRAINS R61 AND K15 BY ANTIGEN-ANTIBODY AFFINITY CHROMATOGRAPHY

A. Marquet, M. Nguyen-Distèche, M. Leyh-Bouille and J. M. Ghuysen

Service de Microbiologie, Faculté de Médicine, Université de Liège, Institut de Botanique, 4000 Sart Tilman, Liège, Belgium

ABSTRACT

The exocellular R61 DD-carboxypeptidase and the lysozyme-releasable K15 DD-carboxypeptidase were purified using anti-exocellular R61 enzyme IgG-Sepharose as immunoadsorbent. The specific activity of R61 enzyme was increased 70 fold (yield 90 %) and that of K15 enzyme was increased 550 fold (yield 50 % of the adsorbed material).

INTRODUCTION

The cellular enzyme system involved in peptide crosslinking during synthesis of the bacterial cell wall peptidoglycan consists of at least 1) a transpeptidase activity which catalyzes peptide bond formation, a reaction through which the carboxyl group of the penultimate D-Ala residue of a peptide donor ending in D-Ala-D-Ala is transferred to the free ω-amino group of another peptide acceptor, and 2) a DD-carboxypeptidase activity which simply hydrolyzes the C-terminal D-Ala-D-Ala sequences of the peptide donors without performing any transfer reaction. Both activities are sensitive to ß-lactams, but the inhibition of either of them may be the cause for cell lysis depending on the species investigated.

The DD-carboxypeptidase-transpeptidase enzyme system in *Streptomyces* sp. is relatively well known (1). In strains K15 und R61, it consists of: 1) a membrane-bound transpeptidase capable of performing low DD-carboxypeptidase activity and 2) a set of DD-carboxypeptidases: a) membrane-bound, b) lysozyme-releasable and c) exocellular, having low transpeptidase activity (2). The relationship between the exocellular, lysozyme-releasable and membrane-bound DD-carboxypeptidase-transpeptidase enzyme system remains to be determined.

The exocellular DD-carboxypeptidase can be isolated from the culture filtrates. At present, the enzyme excreted by strain R61 is the only one which has been purified to homogeneity (3). The procedure makes use of conventional techniques, it is lengthy and the final yield is small.

The lysozyme-releasable DD-carboxypeptidase can be released from the cells through the enzymatic transformation of the mycelium into protoplasts by lysozyme in a sucrose medium (2). This enzyme represents a very small proportion of the total protein present in the original cell extracts and therefore its isolation is not easy. Previous studies have shown that the purified exocellular DD-carboxypeptidase of strain R61 and the lysozyme-releasable DD-carboxypeptidase of this and other Streptomyces strains were immunologically related (4). On this basis, the purification of the lysozyme-releasable enzyme of strain K15 and of the exocellular R61 enzyme was attempted by affinity chromatography using the IgG fraction of an anti-R61 enzyme antiserum as ligand.

PURIFICATION OF THE IgG FRACTION

The antibodies directed against R61 exocellular DD-carboxypeptidase were purified by affinity chromatography using R61 enzyme Sepharose 6B as adsorbent. The Sepharose was activated according to Sundberg and Porath (5). Bisoxyran- rather than CNBr-activated Sepharose was used because very alkaline conditions were required to dissociate the antigen-antibody complex.

Epoxy-activated Sepharose (2 g - 900-1000 μequiv. of epoxide groups/ g dry gel) was added to a solution (4 ml) of the R61 exocellular enzyme (18 mg; i.e. 0.5 μmol of purified enzyme) in 0.5 M carbonate buffer pH 9.5 and the suspension was stirred by rotation for 72 hrs at 28°C. Under these conditions about 30 % of the enzyme was bound to the support. The enzyme-Sepharose complex was successively washed with 1) 250 ml 0.5 M carbonate pH 9.5 - 0.5 M NaCl; 2) 250 ml 0.1 M glycine·HCl pH 3.0 - 0.5 M NaCl; 3) H_2O; 4) 250 ml 30 mM Tris·HCl pH 7.5 - 0.5 M NaCl and 20 mM $CaCl_2$. The remaining free epoxide

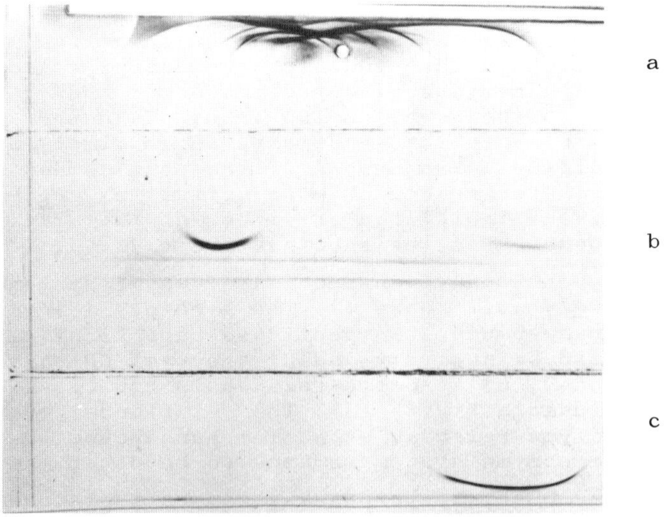

Fig. 1. Immunoelectrophoresis of a) complete exocellular R61 antiserum, b) 0.5 M NaCl eluted material and c) 0.2 M carbonate buffer pH 11.4 eluted IgGs from R61 enzyme Sepharose adsorbent, tested against antiserum to complete rabbit serum.

groups were blocked by reaction with 2 M ethanolamine at pH 9.5 for 8 hrs at 25°C. This adsorbent was washed and finally equilibrated with 10 mM Tris•HCl pH 7.7. 2 ml of complete anti-R61 enzyme antiserum were adsorbed onto a column of the above mentioned adsorbent (1 mg of ligand/ml of gel). The non-adsorbed material had completely lost the capability of inhibiting the DD-carboxypeptidase activity of the R61 exocellular enzyme, which showed that fixation of the specific IgGs had occurred with high efficiency. The column was thoroughly washed with 0.5 M NaCl + 0.01 M Tris•HCl pH 7.7. The IgGs were eluted from the adsorbent with 0.2 M carbonate buffer pH 11.4, rapidly dialysed against 10 mM Tris•HCl buffer pH 7.7 and concentrated to 2 ml by ultrafiltration. Through a series of several runs performed under identical conditions, 30 mg of purified IgG's (Fig. 1) were obtained and used as ligand for the purification of the lysozyme-releasable enzyme.

ENZYME PURIFICATION

Sepharose 6B was activated as described above and the purified IgG fraction was coupled onto it. The reaction was performed in 0.5 M carbonate buffer pH 10.5 for 48 hrs at 34°C. In this step the yield was 84 %. The IgG-Sepharose 6B, washed and treated with ethanolamine as described above, was equilibrated with 30 mM Tris•HCl pH 7.5.

In order to check the behaviour of the immunoadsorbent, the purification of the crude exocellular R61 enzyme was first attempted. The enzyme was applied to the column, eluted with 0.2 M carbonate buffer pH 11.4, dialysed immediately against 10 mM Tris•HCl pH 7.7 and concentrated by ultrafiltration. Measurements of enzyme activity showed a 70 fold increase in specific activity. The yield was 90 % (Table 1).

TABLE 1. Purification of exocellular R61 DD-carboxypeptidase on anti-exocellular R61 DD-carboxypeptidase IgG-Sepharose 6B immunoadsorbent.

	Total activity* (μmoles/min)	Yield	Specific activity** (μ-moles/min/mg)
Crude enzyme	1790	100	0.74
Non-adsorbed material	0	0	
NaCl 0.5 M pH 7.9 eluted material	0	0	
Na_2CO_3 0.2 M pH 11.4 eluted material	1658	92	52

* Substrate: 1.6 mM Ac_2-L-Lys-D-Ala-D-Ala
** Specific activity measured at substrate concentrations of 10 times the Km value (12 mM) as described by J.M. Frère et al. (3).

The same technique was then applied to a crude lysozyme-releasable DD-carboxypeptidase preparation of Streptomyces K15, obtained after ammonium sulphate precipitation. The K15 enzyme was adsorbed on the IgG-Sepharose column, which was then washed with 0.5 M NaCl, 0.2 M carbonate buffer H 10.5. The enzyme was eluted with 0.2 M carbonate pH 11.4, resulting in a 550 fold increase in specific activity with a yield of 50 % of the adsorbed material (Table 2).

TABLE 2. Purification of K15 lysozyme-releasable DD-carboxypeptidase on anti-exocellular R61 DD-carboxypeptidase IgG-sepharose 6B immunoadsorbent.

	Total activity* (μmoles/min)	Yield	Specific activity** (μmoles/min/mg)
Crude enzyme	0.255	100	0.002
Non-adsorbed material	0.139	55	
NaCl 0.5 M pH 10.5 eluted material	0.145	57	
Na_2CO_3 0.2 M pH 11.4 eluted material	0.060	23.5	1.15

* Substrate: 1.6 mM Ac_2-L-Lys-D-Ala-D-Ala

** Specific activity measured at substrate concentration of 10 times the Km value (9 mM) as described by M. Leyh-Bouille *et al.* (2).

CONCLUSION

The anti-exocellular R61 enzyme antiserum was found to inhibit the DD-carboxypeptidase activity of the exocellular and lysozyme-releasable enzymes from other *Streptomyces* strains (4). On this basis, the exocellular R61 DD-carboxypeptidase and lysozyme-releasable K15 DD-carboxypeptidase were purified by using anti-exocellular R61 enzyme IgG-Sepharose 6B as an immunoadsorbent. The results reported here are still preliminary. However, they are very promising and when the technique will be refined, it may well give rise to procedures through which the various enzymes of the Streptomyces peptidoglycan crosslinking enzyme system can be isolated in a specific manner, with high yields and with a minimum of steps involved.

REFERENCES

(1) J.M. Ghuysen, The concept of the penicillin target from 1965 until today, *J. Gen. Microbiol*. 101, 13-33 (1977).

(2) M. Leyh-Bouille, J. Dusart, M. Nguyen-Distèche, J.M. Ghuysen, P.E. Reynolds and H.R. Perkins, The peptidoglycan crosslinking enzyme system in *Streptomyces* strains R61, K15 and *rimosus*. Exocellular, lysozyme-releasable and membrane-bound, *Eur. J. Biochem.*, in press (1977).

(3) J.M. Frère, J.M. Ghuysen, H.R. Perkins and M. Nieto, Molecular weight and amino acid composition of the exocellular DD-carboxypeptidase-transpeptidase of *Streptomyces* R61, *Biochem. J.* 135, 463-468 (1973).

(4) M. Nguyen-Distèche, J.M. Frère, J. Dusart, M. Leyh-Bouille, J.M. Ghuysen, J.J. Pollock and U.J. Iacono, The peptidoglycan crosslinking enzyme system in *Streptomyces* R61, K15 and *rimosus*. Immunological studies. *Eur. J. Biochem*., in press (1977)

(5) L. Sundberg and J. Porath, Preparation of adsorbents for biospecific affinity chromatography. Attachment of group-containing ligands to insoluble polymers by means of bifunctional oxiranes, *J. Chromat*. 90, 87-98 (1974).

ACKNOWLEDGEMENTS

The work was supported in part by the *Fonds de la Recherche Fondamentale Collective*, Brussels, Belgium, the *National Institutes of Health*, Washington, D.C. (contract No 1 R01 AI 13364-01 MBC) and by by an EMBO fellowship to A. Marquet. We also thank *UCB*, Brussels, Belgium, for financial support. M.N.D. is *Chargé de recherches* of the *Fonds National de la Recherche Scientifique*, Brussels, Belgium.

SELECTIVE REMOVAL OF FAST DISSOCIATING ANTIBODIES FROM A HIGH-AFFINITY THYROXINE ANTISERUM TO BE USED IN RADIO-IMMUNOASSAY

Viggo Kruse

National Institute of Animal Science, DK-1958 Copenhagen, Denmark

ABSTRACT

Fast dissociating antibodies were removed in a selective manner from a high-affinity thyroxine antiserum by a method based on the difference in dissociation rate constants between fast and slow dissociating antibodies. The antiserum was first saturated with thyroxine and then incubated with an excess of thyroxine coupled to a Sepharose adsorbent. Antibodies bound to the adsorbent after incubation were removed by centrifugation and discarded. Antibodies remaining in the antiserum after centrifugation as thyroxine-antibody complexes were stripped free of practically all thyroxine. The proportion of slow dissociating antibodies in the treated antiserum increased with treatment time up to 3 d. There was a 46 to 80 % loss of titer units depending on the treatment time. A binding reagent was thus produced in which the slow dissociating antibodies dominated as binders under radioimmunoassay conditions.

INTRODUCTION

It is well known that an antiserum raised against a simple, well-defined hapten contains antibodies with a wide spectrum of affinities. The low-affinity and the high-affinity antibodies directed against the hapten may differ in equilibrium constant (K) by a factor as high as 10,000. It has recently been shown that this heterogeneity in K is mainly caused by variation in the dissociation rate constant. The rate constant for the binding of hapten to antibodies varies remarkably little in contrast to the enormous variations in the rate constant for the dissociation reaction (1).

The dissociation characteristics of a high-affinity thyroxine (T_4) antiserum have been analysed (Kruse, in preparation). The fast, medium and slow dissociating antibodies in the antiserum had dissociation rate constants of 0.5, 0.05 and 0.006 h^{-1}, respectively. It is the presence of fast dissociating antibodies in an antiserum which often makes it difficult to get analytical results of high precision with a radioimmunoassay. This paper describes how to remove the fast dissociating antibodies from an antiserum in a selective manner on the basis of the

* Alexander von Humboldt Fellow from April 1977 at the Institute of Physiology, Technical University of Munich, D-8050 Freising, Federal Republic of Germany

large differences in the dissociation rate constants of the antibodies.

PRINCIPLE

Figure 1 shows the equilibrium between the ligand P, the antibodies Q_1, Q_2 and Q_3, and the fast, medium and slow dissociating ligand-antibody complexes PQ_1, PQ_2 and PQ_3. A P-protein-Sepharose adsorbent is made by coupling P to protein, and by coupling the formed P-protein conjugate to Sepharose. The antiserum is saturated with P so most of the antibodies are present as complexes when the fractionation begins. The saturated antiserum is incubated together with an adsorbent for free antibodies (P-protein-Sepharose; Fig. 1) and an adsorbent for free ligand (activated charcoal; not shown). Dissociated ligand is adsorbed on the charcoal so the binding of ligand to free antibodies is prevented. Practically all dissociated antibodies are bound to the P-protein-Sepharose adsorbent if an excess of the Sepharose adsorbent is used and the concentration of free ligand is kept low.

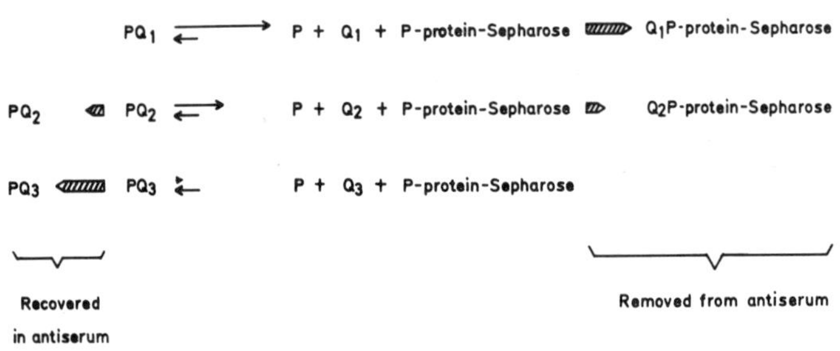

Fig. 1 Principle for the selective removal of fast dissociating antibodies form an antiserum raised against ligand P containing antibodies Q_1, Q_2 and Q_3. The removal is based on the differences in dissociation rate constant between fast (PQ_1), medium (PQ_2) and slow (PQ_3) dissociating ligand-antibody complexes

The P-protein-Sepharose must be made so any dissociated antibody can bind to it. Thus the P-protein part of the adsorbent should be identical to, or at least very similar to the conjugate used for immunization. However, excessive binding of antibodies to the protein, which may occur if the protein part of the immunogen and the adsorbent is identical, can be prevented by using proteins from animals of the same species as the antiserum.

When the fractionation process has proceeded for an appropriate period the Sepharose adsorbent with bound antibody (Fig. 1) and the charcoal with adsorbed ligand are removed from the antiserum by centrifugation. The undissociated ligand-antibody complexes are recovered in the antiserum (Fig. 1). The charcoal is separated from the Sepharose by washing and filtration on a sintered glass funnel. Regeneration of the P-protein-Sepharose adsorbent is done by incubation with an excess of P at high temperature for a long period. Antibodies displaced from the adsorbent by this treatment are removed by filtration and discarded.

The undissociated ligand-antibody complexes recovered in the antiserum (Fig. 1) are stripped free of ligand, and a binding reagent is obtained which contains mainly slow dissociating P antibodies.

It is possible to use other bindable ligands (cross-reacting ligands) for the saturation of the antibodies. The antibodies recovered after fractionation would then be antibodies forming slow dissociating complexes with the particular ligand used for saturation.

One of the main features of the fractionation principle is that antibodies forming slow dissociating complexes with the free ligand we want to measure in radioimmunoassay (f. ex. T_4) are isolated - not antibodies forming slow dissociating complexes with the P-protein conjugate on the Sepharose (f. ex. T_4-protein). Antibodies may have a quite different affinity for the free and the conjugated ligand (2).

MATERIAL AND METHODS

The protein in 5 ml normal rabbit serum was desalted on a column of Sephadex G25-fine equilibrated with 10 mmol/l NaOH. T_4 was coupled to the protein by the carbodiimide method (3). The T_4-protein conjugate was purified by gel filtration and coupled to 50 ml CNBr-activated Sepharose as recommended by the manufacturer (Pharmacia Fine Chemicals). The final product contained 6 mg protein with 70 nmol T_4 per ml Sepharose gel. A pool of T_4 antiserum from rabbit number 8 (Ref. 3) was saturated with 15 nmol T_4 per ml antiserum and incubated overnight at 50°C. Five tubes with 2 ml of saturated antiserum were incubated in barbital buffer at 23° C together with 60 mg methyl cellulose, 100 mg activated charcoal and 4.4 ml T_4-protein-Sepharose in a total volume of 20 ml. Incubation was continued with gentle shaking for 1, 6, 24, 72 or 120 h. The treatment was stopped by chilling the incubates on ice, and gel and charcoal were removed by centrifugation and washed once. The two supernatants were pooled and were regarded as the treated antiserum. A sixth tube with 2 ml of the saturated antiserum had been set aside without treatment and served as control (0 h). All six preparations were stripped free of practically all T_4 simultaneously. It was done as follows: 300 mg methyl cellulose was dissolved in 3 ml ethanol and added to the antiserum under continuous stirring. The volume was adjusted to 33 ml with buffer, and 180 mg dry glutamic acid was added to the solution. The pH was adjusted to 4.4 and 1 g of activated charcoal was added. The stripping suspension was incubated at 50 ° C with gentle shaking for 19 h. The stripping was stopped by removing the charcoal from the antiserum by centrifugation.

The titer of the six preparations plus another control sample made of unsaturated and unstripped antiserum (Unstripped control) was determined as described elsewhere (3). Dissociation curves were made for all seven preparations (3). Briefly, the antiserum preparations were first incubated with labelled T_4 for 2 h at 23° C. The dissociation was then initiated by the addition of a large excess of unlabelled T_4 and followed for a period of 5 d at 23° C.

RESULTS AND DISCUSSION

The antiserum preparations treated by the fractionation method for 0, 1, 6, 24, 72 and 120 h had titers of 100, 54, 50, 24, 26 and 20, respectively, when expressed relative to the titer of the 0 h preparation (absolute titer 16,000). The titer of the unstripped control sample was only 46.

Figure 2 shows that the dissociation curve for the unstripped antiserum (Unstripped control) is steep compared to the curve for the stripped antiserum without treatment by the fractionation method (0 h). This is because a large fraction of the slow dissociating antibodies in the unstripped control is masked by endogenous T_4.

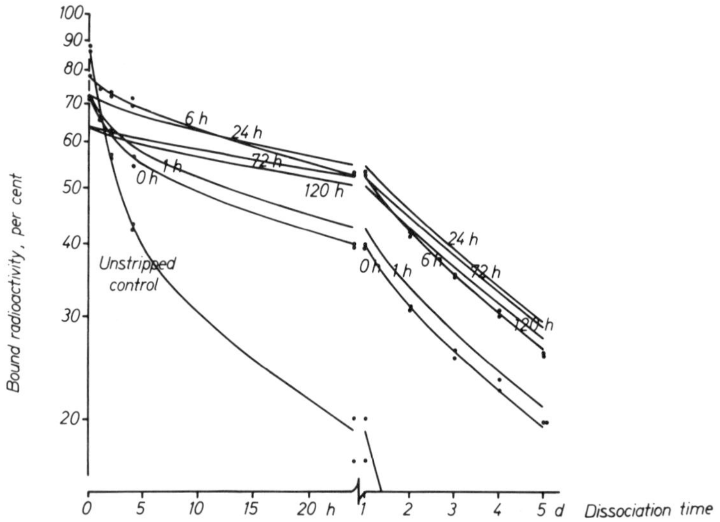

Fig. 2 Dissociation curves for labelled thyroxine-antibody complexes. The complexes were formed by incubating labelled thyroxine and antiserum treated for 0, 1, 6, 24, 72 or 120 h by the fractionation method. A curve for untreated antiserum is also shown (Unstripped control). The filled circles show the experimental results for some of the antiserum preparations, and the curves show the best fit obtained by computer assisted analysis of the experimental results

Such saturated slow dissociating antibodies bind very little labelled T_4 during a short incubation period. The other curves show that the fractionation treatment leads to an almost complete disappearance of the initial steep part of the dissociation curves (Fig. 2). The great majority (92 %) of the T_4-antibody complexes formed after only 2 h of incubation with the antiserum treated for 72 h had a complex half-life of about 5 d (Kruse, in preparation).

REFERENCES

(1) T.W. Smith & K.M. Skubitz, Kinetics of interactions between antibodies and haptens, Biochemistry 14, 1496 (1975)

(2) C. Gomez-Sanchez, L. Milewich & O. B. Holland, Radioiodinated derivatives for steroid radioimmunoassay. Application to the radioimmunoassay of cortisol, J. Lab. Clin. Med. 89, 902 (1977)

(3) V. Kruse, Production and evaluation of high-quality thyroxine antisera for use in radioimmunoassay, Scand. J. clin. Lab. Invest. 36, 95 (1976)

ADSORPTION AND ELUTION CHARACTERISTICS OF *P. AERUGINOSA* TYPE 2 ANTIBODY IN AN AFFINITY CHROMATOGRAPHIC SYSTEM

J. D. Ogle, P. Stace and J. W. Alexander

Departments of Biological Chemistry and Surgery, University of Cincinnati College of Medicine, Cincinnati, Ohio 45267

The effectiveness of immunotherapy against infection by P. aeruginosa with human hyperimmune gamma globulin (HIGG) containing relatively high concentrations of specific anti-Pseudomonas antibodies has been demonstrated in patients with burn injuries (1,2). One disadvantage of this treatment is the large volume required by intramuscular injection to achieve a therapeutic effect. This problem led us to investigate immune affinity chromatography as a possible technique for isolation, purification and concentration of specific anti-Pseudomonas antibody. The isolation of specific antibodies to P. aeruginosa serotype 2 was accomplished by passing human hyperimmune gamma globulin (HIGG) over an affinity column consisting of immobilized cell wall O-antigen followed by elution with water. In order to determine the most effective way to operate such an immunoaffinity column, the patterns of adsorption and desorption of the specific antibody were studied. The affinity matrix consisted of P. aeruginosa serotype two cell wall O-antigen polysaccharide attached by its carboxyl groups to Sepharose 4B containing an amine spacer arm. A H_2O soluble carbodiimide was used to activate the COOH groups. Six tenths of a mg of polysaccharide were immobilized per ml of Sepharose. An affinity chromatographic column was prepared from this matrix. HIGG, containing antibodies to all seven P. aeruginosa serotypes was applied to this column followed by washing with 0.15 M NaCl until no more protein came through. Then the column was washed with water to desorb the adsorbed type 2 antibody. Previously the column had been saturated with type 2 antibody and washed exhaustively with water so that the only binding sites operative were those capable of being eluted with water. The effect of varying the concentration of applied HIGG on the amount of specific antibody adsorbed to the column is shown in Fig. 1. The total number of type 2 units applied was 3200 (600 mg protein) in each application. At very low concentrations of antibody applied to the column, the units of antibody adsorbed increased with increasing concentration but decreased upon continued increase in concentration of applied antibody. This difference from a normal saturation curve may be due to steric hindrance caused by the high concentration of extraneous protein in the HIGG sample. About 1% of the applied protein was type 2 antibody. The decreasing leg of the adsorption curve may actually be biphasic which would implicate different sets of sites on the matrix with different binding constants for the antibody. The accuracy of the assay was not sufficient to determine this. At the optimum concentration of antibody, 9 units/ml only 21% of the antibody molecules were adsorbed. The amount of bound type 2 antibody that could be desorbed with water followed a similar pattern. This is shown in Fig. 2. At the optimum concentration of applied antibody, 61% of the adsorbed antibody was recovered in the water eluant. This corresponds to around 12% yield based on the units

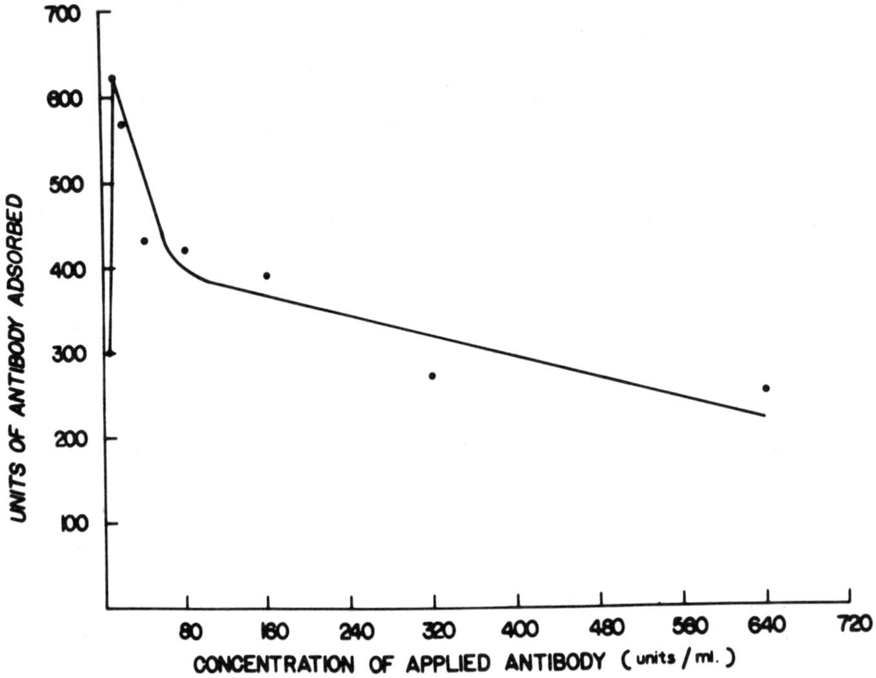

Fig. 1. The effect of varying concentration of applied antiserum on the amount of antibody adsorbed. Column, 1 x 5 cm, flow rate 4 ml/hr.

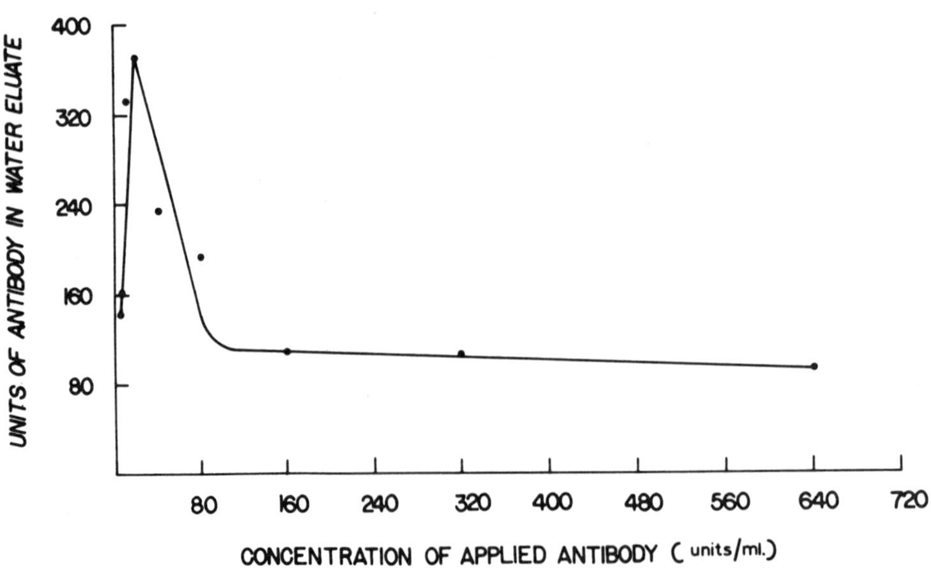

Fig. 2. The effect of varying concentration of applied antiserum on the amount of antibody desorbed with water.

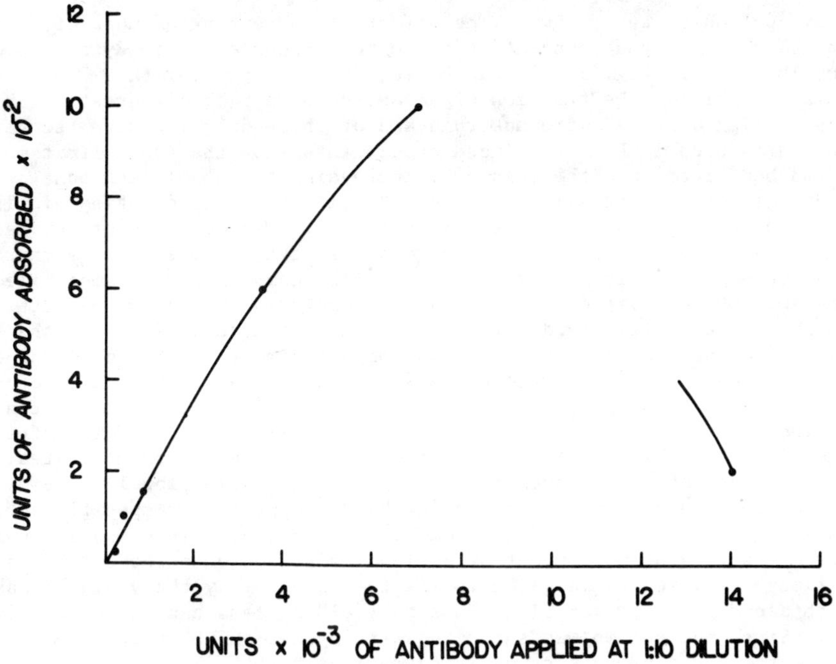

Fig. 3. The effect of amount of applied antiserum on the amount of antibody adsorbed.

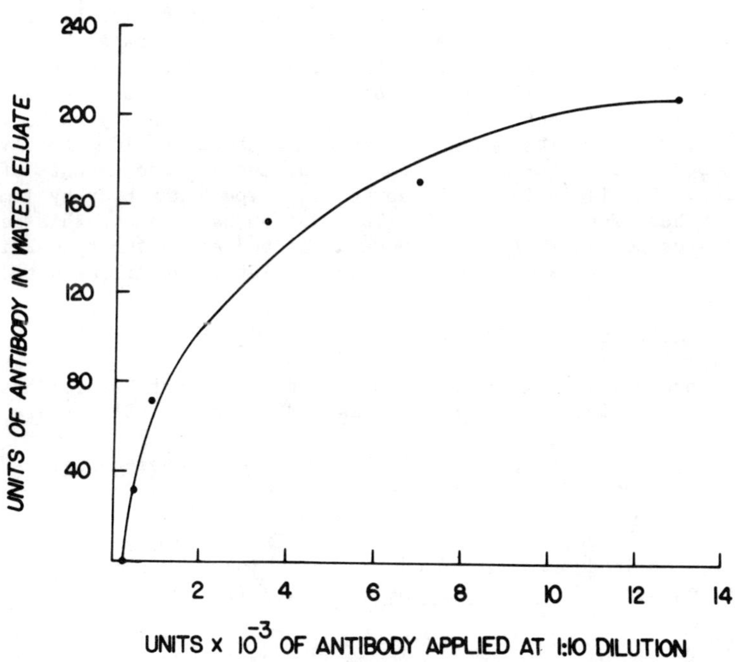

Fig. 4. The effect of amount of applied antiserum on the amount of antibody desorbed with water.

of antibody applied to the column. The effect of increasing amounts of applied type 2 antibody at constant concentration on the amount of antibody adsorbed to the column is shown in Fig. 3. It can be seen that at greater than 700 or 800 antibody units applied, the fraction of adsorbed units fell off drastically. At 7000 units applied only 14% were adsorbed and of these only 17% were recovered by water elution as seen in Fig. 4. These experiments show that to a limit more antibody can be forced onto the column by increasing the absolute amount of applied antibody but not by increasing the concentration of applied antibody. The adsorption and elution properties of a macromolecular, heterogeneous substance such as antibody in an affinity chromatographic system composed also a macromolecular, heterogeneous affinity matrix is bound to be complex. The immobilized antigen MW 300,000 and over consists of a sub-population of molecules with different binding affinities and different degrees of accessibility to the applied antibody, imposed necessarily by the method of attachment to the solid support. The applied antibody (MW 150,000) comprises sub-populations of molecules of different avidities for the antigen, and exists in the gamma globulin sample at a low concentration with respect to nonspecific but similar protein. The effect of differences in accessibility of the attached antigen to antibody molecules is probably emphasized at high concentrations of applied gamma globulin where the extraneous protein molecules may sterically hinder approach of specific antibody molecules to the binding sites of the immobilized antigen. This could result in the observed lowering of the amount of antibody adsorbed onto the column as the concentration of applied gamma globulin was increased. Why the yield of adsorbed antibody obtained in the water eluate was only 61% or less cannot be readily explained. Since the column was saturated with type 2 antibody then washed with water before the start of this study it seems likely that all of the adsorbed antibody was eluted with water in all cases but possibly a certain amount of it was denatured during the desorption procedures. The results of this study indicate that in order to obtain the highest amount of relatively pure specific anti-*Pseudomonas* antibody type 2 by the immune affinity chromatographic system described, one would apply a low concentration of hyperimmune gamma globulin of about 6-7 mg protein/ml (25 units/ml) rather than the 16% solution (ca 600 units/ml) usually available commercially or 6-7% solution such as serum. An amount of gamma globulin 10 times the capacity of the column gave a moderately high yield of purified antibody. Applying the sample in higher amounts did actually yield more purified antibody but at the expense of prohibitively large amounts of starting gamma globulin. The isolation of *P. aeruginosa* type 4 antibody by immunoaffinity chromatography has been reported (3). The elution patterns of this serotype antibody differed considerably from those described above for type 2 insofar as 100% yields of antibody were obtained by water elution at all amounts of applied antiserum.

REFERENCES

(1) J. W. Alexander and M. W. Fisher, Immunization against *Pseudomonas* infection after thermal injury, *J. Infect. Dis.* 130 (Suppl.), S152 (1974).

(2) L. S. Young, R. D. Meyer, and D. Armstrong, *Pseudomonas aeruginosa* vaccine in cancer patients, *Ann. Intern. Med.* 79, 518 (1973).

(3) J. D. Ogle, P. Stace, and J. W. Alexander, Extraction and purification of specific antibody against *Pseudomonas aeruginosa* by immune affinity chromatography, *J. Lab. Clin. Med.* 89, 433 (1977).

BIOSPECIFIC CHROMATOGRAPHY ON NEW DERIVATIVES OF POROUS SILICA BEADS. COUPLING OF GANGLIOSIDE G_{M1} OR ANTICHOLERAGEN ANTIBODY FOR PURIFICATION OF CHOLERA TOXIN

J. L. Tayot, M. Tardy and M. C. Mynard

Institut Merieux, Marcy l'Etoile, 69260 Charbonnières, France

SUMMARY

Porous silica beads "Spherosil" [R] were impregnated with DEAE Dextran which was then cross-linked. These derivatives conferred outstanding properties for ion exchange as well as for immunoaffinity chromatography of proteins. We have now used a very simple procedure to convalently attach various ligands on this support. For instance after the coupling of a ganglioside G_{M1} derivative it was possible to extract choleragen and choleragenoid selectively from several liters of crude culture filtrates of vibrio cholerae. Using this purified preparation as antigen in rabbits monospecific anticholeragen antibodies were obtained and coupled on a silica column for immunoaffinity chromatography. This second column gave an alternate procedure for the purification of cholera toxin. These highly specific methods seem easily applicable to industrial conditions and could be extended to purify other biological macromolecules.

INTRODUCTION

In a recent symposium (1) we described the properties of "Spherosil" [R] porous silica particles impregnated with DEAE Dextran. Like other aminated macromolecules with a positive electric charge it has a very strong affinity for the silica surface. In its presence the strong and irreversible adsorption properties of silica completely disappear and DEAE Dextran gives the support the physico-chemical properties of anion exchangers, comparable to those of corresponding exchangers whose basis is agarose, cellulose or dextran. After cross-linking with bisepoxy reagents, the DEAE Dextran coating is completely stable even in acid or alkaline solutions. Furthermore this new chromatographic support is completely incompressible and easily allows flow-rates from 100 to 400 ml/cm^2 x h under pressure lower than 2 bars. For all these reasons it is particularly well-adapted to the large-scale fractionation of blood proteins by ion exchange (2). It also allows the industrial preparation of purified and concentrated solutions of specific antibodies by immunological affinity after ionic fixation and cross-linkage of their respective antigens (3).

We have now developed a very simple procedure to covalently attach various ligands to the polysaccharide coat of porous silica beads. These supports then allow a very quick and selective extraction of macromolecules presenting biospecific affinity for the chosen ligand. We shall also describe a method of irreversible fixation by first adsorbing and then cross-linking antibodies to the surface of untransformed porous silica beads. The properties and the use of these two types of support with biospecific or immunological affinity will be illustrated in the case of cholera toxin purification.

MATERIALS AND METHODS

Porous silica beads: Spherosil R is manufactured by Rhone-Poulenc. Two different types have been used: XOB 030 - XOC 005. Their particle size varies from 100 to 200 μ. Their mean porous diameter varies from 60 to 5oo nm, their internal surface thus being easily accessible to the macromolecules to be purified and varying from 50 to 10 m^2/g.

Reagents: Apart from DEAE Dextran (Pharmacia) and 1-4 Butanediol-diglycidylether - BDDGE - (Aldrich) the other reagents were obtained from Prolabo. Human albumin and gamma globulin were supplied by Institut Merieux.

Ganglioside G_{M1}: The natural receptor of cholera toxin is prepared according to Svennerholm (4). After an alkaline hydrolysis, where the N-acyl and N-acetyl functions give three amino-groups per molecule (5), we obtained a preparation of lyso G_{M1} which is lyophilized.

Cholera toxin (C.T.): A reference preparation of purified C.T. was purchased from Schwarz/Man, lot GZ 2790, rehydrated to obtain a 1 mg/ml protein concentration. Crude culture filtrates (C.C.F.) are prepared in a fermentor with *vibrio cholerae* INABA 569 B in a modified syncase medium at 32°C or 36°C and pH 7.6.

The biological activity of C.T. is evaluated in guinea pigs by the skin test of Craig (6) and expressed as the highest dilution producing a blue lesion of 6 mm in diameter.

The hemagglutination test (HA) based on agglutination of G_{M1} coated erythrocytes by C.T. has been described in the french patent application n° 77 28164. It is as sensitive as the Craig test and is used to compare the total contents of choleragen and choleragenoid in various preparations. This hemagglutination is specifically inhibited by G_{M1} or lyso G_{M1} (IHA test). This last method is used to titrate specifically these gangliosides in solution or on a carrier.

The preparation of the silica - DEAE Dextran - ligand complex particles is described in detail in the french patent applications n° 76 23176 and 77 28163. Briefly, after cross-linkage with BDDGE, the DEAE Dextran coating is oxidized by $NaIO_4$ and the resulting aldehyde groups react with the amino-groups of the selected ligand (lyso G_{M1} or lysine). The formed Schiff bases are reduced to stable secondary amines with $NaBH_4$. This coupling method has already been proposed by others for agarose (7), Sephadex gel (8), and has given very stable linkages for affinity chromatography. Agarose, however, has few alpha-glycol functions and consequently is not very efficiently derivatized by this method. On the other hand, Sephadex beads are severely degraded or have not the adequate porosity to interact with macromolecules. After a coupling experiment the column is washed with 100 volumes of 0.1 N NaOH, 0.1 N HCl and mixtures of 0.1 N HCl, chloroform and methanol to eliminate traces of non covalently bound lyso G_{M1}. Using 1 g of Spherosil derivatives and 1 mg of lyso G_{M1}, less than 5 % of lyso G_{M1} are detectable in the washing solutions so that we can consider such columns to contain 1 mg ligand per gram of Spherosil.

Anticholeragen antibodies are obtained by immunization of rabbits, either with the reference preparation of C.T. or with our own purified preparations containing 1 mg of $Al(OH)_3$ per ml. As for the CCF the same volume of complete Freund adjuvant is added. Each animal receives three 1 ml subcutaneous injections on D0, D14, D21, and is bled on D28. With CCF and some purified preparations other antibodies than anticholeragen antibodies are also obtained in variable amounts. Specific anticholeragen antibodies are prepared by mixing the antiserum with an adequate volume of CCF freed of choleragen and choleragenoid by filtration through the Spherosil-G_{M1} column (adsorbed CCF). The resulting specific antibodies (gammaglobulins) are then precipitated with 300 g/l ammonium sulfate and purified with the Spherosil-DEAE Dextran anion exchanger (1). After diafiltration in 0.01 M phosphate buffer, pH 6.8, the antibodies are adsorbed into XOC 005 particles until saturation. The excess of non-adsorbed proteins is removed by the same buffer and the protein monolayer is cross-linked with 0.5 % glutaraldehyde. After several washings with 0.1 N glycine·HCl buffer, the column is equilibrated in the chromatographic buffer and ready for use. About 40 mg IgG are fixed per gram of spherosil.

RESULTS AND DISCUSSION

Both types of affinity columns are used in the same way for the C.T. purification starting from C.C.F. The addition of 10 g/l NaCl in the C.C.F. and the phosphate buffer (pH 6.8) entirely suppress the ionic interactions between the support and the macromolecules to be separated. Yet, the biospecific affinity is not hindered under these conditions. 1 g of each type of biospecific support is able to trap at least 3 mg of C.T. For instance with a 40 g column and a C.C.F. containing around 40 ug of choleragen and choleragenoid per ml, there is a 100 % C.T. fixation until 2 l of filtration volume. From 2 l to 4 l, the biological activity (skin test) in the filtrate increases progressively from 0 to 50 %. Then the fixation is negligible. After rinsing with the chromatographic buffer, the C.T. is eluted by 0.05 M Na citrate-citric acid, pH 2.8, and the effluent is continously neutralized. Finally an ultrafiltration step gives a concentrated product as compared to the initial CCF volume. Considering the choleragen and choleragenoid contents before and after purification (HA test), the yield of these purification methods is estimated to be 50 % to 100 %. This is in contrast with the almost irreversible binding and the low yields obtained by Cuatrecasas (9). The beneficial influence either of lyso G_{M1} as a ligand, or of the positive charges of the support remains to be elucidated. Immunoelectrophoretic analysis in agarose (IEP) allows us to identify these preparations. Their concentrations and relative purities are compared with the reference C.T.s (Fig. 1). These preparations contain high levels of toxic activity, at least in the skin test of Craig and are completely identical with reference cholera toxin in double immunodiffusion.

In parallel experiments we have demonstrated that other biospecific columns (Spherosil - lysine or Spherosil-non immunoglobulins), made and used in the same way, did not trap any traces of cholera toxin. This proves the specificity of the observed interactions.

Without NaCl in the chromatographic buffer, the Spherosil - G_{M1} column can be used simultaneously as an ion exchange as well as an affinity column. This is shown by the complete separation of an artificial sample containing three different proteins (Fig. 2).

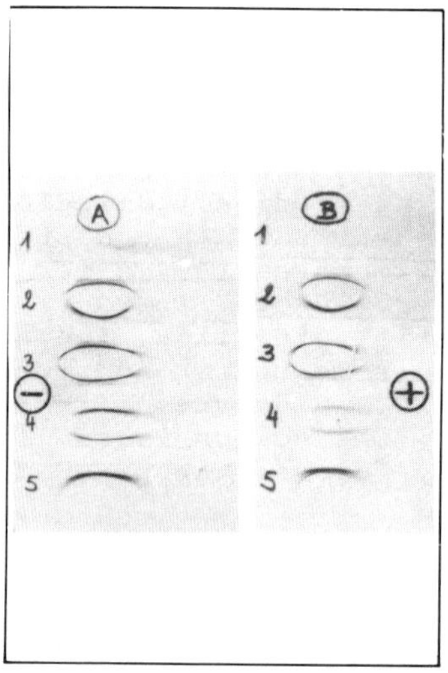

Fig. 1 Immunoelectrophoretic analysis of purified cholera toxin preparations
1 = C.C.F. (7 x c)
2 = reference cholera toxin (1 mg/ml)
3 = cholera toxin (150 x c) from spherosil-anti C.T. column
4-5 = cholera toxin (30 x c) from spherosil-lyso GM1 column
A = anti C.C.F. serum
B = specific anti cholera toxin antibodies

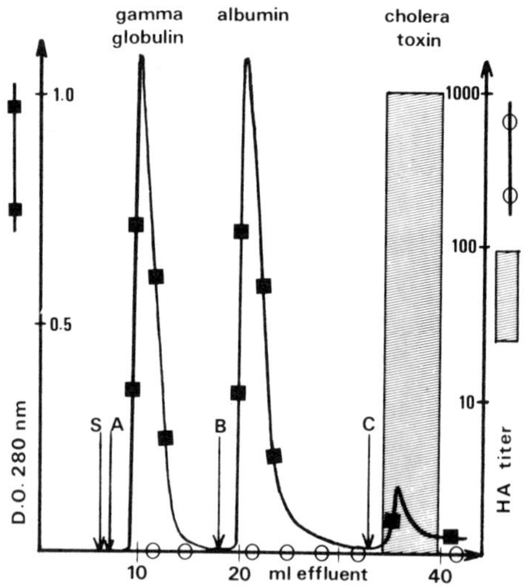

Fig. 2 Association of ion exchange and affinity chromatography on 1 g of Spherosil-DEAE-Dextran-lyso G_{M1}
S = sample (Gamma Globulin 6 mg - Albumin 12 mg reference cholera toxin 0.1 mg in 1 ml of buffer A
A = phosphate buffer 0.01 M pH 6.8
B = NaCl 10 g/l in buffer A
C = Na citrate - citric acid 0.05 M pH 2.8

The electropositive gamma globulin is not adsorbed on the modified DEAE Dextran layer, whereas the negatively charged albumin is fixed and then eluted by NaCl addition. The low ionic strength of the buffer and the numerous positive charges of the support do not hinder the affinity between lyso G_{M1} and C.T. which is completely fixed. C.T. is only eluted by the citrate buffer, but the yield in this case is very low (5 %). We have repeatedly observed such low yields when the column is not used at its full capacity. C.T. is probably denatured at this pH when its concentration is too low. Under these conditions, each protein is quantitatively recovered in almost symmetrical peaks. This also illustrates the specificity of the interactions with these new derivatives of porous silica. These experiments show that electrostatic interactions are not troublesome in affinity chromatography, at least in this case. Furthermore, they confer some new properties which can be used with benefit. Other experiments are in progress with other ligands to extend these observations.
The IHA test has been used to determine that 1 mg of lyso G_{M1} in aqueous solution can neutralize around 4 mg of C.T. This is very close to the fixation capacity of 1 g of Spherosil-G_{M1} which also contains 1 mg of lyso G_{M1}. This result underlines that lyso-G_{M1} is an interesting ligand and that porous silica derivatives are widely accessible to macromolecules.

Repeated rinsing with 0.1 N HCl, even between each purification cycle, do not alter the specificity nor the capacity of these affinity columns. This indicates an excellent stability of the biospecific surface coating the porous silica beads. This stability even in acid solutions contrasts with the known lability of G_{M1} which is quickly transformed into G_{AI} in these conditions, at least in aqueous solutions. If it would be also the case for coupled lyso G_{M1}, this might be interpreted as a proof of affinity between G_{AI} and C.T. This hypothesis remains to be further studied.

In conclusion these new derivatives of porous silica appear to have numerous qualities such as specificity, stability, incompressibility, permeability, large pores widely accessible, which are needed for biospecific chromatography. Large scale applications of these affinity columns seem now possible, particularly in the field of purified enzymes, blood proteins and vaccines.

ACKNOWLEDGEMENTS

We thank Prof. L. Svennerholm for his interest and for providing some samples of pure lyso G_{M1}.
The technical assistance of Mrs. G. Favre and Miss C. Legoff has been appreciated.

REFERENCES

(1) Advances in the chromatographic fractionation of macromolecules Birmingham Symposium (1976). *Chromatography of synthetic and biopolymers*, Ellis Horwood, Chichester, U.K., in press.

(2) Tayot, J.L., Tardy, M., Gattel, P., Plan, R., and Roumiantzeff, M. (see 1)

(3) Tardy, M., Tayot, J.L., Roumiantzeff, M., and Plan, R. (see 1)

(4) Svennerholm, L. (1972), *Methods in carbohydrate chemistry VI*, 464-474, Whistler - Bemiller, Acad. Press.

(5) Holmgren, J., Mänsson, J.E., and Svennerholm, L., *Medical Biology* 52, 229-233 (1974).

(6) Craig, J.P., *Nature* 207, 4997, 614-616 (1965).

(7) Parikh, I., March, S., and Cuatrecasas, P. (1974). *Methods in Enzymology* XXXIV, 77-102. Jakoby - Wilchek, Acad. Press.

(8) Wilson, M.B., and Nakane, P.K., *J. of Immunol. Methods* 12, 1-2, 171-181 (1976).

(9) Cuatrecasas, P., *Biochemistry* 12, 21, 4253-4264 (1973).

PART V

MATRIX-BOUND LECTINS AND AFFINITY CHROMATOGRAPHY OF CELLS

FRACTIONATION OF LYMPHOCYTES ON INSOLUBILIZED *HELIX POMATIA* A HEMAGGLUTININ AND WHEAT GERM AGGLUTININ

Sten Hammarström, Ulla Hellström, Marie-Louise Dillner, Peter Perlmann, Hedvig Perlmann, Bernt Axelsson and Eva-Stina Robertsson

Department of Immunology, University of Stockholm, S-10691 Stockholm, Sweden

INTRODUCTION

The immune system constitutes an interacting multicellular system in which two major classes of cells (T- and B-lymphocytes) participate. However, within both cell-classes a further cellular heterogeneity with respect to function, degree of differentiation and surface marker distribution has been recognized. For the understanding of the biological processes involved in the immune response it is essential to develop methods for the separation of the interacting cell types. In this communication we describe the use of two carbohydrate binding proteins *Helix pomatia* A hemagglutinin (HP) and wheat germ agglutinin (WGA) for the fractionation and characterization of lymphocytes. For fractionation, HP or WGA are covalently bound to Sepharose 6MB and packed into columns. Cells with no receptors or with weakly binding receptors for these proteins pass through the columns while cells with receptors are bound to the column. These latter cells can then be eluted with the corresponding competitive sugar hapten, thus rendering both bound and non-bound cells from the same cell population. It is shown that T- and B--cells can be separated from each other on HP-Sepharose columns and that T--cells can be subdivided into two functionally different subpopulations on WGA-Sepharose columns.

MATERIALS AND METHODS

Human Lymphocytes

Peripheral blood lymphocytes (PBL) from healthy donors, patients with chronic lymphocytic leukaemia (CLL) and multiple myeloma, most of them of blood group O and B, were used. In the fractionation experiments with HP blood group A donors were avoided since HP reacts with contaminating erythrocytes in the lymphocyte preparations from such donors (1). The cells were purified from defibrinated venous blood by gelatin sedimentation, treatment with carbonyl iron and Ficoll-Isopaque centrifugation as previously described (2). Differential white cell counts on smears stained with May Grünwald-Giemsa stain showed that ≥98% of the white cells were lymphocytes. Cell viability was ≥97% as assessed by trypan blue exclusion. Erythrocyte contamination was 5-10%. In some experiments the purified lymphocyte preparations were subjected to an adherence step in plastic dishes to remove remaining monocytes (3). Lymphocytes were also purified from cord blood by the procedure described above except that heparinized blood was used. ≥ 90 % of the cells were lymphocytes. Approximately 6% were larger blastlike cells of an immature type.

Animal Lymphocytes

Lymphocytes were obtained from the following sources: peripheral blood of cattle; mouse spleen (CBA, C57/bl); rat spleen, lymph node and thymus (DA). Heparinized cattle blood was separated by the Ficoll-Isopaque method of Bøjum (4). 95% of the nucleated cells were mononuclear and the cell viability was 97-99%. Single cell suspensions were prepared from spleen, lymphnode and thymus. Spleen lymphocytes were purified by the Ficoll-Isopaque gradient centrifugation method (4) and incubated on tissue culture grade plastic flasks to remove adherent cells. 95% or more of the cells were lymphocytes. The cell viability was ≥ 98%.

Carbohydrate Binding Proteins

Helix pomatia A hemagglutinin (HP) was prepared on insolubilized hog A+H blood group substance as described earlier (5). HP is a hexameric glycoprotein containing six homogeneous carbohydrate binding sites (molecular weight 79,000 ± 4,000) (6,7). The specificity range of HP is α-D-GalNAc*p* > α-D-GlcNAc*p* ≃ β-D-GalNAc*p* > β-D-GlcNAc*p* > α-D-Gal*p* (8). HP does not induce DNA-synthesis in neuraminidase treated or untreated human peripheral blood lymphocytes (9).

Highly purified wheat germ agglutinin (WGA) was obtained from Pharmacia Fine Chemicals AB, Uppsala, Sweden. It was purified by affinity chromatography using D-GlcNAc coupled to Epoxy-activated Sepharose. WGA is a dimeric non-glycosyl-containing protein containing two or four carbohydrate binding sites (molecular weight ~ 36,000) (10,11). The binding site is complementary to a sequence of three β-(1→4)-linked D-GlcNAc units (N,N',N''-triacetyl chitotriose) (10,12). WGA is non-mitogenic (reference 13 and Hammarström *et al.*, unpublished observations).

Surface Receptors for HP and WGA

Binding of HP and WGA to lymphocytes was assessed either by immunofluorescence on living cells using fluorochrome labelled HP or WGA or by cell binding experiments with 125I-labelled reagents. The details of fluorochrome and radioactive labelling of the carbohydrate binding proteins as well as of the immunofluorescence and cell binding experiments have been published (1,14). WGA binds to untreated human lymphocytes as well as to neuraminidase treated cells. To create surface receptors for HP, neuraminidase (*Cl. perfringens* or *Vibrio cholera*) treatment of the cells is necessary. Optimal conditions for human and rat lymphocytes were 10 μg neuraminidase *Cl. perfringens* type VI 1-3 U/mg, NAN-lactose substrate (Sigma Chemical Company, St. Louis, Mo., USA) / 25x10^6 lymphocytes in 2 ml Tris buffered Hank's solution, pH 7.4 (TH) at 37°C for 45 minutes (15,16). For mouse and cow lymphocytes 25 μg of neuraminidase / 25x10^6 cells was found to be optimal (Hammarström *et al.*, unpublished observations). After enzyme treatment the cells were washed twice in TH containing 0.2% human serum albumin (HSA).

HP-binding Surface Glycoproteins

Purified human T-cells were labelled by lactoperoxidase catalysed surface iodination (17) and solubilized in 0.5% Nonidet P-40. The solubilized components were then treated with excess HP-Sepharose and unbound material was removed by washing. Bound components were eluted with D-GalNAc, precipitated with aceton and dissolved in 2% SDS containing 0.2 M dithiothreitol.

The material was then subjected to gradient polyacrylamide gel electrophoresis (7.5-15%) using a discontinuous buffer system. After fixation and staining with Coomassie blue labelled components were developed by autoradiography.

Characterization of lymphocytes

Surface immunoglobulin positive (SIg^+) cells were detected by direct or indirect immunofluorescence on living cells. The following reagents were used: 1) human lymphocytes, FITC-labelled $F(ab')_2$-fragments of immunoadsorbent purified rabbit antibodies specific for human $F(ab')_2$ [= a-$F(ab')_2$]; FITC-labelled $F(ab')_2$-fragments of immunoadsorbent purified rabbit antibodies specific for human γ-chain [= a-γ] ; FITC-labelled $F(ab')_2$-fragments of immunoadsorbent purified goat antibodies specific for human δ-chain [= a-δ]; FITC-labelled $F(ab')_2$-fragments of rabbit IgG specific for human μ-chain [= a-μ] ; immunoadsorbent purified rabbit antibodies against human IgG (antibodies precipitating κ- and λ-light chains and γ-heavy chains) and FITC-labelled IgG from sheep anti rabbit immunoglobulin. The detailed purification procedures for these antibodies, their F/P molar ratios, and the conditions for the immunofluorescence tests have been described (18,19). 2) Bovine lymphocytes, FITC-labelled IgG of rabbit anti bovine IgG (antiserum from National Bacteriological Laboratories, Stockholm, Sweden, F/P molar ratio = 3.4). 3) Rat lymphocytes, rabbit anti rat IgG antiserum and FITC-labelled IgG from sheep anti rabbit immunoglobulin (16). 4) Mouse lymphocytes, rabbit anti mouse IgG antiserum and FITC-labelled IgG from sheep anti rabbit immunoglobulin.

Lymphocytes forming spontaneous rosettes (E_s^+ cells) with sheep erythrocytes were determined by the method of Jondal et al. (20) as modified by Yata et al. (21). Rosettes were allowed to form in concentrated fetal bovine serum by incubation at 4°C over night.

Fc-receptor bearing cells (E_bA^+ cells) were determined by a rosette assay using bovine erythrocytes coated with human IgG as described by Hallberg et al. (22).

Complement receptor bearing cells (E_sAC^+ cells) were assayed as described by Perlmann et al. (23). Briefly, sheep erythrocytes (E_s) were sensitized with the IgM fraction of a rabbit antiserum against sheep stromata (EA) followed by treatment with C6-deficient rabbit serum to form E_sAC.

Response to mitogens. Incorporation of ^{14}C-thymidine into DNA of unfractionated and fractionated lymphocytes was determined after exposure of the cells to leucoagglutinin from *Phaseolus vulgaris* (La), a purified mitogen from PHA (24), concanavalin A (Con A), and *Staph. aureus* Cowan I bacteria. The cells were cultivated for 72 hrs in the presence of different doses of mitogen as described elsewhere (9). In the experiments with *Staph. aureus* Cowan I bacteria mercaptoethanol of a final concentration of 0.7 mM was included in the culture medium (25).

Antibody dependent cytolytic potential of the lymphocytes was studied with ^{51}Cr-labelled chicken erythrocytes as target cells in the presence of high dilutions of rabbit anti-chicken erythrocyte serum. For details see Perlmann et al. (3).

Preparation of HP- and WGA-Sepharose

HP was conjugated to cyanogen-bromide-activated Sepharose 6MB (Pharmacia Fine Chemicals, Uppsala, Sweden). These Sepharose particles were specially developed for cell fractionation and have large, relatively uniform particle size (250-315 μ). Conjugation was performed under standard conditions as described in detail by Hellström et al. (18). HP-Sepharose gels containing 1 mg HP/ml swollen gel was found to be optimal for lymphocyte fractionation. Three to four ml HP-Sepharose gel, substituted as indicated above, was found to be sufficient to retain practically all HP^+ cells present in $80-100x10^6$ neuraminidase treated human lymphocytes. Lower degree of substitution gave incomplete retention of binding cells. Higher degree of substitutions made it difficult to elute bound cells with the sugar hapten concentrations chosen (e.g. 1 mg/ml of D-GalNAc). Moderately oversubstituted gels 1.5-2 mg HP/ml swollen gel could, however, be used if either higher sugar concentrations were used for elution or if the beads were removed from the column, resuspended in the sugar hapten solution and allowed to settle. Remaining lymphocytes could then be recovered from the supernatant.

WGA-Sepharose 6MB was prepared by the same procedure as described for HP. The degree of substitution used was 5 mg WGA/ml swollen gel. Up to $50x10^6$ human T-lymphocytes could be successfully fractionated on 1-2 ml WGA-Sepharose.

Fractionation Procedure

HP-Sepharose or WGA-Sepharose gel (3-4 ml and 1-2 ml, respectively) was packed in a small column (0.9x15 cm, K9/15 Pharmacia Fine Chemicals) equipped with a 85 μ nylon net (Nytal Seidengazefabrik A.G., Thal, Switzerland). Lymphocytes ($\leq 100x10^6$ neuraminidase treated human or cow blood lymphocytes or neuraminidase treated rat or mouse spleen cells - HP-Sepharose or $\leq 50x10^6$ human T-cells - WGA-Sepharose) in 1-1.5 ml TH containing 0.2% HSA and 0.02% NaN_3 were applied to the column and allowed to penetrate the gel. The cells were then incubated on the gel (15 min - HP-Sepharose, 3 min - WGA-Sepharose). Cells not retained on the column were eluted with 100-150 ml buffer at a flow rate of 6-10 ml/min. Retained cells on the HP-Sepharose column were eluted in two steps. First with 80 ml 0.1 mg/ml D-GalNAc in buffer and then with 80 ml 1 mg/ml D-GalNAc in buffer. The retained cells on the WGA-Sepharose column were eluted with 100 ml buffer containing 25 mg D-GlcNAc/ml. During elution with high sugar concentrations the flow was stopped for 5-10 min in order to increase the efficiency of dissociation. The passed and eluted cells were then concentrated by centrifugation and washed with TH containing 0.2% HSA to remove the hapten.

RESULTS

HP-receptors, Cellular Distribution, Number and Chemical Nature

Human peripheral blood lymphocytes (PBL) do not stain with FITC-HP. It was, however, found that neuraminidase treatment uncovered binding sites for HP on a fraction of PBL. Between 60 and 80% of the lymphocytes thus became HP-positive (Fig. 1 and reference 15). Increase of the enzyme concentration above 10 μg/2 ml and $25x10^6$ cells did not increase the number of HP^+ cells (15). Binding of FITC-HP to the cells was specific since it could be inhibited by D-GalNAc but not with D-Glc or L-Fuc (15). Analysis of bovine PBL, mouse and rat spleen cells, rat lymph node cells and rat thymus cells similarly demonstrated that neuraminidase treatment exposed HP-receptors on a fraction of the cells from all these sources. With bovine and mouse lymphocytes slightly higher neuraminidase concentrations were needed to reach a constant fraction of stained cells (see methods). As can be seen from Fig. 1 40-50% of bovine PBL, 30-40% of rat spleen cells and 40-50% of mouse spleen cells were HP^+ using FITC-HP. Moreover about 50% of rat lymph node lymphocytes and 80-95% of rat thymus lymphocytes were HP^+ after neuraminidase treatment (16).

To find out whether the fraction of HP^+ cells was increased if the more sensitive indirect immunofluorescence technique (HP and FITC-labelled immunoadsorbent purified rabbit antibodies against HP) was used we have also analysed human and cow PBL by this technique. Only slightly higher values were obtained for human PBL (i.e. 80-85%) while for bovine PBL the percentage of HP^+ cells was increased to about 60%.

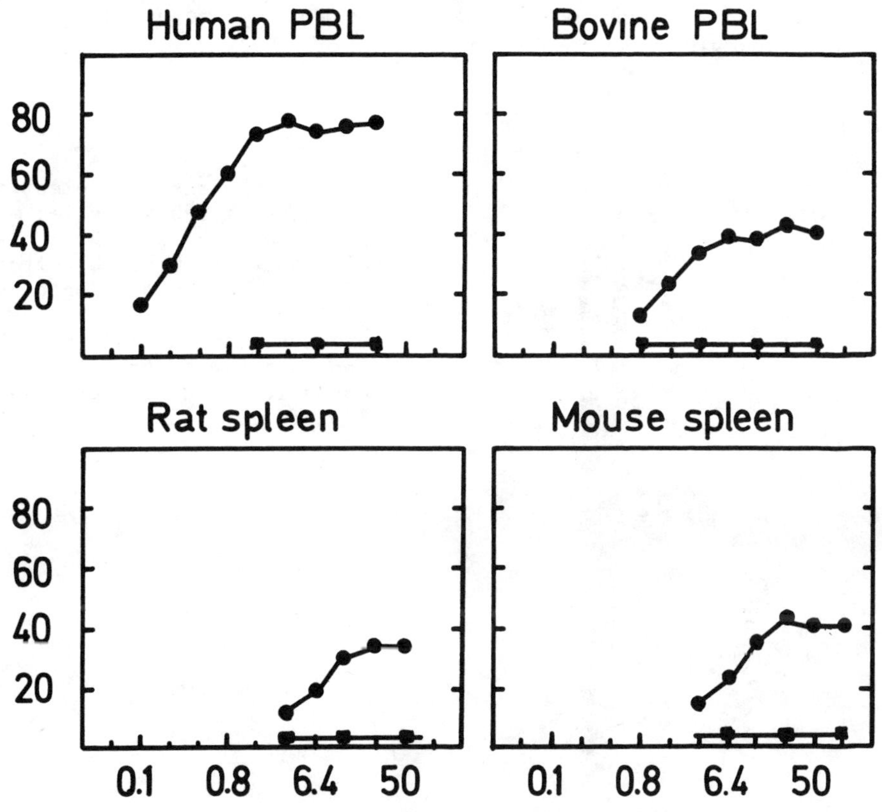

Fig. 1. Surface staining of neuraminidase treated (●) or untreated (■) lymphocytes for HP-receptors. Ordinate: % stained cells; abscissa: μg FITC-HP added.

Human PBL bound $0.6-1.1 \times 10^6$ molecules of intact HP and 1.7×10^6 molecules of divalent HP at saturation as determined by cell binding experiments using 125I-labelled intact or partially reduced HP (14). The data are calculated on the number of HP^+ cells which was 75%. Similar analysis of cow PBL and mouse spleen cells gave slightly lower numbers, e.g. $0.5-0.6 \times 10^6$ and 0.5-

-0.8x10^6, respectively.

It was of interest to find out whether HP recognizes many or only a few glycoproteins on the surface of neuraminidase treated lymphocytes. We therefore subjected enzyme treated purified T-cells from human PBL to lactoperoxidase catalyzed surface iodinction followed by detergent extraction and absorbtion of the labelled solubilized components to insolubilized HP followed by specific elution with D-GalNAc. The eluate was then analysed on SDS-PAGE and the banding pattern developed by autoradiography. Fig. 2 shows a representative experiment. A maximum of 3 labelled components was detected. However, only one component, apparent molecular weight 150,000, was strongly labelled. It would appear that this is the major HP-binding glycoprotein on the lymphocyte surface since this was the only band detected when 125I-HP was used to stain an electrophoretogram of unlabelled solubilized lymphocyte membrane components (Axelsson et al. unpublished observations). Similar studies on mouse spleen cells indicate that mouse lymphocytes also contain a major HP-binding glycoprotein with a molecular weight of about 150,000 (Kimura et al. unpublished observations).

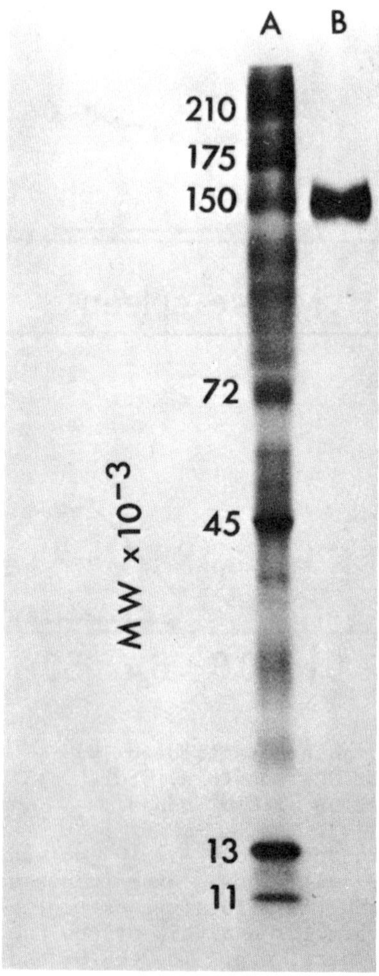

Fig. 2. SDS slab gel electrophoresis of 125I-labelled membrane components of human T-lymphocytes. Banding pattern developed by autoradiography. (A) unfractionated labelled membrane components. (B) labelled components bound to HP-Sepharose and subsequently eluted with D-GalNAc.

The presence of HP-receptors on the majority of the lymphocytes in human PBL and on most rat thymus cells but only on about 30-50% of the lymphocytes in mouse and rat spleen indicated that HP was a T-cell marker. Immunofluorescence studies with FITC-HP on T-cell enriched cell populations from human PBL showed that this indeed was the case. Thus, 90-95% of the cells obtained after passage of PBL over an Ig/anti Ig complex column or of the cells obtained by sheep erythrocyte rosette separation (3) were HP^+ (reference 15 and Hammarström et al. unpublished observations).

From the results above it may be inferred that most neuraminidase treated B-cells are HP-receptor negative. To study this question in more detail we have recently performed double marker experiments with fluorochrome labelled anti Ig and HP on human PBL (19). It was found that a mean of 15% (range 3-24%, n = 7) of the B-cells in PBL as detected with the a-$F(ab')_2$ reagent also were HP^+. Similar analysis with the a-μ and a-δ reagents showed that 10% (range 0-21%, n = 8) and 10% (range 2-20%, n = 6) respectively were HP^+ (19). These results are of considerable interest when taken together with the results of a similar study performed on lymphocytes from cord blood. In cord blood the proportion of B-cells is considerably higher (about 20% as compared to 5-8% for adult blood). Of these B-cells a mean of 83% (range 80-86%, n = 7) were HP^+ [a-$F(ab')_2$-reagent] (19). Similar results were obtained with a-μ and a-δ reagents. Analysis of neuraminidase treated PBL from patients with chronic lymphocytic leukaemia (CLL) and multiple myeloma (reference 26 and Hellström et al. unpublished observations) showed that the malignant cells in CLL also were HP^+ while malignant B-cells from multiple myeloma generally were HP^-. The malignant cells in CLL are believed to be less differentiated B-cells than the corresponding malignant cells in myeloma (27). It would therefore appear as if HP is a differentiation marker for human lymphocytes present both on T-cells and on immature B-cells.

Cell Fractionation on HP-Sepharose

HP^+ lymphocytes could be successfully separated from HP^- lymphocytes on columns charged with HP covalently bound to Sepharose 6MB particles. Fig. 3 shows 4 representative experiments in which human PBL, bovine PBL, rat spleen cells and mouse spleen cells, respectively, were fractionated on HP--Sepharose after treatment of the cells with neuraminidase. The total cell yield was typically between 60-70%. With human PBL between 5-10% of the cells were recovered in the passed fraction, about 15% in the 0.1 mg/ml D-GalNAc fraction and about 45% in the 1 mg/ml D-GalNAc fraction (18). Analysis of the recovery of cells with different surface markers (HP^+, E_s^+, SIg^+, E_sAC^+) furthermore demonstrated that there was no selective loss of cells with any of these surface markers (18). In order to ensure the specificity of our fractionation procedure two types of control experiments were performed 1) column fractionation on CNBr-activated and deactivated Sepharose 6MB particles and 2) column fractionation experiments in which elution was attempted with a non-competitive sugar hapten (e.g. D-Man). In the former series of experiments the cell recovery of human PBL was 80±11% (n = 8). Most importantly, however, there was no selective loss of cells with any of the four surface markers (see above) (18). In the second type of experiments no cells could be eluted demonstrating the specificity of the interaction.

As can be seen from Fig. 3, HP^+ cells were strongly depleted in the passed cell fraction while the fraction eluted with high sugar concentration (1 mg/ml of D-GalNAc) was enriched in HP^+ cells as compared with the original cell population. In contrast SIg^+ cells were enriched in the passed fraction and strongly depleted or absent in the 1 mg/ml D-GalNAc eluted fraction. This was true for neuraminidase treated lymphocytes from all four species. It should, however, be noted that only in the case of human PBL were the conditions such that all SIg$^+$ cells recovered were true B-cells (i.e. the cells were incubated at 37°C for 1 hr in order to remove SIg bound to Fc-receptors (28) and FITC-labelled F(ab')$_2$-fragments of rabbit a-F(ab')$_2$ antibodies were used). It is noteworthy that the fraction eluted with high sugar concentration was completely devoid of B-cells (compare also reference 18). It is furthermore of interest that all HP$^+$ B-cells (~15% of all B-cells, see above) were recovered in the 0.1 mg/ml D-GalNAc fraction perhaps indicating that the HP-receptors on these cells are quantitatively or qualitatively different from those on T-cells. In the other experiments membrane bound Ig as well as Ig bound to Fc-receptors and possibly also free Fc-receptors may contribute to the positive staining with the anti Ig reagents (29). In analogy with our earlier studies on human PBL (18) it is in fact probable that most of the positively stained cells in the 1 mg/ml D-GalNAc fraction are Fc-receptor positive cells with absorbed IgG.

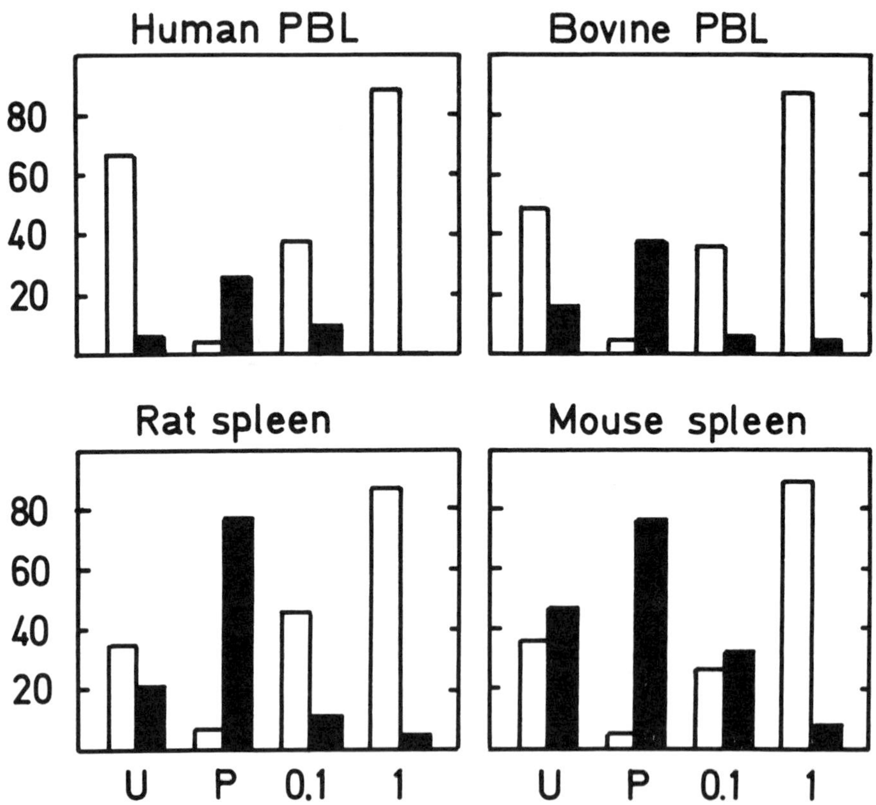

Fig. 3. Fractionation of neuraminidase treated lymphocytes on HP-Sepharose columns. Ordinate: % of cells with surface marker. Blank columns: HP-receptor positive cells. Filled columns: SIg$^+$ cells. (U) Unfractionated cells. (P) Passed cells. (0.1) Eluted cells, 0.1 mg/ml D-GalNAc. (1) Eluted cells, 1 mg/ml D-GalNAc.

TABLE 1 Fractionation of NANAase Treated Human Lymphocytes on HP-Sepharose

Expt. no.		Before fract.	Passed	Eluted 0.1	Eluted 1.0
1	Yield(%)	-	6	27	48
	Surface markers(%)				
	HP^+	71	3	56	92
	E_s^+	78	6	69	94
	E_s^sAC	22	70	24	10
2	Yield(%)	-	13	15	38
	Surface markers(%)				
	HP^+	67	10	42	87
	E_s^+	78	19	65	95
	E_b^sA	21	32	26	9

Human PBL fractionated on HP-Sepharose were also analysed for the distribution of cells with receptors for sheep erythrocytes (T-cell marker), the Fc-portion of IgG and for C3 receptors using three different rosette assays. Table 1 shows a typical experiment. As can be seen E-rosetting cells were strongly depleted from the passed fraction and enriched in the fraction eluted with high sugar concentration thereby confirming that the great majority of HP^+-cells are T-cells. Cells with Fc-receptors and C3 receptors showed essentially the opposite distribution. Such cells were enriched in the passed fraction, slightly enriched in the 0.1 mg/ml D-GalNAc fraction and partially depleted from the 1 mg/ml D-GalNAc fraction. However, these latter results also show that there are both HP^+, Fc^+ cells and HP^-, Fc^+ cells as well as HP^+, EAC^+ cells and HP^-, EAC^+ cells. The former result has also been confirmed by double marker experiments (30).

In some experiments with mouse lymphocytes, the cells were also characterized by means of the θ-antigen, a T-cell marker. It was found that θ^+ cells closely parallelled the distribution of HP^+ cells on column fractionation of neuraminidase treated mouse spleen cells. Similar studies on rat and cow lymphocytes could not be performed because of lack of established T-cell markers.

The functional integrity of the lymphocytes after fractionation and the efficiency of the fractionation procedure was investigated by 1) measuring response to the mitogens La, Con A and Staph. aureus Cowan 1 bacteria and 2) antibody dependent cell mediated cytotoxicity (K-cell activity to ^{51}Cr--labelled chicken erythrocytes). In human PBL, cells responding to the T--cell mitogens La and Con A were enriched in the fraction eluted with high sugar concentration and almost completely depleted from the passed fraction (Table 2, Exp. 1 and reference 18). However, the reverse was true when the fractions were stimulated with Staph. aureus Cowan 1 bacteria (Table 2, Exp. 1). These bacteria stimulate B-cells presumably via protein A present on the bacterial surface and interacting with SIg on B-cells (25). K-cell activity was enhanced in the passed and generally also in the 0.1 mg/ml D-GalNAc fraction. It was always low but significant in the cell fraction eluted with high sugar concentration (Table 2, Exp. 2 and references 18, 30). Taken together these data show that the HP-column fractionation procedure can be used to enrich or deplete subpopulations of lymphocytes associated with different functional activities. The results also show that the fractionation procedure does not adversly affect the functional integrity of the cells and finally that column passage does not activate the lymphocytes.

TABLE 2 Fractionation of NANAase Treated Human Lymphocytes on HP-Sepharose

Expt. no.		Before fract.	Passed	Eluted 0.1	Eluted 1.0
1	Yield(%)	-	5	8	41
	Surface marker (%)				
	HP^+	80	3	52	92
	Stimulation ($CPM\cdot 10^{-3}$)				
	La*	7.2	2.6	4.0	11.8
	Cowan I**	7.4	15.2	6.7	0.9
2	Yield(%)	-	14	23	47
	Surface markers (%)				
	HP^+	70	17	55	98
	^{51}Cr-release(%)***	63	77	67	15

* 8 µg La/ml=optimal dose; $CPM/0.2\cdot 10^6$ lymphocytes

** $4\cdot 10^7$ Staph. aureus Cowan I bacteria/ml=optimal dose; $CPM/0.2\cdot 10^6$ lymphocytes

*** ADCC with ^{51}Cr-labelled chicken erythrocytes

WGA-receptors. Number and Cellular Distribution

WGA binds to all lymphocytes in human peripheral blood irrespectively of whether they are treated with neuraminidase or not. However, titration experiments with FITC-WGA showed that the curve was biphasic with only a fraction of the cells (20-30%) stained at low lectin concentrations (31). Similarly a biphasic binding curve was obtained when PBL or purified T--cells [< 1% SIg^+ cells and > 90% E^+ and HP^+ cells] were titrated with increasing concentrations of 125I-WGA ($K_{1\ app.} = 7\times10^7\ M^{-1}$, $n_1 = 6\times10^6$; $K_{2\ app.} = 5\times10^6\ M^{-1}$, $n_2 = 15\times10^6$) (31). The difference in apparent binding constants was approximately 14 fold. These data indicate that there are two types of WGA-receptors on human T-cells - strong and weak binding receptors - and that they at least partly are distributed on different cells.

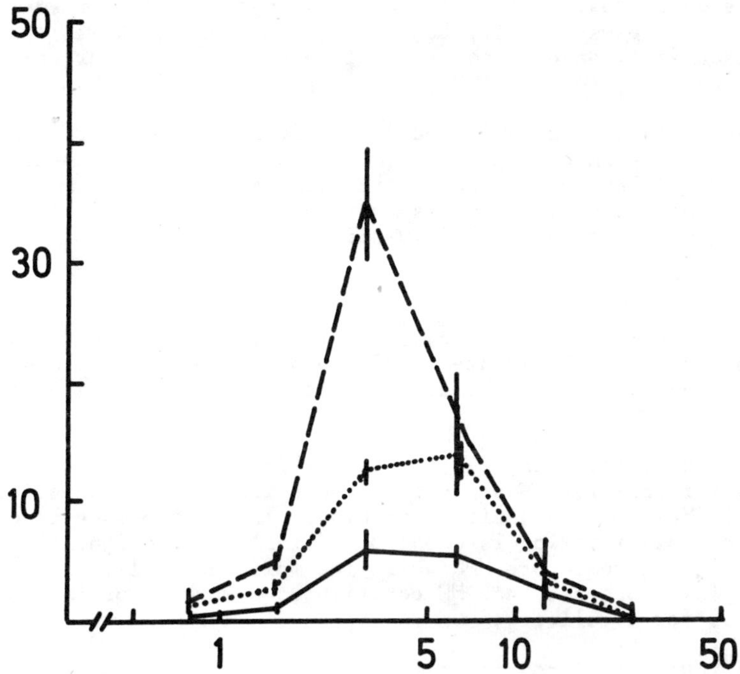

Fig. 4. Dose response curve for stimulation of DNA synthesis with Con A of human T-lymphocytes (dotted line), passed T-lymphocyte fraction from the WGA--Sepharose column (solid line) and eluted T-lymphocyte fraction from the WGA-Sepharose column (dashed line). Cells were activated for 72 hrs and thereafter pulsed for 16 hrs with [14C]-thymidine. The vertical bars show mean ± SD of triplicates. Ordinate: (CPM/0.2×10^6 lymphocytes) x 10^{-3}. Abscissa μg Con A/ml and 10^6 cells.

Cell Fractionation on WGA-Sepharose

Using WGA covalently bound to Sepharose 6MB particles T-cells with strong and weak WGA-binding receptors could be separated from each other. Cells with weak WGA-binding receptors passed the column while the more tightly bound cells were eluted with the competitive sugar hapten D-GlcNAc. 43±15% (mean ± SD) of the cells passed the column and 22±11% were eluted with the sugar (eleven experiments) (31). The total cell recovery was 65±19%. Cell viability was not affected by column passage. The cells were furthermore not stimulated by passage over the column.

Figure 4 shows that the two T-cell populations obtained by the column fractionation procedure were functionally different with respect to their ability to respond to Con A. The passed fraction was depleted of Con A responder cells (or accessory cells needed for Con A response) while the eluted cell fraction showed an enhanced Con A response. A similar study performed with La gave the same result (31). In addition it was demonstrated that the response to La at an optimal dose decreased linearly with the dilution of the eluted cells by the passed cells. These results indicate that the increased responsiveness of the eluted cells in the La-system was not due to removal of a suppressor cell. Conversely the lack of La response found in the passed fraction does not reflect removal of a helper cell in the eluted fraction.

Passed and eluted cells were also tested for their ability to act as responder cells in mixed lymphocyte culture (MLC). In several experiments responder cells were enriched in the eluted fraction. In other experiments there was no clear difference in MLC-responsiveness of the two fractions. Further work is needed to resolve this question.

CONCLUSIONS

1. Neuraminidase treatment of lymphocytes from man, rat, mouse and cow uncovers receptors for the carbohydrate binding protein *Helix pomatia* A hemagglutinin (HP) on a fraction of the cells. Untreated lymphocytes do not bind HP.

2. All (or almost all) neuraminidase treated T-lymphocytes from these species are HP-receptor positive. Most neuraminidase treated B-lymphocytes lack HP-receptors. However, in adult human blood 15% of the B-cells are HP^+. In cord blood this fraction is about 80%. Similarly, malignant cells in chronic lymphocytic leukaemia are HP^+ while the malignant cells in multiple myeloma are HP^-. This suggests that HP constitutes a marker for immature B-cells as well as for T-cells.

3. A single cell surface glycoprotein (mol. weight ~150,000) appears to be responsible for almost all HP-binding to neuraminidase treated human or mouse T-cells.

4. Fractionation of neuraminidase treated lymphocytes from all four species on columns charged with HP covalently bound to Sepharose 6MB particles yields three cell fractions: a passed fraction, a fraction eluted with low concentrations of the competitive sugar hapten N-acetyl-D-galactosamine (D-GalNAc) and a fraction eluted with high D-GalNAc concentrations. HP^+ cells are depleted from the passed fraction and enriched in the fraction eluted with high D-GalNAc concentrations. Column fractionation can be performed without selective loss of any of the major lymphocyte categories. Fractionation does not affect viability or functional activity of the cells. Column passage does not activate the lymphocytes.

The HP-column passed cell fraction is enriched in B-cells and depleted of T-cells as assessed by surface markers and functional tests. The fraction eluted with high sugar concentration is almost exclusively made up of T-cells. The fraction eluted with low sugar concentration is a heterogenous mixture containing both T-cells and B-cells which bind more weakly to the HP-column. Cells with Fc-receptors for IgG and K-cell activity are also relatively abundant in this fraction.

5. T-cells from human peripheral blood are heterogeneous with respect to binding of wheat germ agglutinin (WGA). About 25% of the peripheral T-cells have "strong" WGA-binding receptors while the remaining cells have "weak" WGA-binding receptors. This difference in WGA-binding can be used for fractionation of T-cells on WGA-Sepharose 6MB columns. Two cell fractions are obtained: a passed fraction and a fraction eluted with the sugar hapten N-acetyl-D-glucosamine (D-GlcNAc). The eluted fraction is enriched in cells

responding to the mitogens leucoagglutinin from Phaseolus vulgaris (La) and concanavalin A (Con A). The passed fraction is significantly depleted of cells responding to these mitogens.

ACKNOWLEDGEMENTS

Supported by grants no. B 3485-008 and B 2032-047 from the Swedish Natural Science Research Council and United States NIH-NCI Contract no. 1-CB-43883.

REFERENCES

(1) S. Hammarström, Binding of Helix pomatia A hemagglutinin to human erythrocytes and other cells. Influence of multivalent interaction on affinity, Scand. J. Immunol. 2, 53 (1973).

(2) P. Perlmann, H. Perlmann and H. Wigzell, Lymphocyte mediated cytotoxicity in vitro. Induction and inhibition by humoral antibody and nature of effector cells, Transplant. Rev. 13, 91 (1972).

(3) H. Perlmann, P. Perlmann, G. R. Pape and G. Halldén, Purification, fractionation and assay of antibody-dependent lymphocytic effector cells (K-cells) in human blood, Scand. J. Immunol. 5, Suppl. 5, 57 (1976).

(4) A. Böyum, Separation of lymphocytes and erythrocytes by centrifugation, Scand. J. clin. Invest. 97, Suppl. 21, 77 (1968).

(5) S. Hammarström and E. A. Kabat, Purification and characterization of a blood group A reactive hemagglutinin from the snail Helix pomatia and a study of its combining site, Biochemistry 8, 2696 (1969).

(6) S. Hammarström and E. A. Kabat, Studies on specificity and binding properties of the blood group A reactive hemagglutinin from Helix pomatia, Biochemistry 10, 1684 (1971).

(7) S. Hammarström, A. Westöö and I. Björk, Subunit structure of Helix pomatia A hemagglutinin, Scand. J. Immunol. 1, 295 (1972).

(8) S. Hammarström, L. A. Murphy, I. J. Goldstein and M. E. Etzler, Carbohydrate binding specificity of four N-acetyl-D-galactosamine - "specific" lectins: Helix pomatia A hemagglutinin, soy bean agglutinin, lima bean lectin and Dolichos biflorus lectin, Biochemistry 16, 2750 (1977).

(9) M.-L. Dillner, S. Hammarström and P. Perlmann, The lack of mitogenic response of neuraminidase treated and untreated human blood lymphocytes to divalent, hexavalent or insoluble Helix pomatia A hemagglutinin, Exp. Cell Res. 96, 374 (1975).

(10) Y. Nagata and M. M. Burger, Wheat germ agglutinin: molecular characteristics and specificity of sugar binding, J. Biol. Chem. 249, 3116 (1974).

(11) R. H. Rice and M. E. Etzler, Subunit structure of wheat germ agglutinin, Biochem. Biophys. Res. Commun. 59, 414 (1974).

(12) I. J. Goldstein, S. Hammarström and G. Sundblad, Precipitation and carbohydrate-binding specificity studies on wheat germ agglutinin, Biochim. Biophys. Acta 405, 53 (1975).

(13) D. H. Boldt, R. P. Mac Dermott and E.P. Jorolan, Interaction of plant lectins with purified human lymphocyte populations: binding characteristics and kinetics of proliferation, J. Immunol. 114, 1532 (1975).

(14) U. Hellström, M.-L. Dillner, S. Hammarström and P. Perlmann, The interaction of non-mitogenic and mitogenic lectins with T lymphocytes: association of cellular receptor sites, Scand. J. Immunol. 5, 45 (1976).

(15) S. Hammarström, U. Hellström, P. Perlmann and M.-L. Dillner, A new surface marker on T lymphocytes of human peripheral blood, J. Exp. Med. 138, 1270 (1973).

(16) R. H. Swanborg, U. Hellström, H. Perlmann, S. Hammarström and P. Perlmann, A new rat lymphocyte surface marker: characterization and separation of cells with receptors for Helix pomatia A hemagglutinin, Scand. J. Immunol. 6, 235 (1977).

(17) J. J. Marchalonis, R. E. Cone and V. Santer, Enzymic iodination. A probe for accessible surface proteins of normal and neoplastic lymphocytes, Biochem. J. 124, 921 (1971).

(18) U. Hellström, S. Hammarström, M.-L. Dillner, H. Perlmann and P. Perlmann, Fractionation of human blood lymphocytes on Helix pomatia A hemagglutinin coupled to Sepharose beads, Scand. J. Immunol. 5, Suppl. 5, 45 (1976).

(19) U. Hellström, P. Perlmann, E. S. Robertsson and S. Hammarström, Receptors for Helix pomatia A hemagglutinin (HP) on a subpopulation of human B cells, Scand. J. Immunol. (1977) in press.

(20) M. Jondal, G. Holm and H. Wigzell, Surface markers on human T and B lymphocytes. I. A large population of lymphocytes forming non-immune rosettes with sheep red blood cells, J. Exp. Med. 136, 207 (1972).

(21) J. Yata, C. Desgranges, T. Tachibana and G. de-Thé, Separation of human lymphocytes forming spontaneous rosettes with sheep erythrocytes, Biomedicine 19, 475 (1973).

(22) T. Hallberg, B. Gurner and R. R. Coombs, Opsonic adherence of sensitized ox red cells to human lymphocytes as measured by rosette formation, Int. Arch. Allergy 44, 500 (1973).

(23) P. Perlmann, H. Perlmann and H. Müller-Eberhard, Cytolytic lymphocytic cells with complement receptor in human blood. Induction of cytolysis by IgG antibody but not by target cell bound C3, J. Exp. Med. 141, 287 (1975).

(24) T. Weber, Isolation and characterization of a lymphocyte stimulating leucoagglutinin from red kidney beans (Phaseolus vulgaris), Scand. J. clin. Lab. Invest. Suppl. 111, 33 (1969).

(25) A. Forsgren, A. Svedjelund and H. Wigzell, Lymphocyte stimulation by protein A of Staphylococcus aureus, Eur. J. Immunol. 6, 207 (1976).

(26) U. Hellström, H. Mellstedt, P. Perlmann, G. Holm and D. Pettersson, Receptors for Helix pomatia A hemagglutinin on leukaemic lymphocytes from patients with chronic lymphocytic leukaemia (CLL), Clin. Exp. Immunol. 26, 196 (1976).

(27) G. Holm, H. Mellstedt, D. Pettersson and P. Biberfeld, Idiotypic immunoglobulin structures on blood lymphocytes in human plasma cell myeloma, Transpl. Rev. 34, 139 (1977).

(28) P. I. Lobo, F. B. Westerveld and D. A. Horwitz, Identification of two populations of immunoglobulin-bearing lymphocytes in man, J. Immunol. 114, 116 (1975).

(29) R. J. Winchester, S. M. Fu, T. Hoffman and H. G. Kunkel, IgG on lymphocyte surfaces; technical problems and the significance of a third cell population, J. Immunol. 114, 1210 (1975).

(30) P. Perlmann, H. Perlmann, B. Wåhlin and S. Hammarström, Quantitation, fractionation and surface marker analysis of IgG- and IgM-dependent cytolytic lymphocytes (K-cells) in human blood, Immunopathology, VIIth International Symposium (P. Miescher ed) p. 321 (1976) Schwabe, Basel.

(31) U. Hellström, M.-L. Dillner, S. Hammarström and P. Perlmann, Fractionation of human T lymphocytes on wheat germ agglutinin-Sepharose, J. Exp. Med. 144, 1381 (1976).

INVESTIGATION OF THE OLIGOSACCHARIDE HETEROGENEITY OF HUMAN ENTEROKINASE BY AFFINITY CHROMATOGRAPHY ON IMMOBILISED LECTINS

Anthony I. Magee, David A. W. Grant and John Hermon-Taylor

Department of Surgery, St. George's Hospital Medical School, London SW17 ORE

INTRODUCTION

Human enterokinase has been shown to be a glycoprotein enzyme of apparent molecular weight 296,000 containing 57% carbohydrate by weight (1, 2). Fucose, mannose, galactose and N-acetylglucosamine are present in the approximate molar ratio 7 : 4 : 4 : 3; N-acetylgalactosamine, glucose and sialic acid were not detected (1). The present study investigates the peripherally available sugars in the oligosaccharide moiety of human enterokinase, maltase and intestinal aminopeptidase by affinity chromatography on a variety of immobilised lectins.

MATERIALS AND METHODS

Human duodenal fluid, after secretin / pancreozymin stimulation, was obtained during routine clinical testing of pancreatic secretion. Enterokinase was partly purified about 100-fold from the pooled starting material by ion exchange chromatography and affinity chromatography on glycylglycine-p-aminobenzamidine-substituted Sepharose 4B as previously described (1); intestinal aminopeptidase and maltase are present in this partly purified preparation.

Lotus tetragonolobus lectin (LTL) was obtained from Miles, Slough, U.K. Soybean and wheat germ lectins (WGL), Concanavalin A-Sepharose (ConA-Seph.) and Sepharose 4B were from Pharmacia G.B., London. Cyanogen bromide (CNBr) was from Koch-Light, U.K. and DE52-cellulose from Whatman, U.K. Sugars were supplied by Sigma, London and all other chemicals were from BDH. Lectins were coupled to Sepharose 4B using CNBr by the method of Cuatrecasas (5) and chromatography was carried out at 4°C in 10mM sodium phosphate buffer pH 7.6 containing 0.1M NaCl (operational buffer). Elution was with 50 - 250mM sugar in this buffer. Enterokinase, intestinal aminopeptidase and maltase were assayed as previously described (1). Between runs all columns were washed extensively with 50mM sodium phosphate buffer pH 7.6 + 1M NaCl and 50mM sodium citrate buffer pH 3.8 + 1M NaCl and re-equilibrated in operational buffer. The columns could be re-used many times over a period of six months without deterioration.

ConA-Sepharose (substitution ∿ 10mg / g gel) bound only 4.7% of applied enterokinase and 15.6% of the maltase; 96% of the aminopeptidase activity was retained. Enterokinase and maltase were eluted by 100mM α-Methyl mannoside + 1M NaCl in the operational buffer; no aminopeptidase was displaced. Soybean lectin-Sepharose (substitution ∿ 1.2mg/g) did not bind any of the enzymes under the conditions tested.

Figure 1 shows the binding of the three enzymes to LTL-Sepharose (substitution ∿ 1.8mg/g). 58% of the enterokinase and 32% of the maltase applied bound and was subsequently eluted with 250mM Fucose. Recovery of enterokinase was about 80%. No aminopeptidase was retained. Unbound and bound enterokinase fractions were concentrated separately on small columns of DEAE-cellulose and re-applied to the lectin column. 91% of the unbound fraction eluted straight through the column; all the bound fraction was retained and eluted with 250mM Fucose.

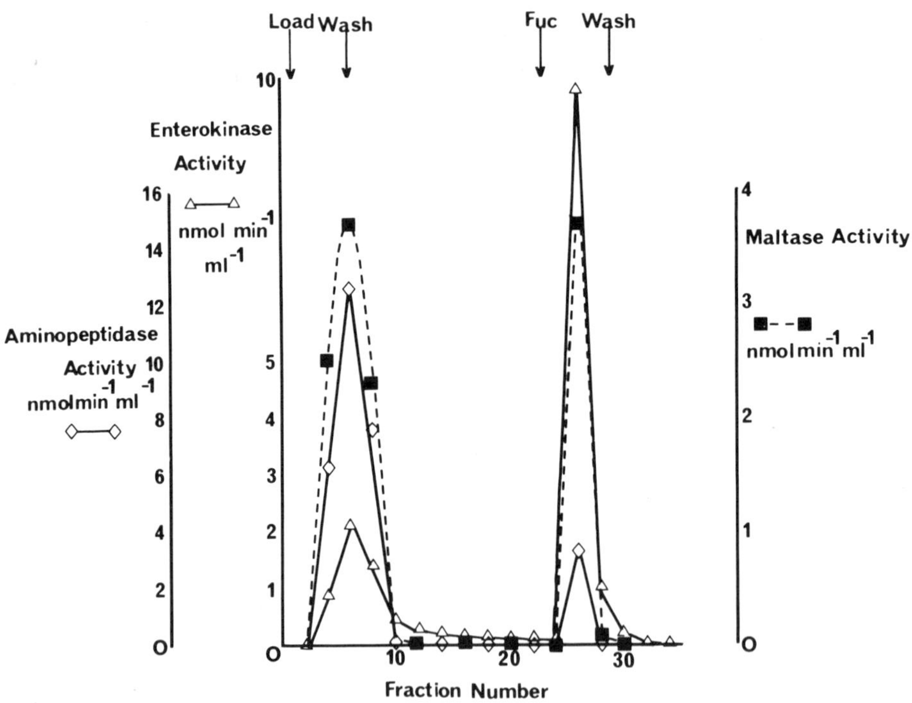

Fig. 1. Separation of human duodenal enzymes on LTL-Sepharose

These two distinct enterokinase populations distinguished by the peripheral availability of fucose, were now tested separately on N-acetylglucosamine (GlcNAc) specific wheat germ lectin (WGL-Sepharose). Enterokinase containing fractions were concentrated and re-applied as before. Each population was consistently subdivided into two fractions as summarised in Table 1.

TABLE 1 Separation of Human Intestinal Enzymes on LTL-Seph. and WGL-Seph. Adsorbants

Binding To:		% Total Activity		
LTL-Seph.	WGL-Seph.	Entero-Kinase	Amino-peptidase	Maltase
+	-	55	0	32
-	-	22	100	68
-	+	20	0	0
+	+	3	0	0

The study was repeated in a similar manner with another batch of enzyme reversing the order of the separations so that chromatography on WGL-Sepharose was performed first (Fig. 2.)

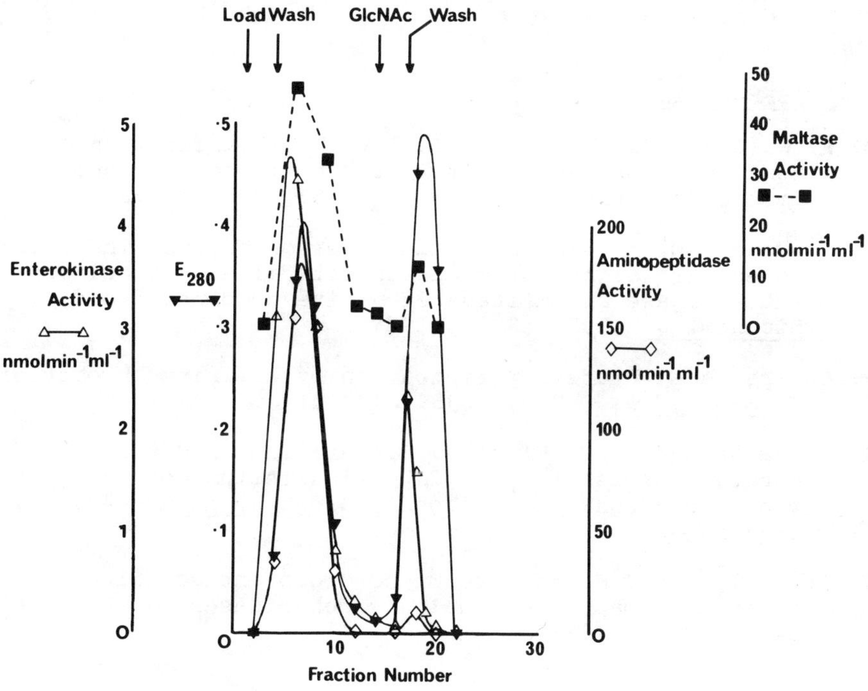

Fig. 2. Separation of human duodenal enzymes on WGL-Sepharose

There was close agreement in the proportions of enterokinase in each of the four fractions.

It appears that there are at least four types of human enterokinase distinguished by differences in their peripherally available sugars. The major sub-population (55%) has terminal non-reducing fucose residues only; 20% of the enzyme molecules have terminally available GlcNac only, while 3% have both residues available for binding. 22% of the enzyme molecules bound to none of the lectins tested. Human intestinal aminopeptidase has terminal glucose or mannose similar to the porcine enzyme (4) and maltase has peripheral fucose, glucose or mannose in agreement with the previously described carbohydrate content of this enzyme (5). These differences in peripherally available sugars on enzymes synthesised by the intestinal epithelial cells, exposed to a similar environment in the duodenal lumen, and prepared under identical conditions, are more likely to reflect differences in enzymic glycosylation during biosynthesis than susceptibility to oligosaccharride degradation.

REFERENCES

(1) D. A. W. Grant and J. Hermon-Taylor, The Purification of Human Enterokinase by Affinity Chromatography and Immunoadsorption, Biochem. J. 155, 243 (1976)

(2) A. I. Magee, D. A. W. Grant and J. Hermon-Taylor, The apparent molecular weights of human intestinal aminopeptidase, enterokinase and maltase in native duodenal fluid, Biochem. J. 165, 583 (1977)

(3) P. Cuatrecasas, Protein purification by affinity chromatography J. Biol. Chem. 245, 3059 (1970)

(4) R. J. Barns and R. G. Elmslie, Preparation of Porcine Enteropeptidase free of Intestinal Aminopeptidase activity with ConA-Sepharose, Biochim. Biophys. Acta 445, 815 (1976)

(5) J. J. Kelly and D.H. Alpers, Blood group antigenicity of purified human intestinal disaccharidases, J. Biol. Chem. 248, 8216 (1973)

ON THE BINDING SPECIFICITY OF WHEAT GERM LECTIN-SEPHAROSE TOWARDS CARBOHYDRATE RESIDUES IN GLYCOPROTEINS

Per Vretblad and Rolf Hjorth

Pharmacia Fine Chemicals AB, Box 175, S-75104 Uppsala, Sweden

INTRODUCTION

Lectins are proteins which have proved extremely useful as analytical tools in biochemistry and cell biology. Their usefulness depends on their property to interact specifically with one or a very limited number of monosaccharides.

Wheat germ contains one lectin which has been purified and studied in several laboratories. The molecular weight of this protein is about 35 000 (1) and it consists of two polypeptide chains of similar size, each of which carries two binding sites (1,2). Many authors agree, from inhibition studies (3-5), equilibrium dialysis experiments (1,2) and fluorescence studies (6,7) that wheat germ lectin (WGL) is specific for N-acetyl-D-glucosamine (GlcN) and that it binds more efficiently to the dimer and trimer of this sugar. The combining site can accommodate about three monosaccharide residues (4,7). According to Allen et al. (4) the binding site contains three or four subsites with slightly different structural requirements. The lectin can bind to GlcN residues in a macromolecule which are not in a terminal position (5).

Recently, however, some papers have appeared, the results of which indicate that N-acetylneuraminic acid (NANA) can also bind to WGL (8-12). Association constants for the interaction between the free lectin and GlcNAc or NANA have been determined under various conditions. The values (in M^{-1}) are about 10^3 (refs. 1,7,12) for GlcNAc, in the range of 10^4 for its dimer and trimer (1,6,7) and 560 for NANA (12).

Immobilized WGL has for some years been used as an affinity chromatography medium for the fractionation of glycoproteins, membrane components and cells. It is therefore of interest to establish the binding specificity of the immobilized lectin. Most of the work on the specificity of WGL has been carried out with the protein in free solution and with low molecular weight ligands, and the results of these studies may not be applicable to affinity chromatography systems. If the association constants given above were also true for the interaction between immobilized WGL and sugar residues in a glycoprotein, it would not be possible to obtain significant binding of glycoproteins containing single GlcN or NANA residues. A K_{ass} of about 10^4 is necessary in affinity chromatography (13). However, in glycopeptides both GlcN and NANA occur in varying amounts and the effects of interaction between different glycopeptide chains and of steric blocking are difficult to predict.

We have studied the interaction between Wheat germ Lectin-Sepharose[R] 6MB and a range of glycoproteins of different sugar composition.

EXPERIMENTAL

Fresh human serum was obtained from the Academic Hospital,Uppsala, Sweden. Bovine submaxillary mucin, ovomucoid, ovalbumin, avidin A, NANA and neuraminidase(*Clostridium perfringens*) were products of Sigma. GlcN was from Fluka, glucose oxidase from Worthington and rabbit antisera to human proteins from Hoechst. Wheat germ Lectin-Sepharose 6MB (WGL-Sepharose) and Polyacrylamide Gradient Gels PAA 4/30 were products of Pharmacia Fine Chemicals. Hemopexin was purified as described elsewhere(14).

In the standard experiment, WGL-Sepharose (4-5 ml) was equilibrated with 0.01 M sodium phosphate buffer, containing 0.02% sodium azide, pH 7.0 (starting buffer) in a K 16/20 chromatographic column. A protein sample (1 mg/ml, 1-2 ml) in starting buffer or human serum (2 ml), dialyzed against starting buffer, was applied to the column at a flow rate of 6 ml/h. The column was developed with starting buffer to remove those proteins which did not interact with the gel. Some experiments were also run in 0.05 M Na phosphate buffer containing 0.2 M NaCl and 0.02% NaN_3, pH 7.0.

Several different procedures were used to remove bound proteins. Pulses of GlcN (2.5 ml of starting buffer containing 100 mg/ml) or NANA (2.5 ml of buffer containing 45 mg/ml) were applied. In the latter case, the solution was adjusted to pH 7 with sodium hydroxide. Gradients of NaCl (0-2 M) in starting buffer were also used. In some experiments, 4 ml of starting buffer containing neuraminidase (1.5-3 U) was incubated on the column for 5 h at 23° C. The column was then developed with starting buffer. Different gel columns were used for each model protein. The fractions desorbed from the lectin columns in the serum experiments were analyzed by immunodiffusion and by electrophoresis at pH 8.35 in polyacrylamide gradient gels.

RESULTS AND DISCUSSION

Three of the model glycoproteins, namely ovalbumin, glucose oxidase, and avidin, contain GlcN but no NANA. Ovomucoid contains much GlcN (about 15% per weight) but also traces of NANA. Of these, only avidin and ovomucoid bound to WGL-Sepharose under the conditions used. The binding thus is correlated to the GlcN content of the proteins, which is about 4.3% in avidin, 2% in glucose oxidase, and 1.2% in ovalbumin. Obviously, however, the content of sugar residue per unit weight of glycoprotein cannot be the single determinant factor for binding since the carbohydrate residues must also be accessible to the lectin.

Avidin could be released by a salt gradient (at 0.35 M NaCl) but not by GlcN, which indicates that electrostatic interaction might be the most important element in the binding. However, this experiment was carried out at low ionic strength which strongly favours electrostatic interactions. When avidin was instead applied in the presence of 0.05 M Na phosphate, 0.2 M NaCl, it could be partly released by GlcN but not by galactose. These results show that the nature of the interaction between lectin and glycoprotein is dependent on the experimental conditions and that it is difficult to distinguish one type of interaction from another. Ovomucoid showed the same pattern as avidin in that it was partly desorbed by GlcN when applied in 0.2 M salt but not when applied in 0.01 M buffer.

Bovine submaxillary mucin, which is a protein with an extremely high content of NANA (about 30% by weight) and rather few GlcN residues, did not bind to WGL-Sepharose. It is known that NANA residues are exposed on the surface of the protein, and this result therefore suggests that interaction between the lectin and NANA alone is not sufficient for binding.

Hemopexin is a protein which contains about 7% of each of NANA and GlcN. This protein bound well to WGL-Sepharose and could be eluted by GlcN, whereas it was not eluted by salt or by NANA. Neither could hemopexin be desorbed from WGL-Sepharose by neuraminidase. However, desialyzed hemopexin bound to WGL-Sepharose and could be desorbed by GlcN. Thus, in the case of hemopexin there is a clear indication that GlcN, and not NANA, is responsible for the interaction with the lectin.

Human serum comprises a much more complex model system. Fractionation of human serum on WGL-Sepharose has been reported earlier (15). It was shown that apart from hemopexin, α_1-acid glycoprotein, ceruloplasmin, haptoglobin, α_2-macroglobulin, antithrombin III, α_1-antichymotrypsin, and small amounts of IgG, IgA and IgM were bound in the presence of 0.2 M NaCl and could be eluted by GlcN. This experiment was now repeated in the presence of 0.01 M Na phosphate and gave the same result with the exceptions that also α_2-HS-glycoprotein and traces of α_1-antitrypsin were reversibly bound. The composition of the fraction desorbed by GlcN is shown in Fig. 1.

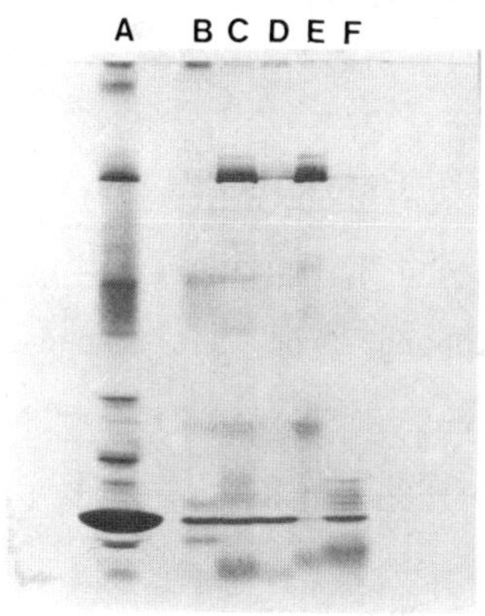

Fig. 1. Gradient gel electrophoresis of human serum proteins reversibly bound to Wheat germ Lectin-Sepharose 6MB in 0.01 M Na phosphate buffer, pH 7.0. A, unfractionated serum. The other samples are fractions desorbed with the following agents: B = NaCl gradient; C = N-acetyl-D-glucosamine (GlcN); D = N-acetylneuraminic acid (NANA); E = GlcN after desorption with NANA; F = neuraminidase.

Similar experiments were performed in which NANA was used as desorbing agent. In the presence of 0.01 M Na phosphate NANA could desorb the same proteins as GlcN except for antithrombin III, ceruloplasmin and hemopexin. However, the amounts were small and the fraction contained a substantial amount of albumin, which is not a glycoprotein. GlcN is clearly a more effective desorbing agent than NANA since about an equal amount of protein (A_{280} units) could be eluted by GlcN after NANA desorption (Fig. 1). In this GlcN fraction, all ceruloplasmin and hemopexin were recovered. When serum was applied to a WGL-Sepharose column in 0.2 M NaCl and desorption was carried out with NANA, only a few proteins could be detected in small amounts, namely: IgG, IgA, haptoglobin, albumin and α_1-antitrypsin. Also, when serum proteins were bound to the gel in 0.01 M Na phosphate and desorption was carried out by a salt gradient, the protein pattern was similar to that obtained by NANA desorption but contained a few more proteins, e.g. α_1-antitrypsin. This indicates that NANA desorbs proteins in a non-specific manner, acting mainly as an electrolyte. The fact that WGL is positively charged under the experimental conditions (pI = 8.5) supports this hypothesis.

Neuraminidase treatment of WGL-Sepharose columns releases small amounts of most bound serum proteins. These proteins were identified by immunodiffusion experiments. Electrophoresis patterns (Fig. 1) are difficult to interpret since the molecular size and conformation of proteins will be altered by desialylation.

In summary the experimental results indicate that GlcN is the sugar residue which is responsible for lectin binding. Most glycoproteins contain GlcN but all of these do not bind to WGL-Sepharose. It is therefore interesting to speculate on the more precise structural requirements of the binding site of the immobilized lectin. Transferrin does not bind to WGL, whereas α_1-acid glycoprotein does. The structures of their glycopeptides have been reported (16,17) and they are quite similar. However, there is one structural difference between the two glycopeptides. The sequence GlcN-Man-GlcN only appears in α_1-acid glycoprotein and would fit well into the active site of the lectin according to the subsite theory of Allen et al. (4). Their results indicated that two GlcN separated by a different residue - in their case muramic acid - would fit even better in the active site than a GlcN-GlcN sequence. The GlcN-Man-GlcN sequence is reported to be present also in ovomucoid, which binds well to WGL; it is absent from α_1-antitrypsin and probably from ovalbumin, which do not bind.

REFERENCES

1. Y. Nagata and M. M. Burger. J. Biol. Chem. 249, 3116 (1974).

2. J. P. Privat, F. Delmotte and M. Monsigny, FEBS Lett. 46, 224 (1974).

3. M. M. Burger and A. R. Goldberg, Proc. Natl. Acad Sci. U.S. 57, 359 (1967).

4. A. K. Allen, A. Neuberger and N. Sharon, Biochem. J. 131, 155 (1973).

5. I. J. Goldstein, S. Hammarström and G. Sundblad, Biochim. Biophys. Acta 405, 53 (1975).

6. R. Lotan and N. Sharon, Biochem. Biophys. Res. Commun. 55, 1340 (1973).

7. J. P. Privat, F. Delmotte, G. Mialonier, P. Bouchard and M. Monsigny, Eur. J. Biochem. 47, 5 (1974).

8. P. J. Greenaway and D. LeVine, Nature New Biol. 241, 191 (1973).

9. L. Winquist, L. Eriksson, G. Dallner and B. Ersson, Biochem. Biophys. Res. Commun. 68, 1020 (1976).

10. W. R. Redwood and T. G. Polefka, Biochim. Biophys. Acta 455, 631 (1976).

11. D. H. Boldt, S. F. Speckart, R. L. Richards and C. R. Alving, Biochem. Biophys. Res. Commun. 74, 208 (1977).

12. F. Jordan, E. Bassett and W. R. Redwood, Biochem. Biophys. Res. Commun. 75, 1015 (1977).

13. C. R. Lowe, Int. J. Biochemistry 8, 177 (1977).

14. P. Vretblad and R. Hjorth, Biochem. J., in the press.

15. R. Hjorth and P. Vretblad in Advances in the Chromatographic Fractionation of Macromolecules (Epton, R., ed.,) E. Horwood, Chichester, U.K., in the press.

16. G. Spik, B. Bayard, B. Fournet, G. Strecker, S. Bouquelet and J. Montreuil, FEBS Lett. 50, 293 (1975).

17. R. Kornfeld and S. Kornfeld, Annu. Rev. Biochem. 45, 217 (1976).

BINDING SPECIFICITY AND PURIFICATION OF *MEDICAGO SATIVA* LECTIN

Werner Kamberger

Institut für Mikrobiologie der Universität Erlangen-Nürnberg, Friedrichstr. 33, D-8520 Erlangen GFR

INTRODUCTION

Plant lectins are a group of proteins, which share the ability to selectively bind to specific carbohydrate receptors. Concerning the physiological function, it is suggested that an interaction between legume lectins and polysaccharides (PS) on the surface of Rhizobium cells either of capsular origin or lipopolysaccharide (LPS), may account for the specificity expressed between Rhizobia and their host plants in the initiation of the root nodule symbiosis (Ref. 3). The present paper describes the purification of a new galactose binding phytolectin on three different affinity adsorbents. The interaction of this lectin with the exopolysaccharides (EPS) of a Rhizobium meliloti strain, which produces nitrogen fixing root nodules on Medicago sativa plants, is demonstrated.

METHODS

Preparation of M. sativa extracts. A fine ground powder (50g) of M. sativa seeds was mixed with 200 ml phosphate buffer (0.05M) containing sodium ascorbate (0.05M), NaCl (0.15M) and $MgSO_4$ (0.001M). Washed, insoluble polyvinylpyrrolidone (10g) was added and the pH adjusted to 4.0. The slurry was stirred for 4h at 4°C, centrifuged at 25.000xg for 1h and the pH of the clear supernatant adjusted to 7.0 with 1N NaOH. This seed extract was dialysed against phosphate buffered saline, pH 7.2 (PBS), lyophilized, tested for agglutinating activity and stored at -20°C.

Haemagglutination and haemagglutination-inhibition assays. Rabbit erythrocytes were washed three times in saline and diluted to a suspension of 2%. Aliquots of this suspension (25µl) were added to equal volumes of serial dilutions of the seed extract in saline. The stock solution contained 5% seed extract. In the inhibition test, the serial dilutions of the extract were performed in 0.05M solutions of various sugars. After an incubation time of 20 min at $30^{o}C$ the erythrocytes were added.

Determination of the sugar components of the R.meliloti 2011 EPS. The purified EPS were treated with 1N H_2SO_4 at $100^{o}C$ for 6h and the hydrolysate analysed by thin-layer chromatography (TLC) on aluminium sheets (20x20cm) precoated with cellulose F (0.1mm). As developing solvent pyridine-butanol-water (4:6:3) was used.

Synthesis of biospecific adsorbents. N-caproyl-D-galactosamine-Sepharose was prepared by coupling of D-galactosamine•HCl to CH-Sepharose 4B according to the method of Allen and Neuberger (Ref. 2). N-acetyl-D-galactosamine-Sepharose was synthesized by coupling of N-acetyl-D-galactosamine to Epoxy-activated Sepharose 6B. The coupling procedure was basically that described by Vretblad (Ref. 4) with the exception that the pH of the coupling solution was 10.5. Defined acid-hydrolysed agarose was prepared by a 2h treatment of Sepharose 6B with 0.2M HCl at $50^{o}C$. By this procedure certain types of D-galactose oligosaccharide chains are made accessible to the binding of the lectin (Ref. 1).

RESULTS AND DISCUSSION

The EPS of R. meliloti 2011 were composed only of acidic PS. No aminosugars were detected. The complete composition of the EPS is given in Table 1. The relative amount of the sugar components was calculated densitometrically from the TLC.

Unlike most other isolated phytolectins M. sativa lectin is relatively unstable. Sodium ascorbate and insoluble polyvinylpyrrolidone was added during the extraction procedure to protect the lectin against autoxidation.A fractionted $(NH_4)_2SO_4$ precipitation resulted in a complete inactivation of the haemagglutinating activity.

M. sativa seed extract (200mg in 2ml PBS) was applied to a

TABLE 1 Composition of R. meliloti 2011 EPS

Glucose	(50% ± 10%)
Galactose	(15% ± 3%)
Mannose	(6% ± 2%)
Xylose	(8% ± 2%)
Uronic acids	(20% ± 5%)

column (80x9mm) containing N-caproyl-D-galactosamine-Sepharose, N-acetyl-D-galactosamine-Sepharose or acid-hydrolysed agarose. At 4°C the gel was washed with PBS until the A_{280} absorption of the effluent reached zero. Fractions (2ml) were collected and the lectin eluted with 0.2M galactose in PBS (Fig. 1). The peak fraction was dialysed against 1:100 diluted PBS and lyophilized. The void volume peak was collected, lyophilized and assayed for haemagglutinating activity. All agglutinating activity was found in the biospecific desorbed material. The purification factor was about the same for all three affinity adsorbents and reached 512 according to the agglutination titer.

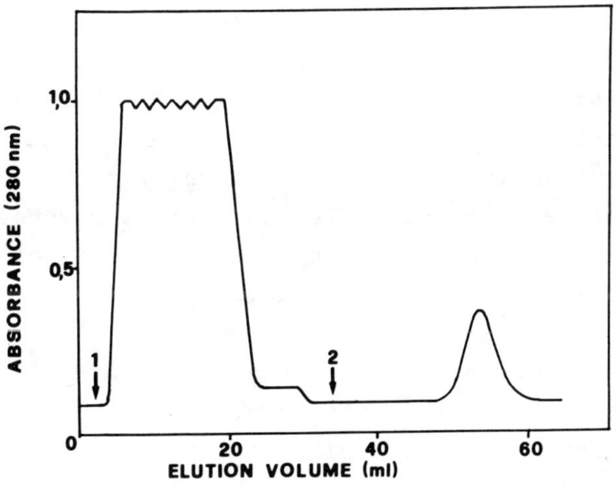

Fig. 1. Affinity chromatography of M. sativa lectin on N-caproyl-D-galactosamine-Sepharose. Arrow 1: Application of the sample in PBS. Arrow 2: Start of elution with 0.2M galactose in PBS.

Lectin elution by a 0.01M glycine•HCl buffer pH 2.0 (Fig. 2) resulted in a purification factor of 64 to 128 according to the agglutination titer. The lower activity could be due to a more specific

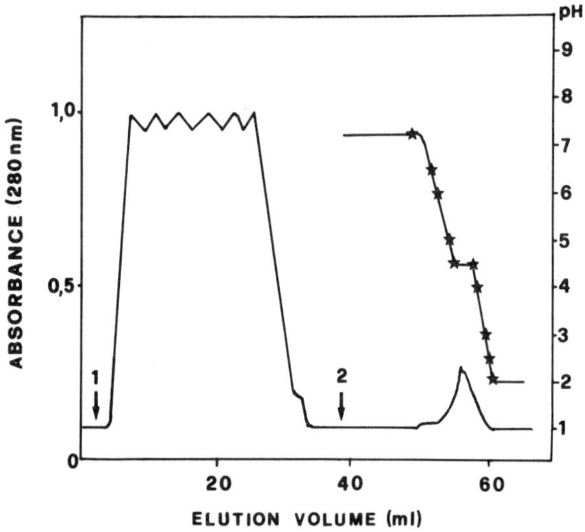

Fig. 2. Affinity chromatography of M. sativa lectin on acid treated Sepharose 6B. Arrow 1: Application of the sample in PBS. Arrow 2: Start of elution with glycine•HCl buffer pH 2.0. The line marked with * shows the pH of the fractions.

desorption by galactose. A partly irreversible inactivation of the lectin by the low pH buffer, however, could not be ruled out.

A direct interaction between purified D-galactose binding M. sativa lectin and D-galactose containing capsular material was indicated by agglutination of cells of R. meliloti 2011. These results provide further evidence that legume lectins may have a function in the recognition mechanism leading to host specificity in the root nodule symbiosis of legumes with Rhizobia.

The presented procedure offers a route for the isolation of a new, relatively unstable lectin in a single step by affinity chromatography on biospecific adsorbents, thus affording sufficient material for further characterization of the chemical and physiological properties.

REFERENCES

(1) H.J.Allen and E.A.Johnson, Carbohydr. Res. 50, 121 (1976)
(2) A.K.Allen and A.Neuberger, FEBS Let. 50, 362 (1974)
(3) F.B.Dazzo and D.Hubbel, Appl. Microbiol. 30, 1017 (1975)
(4) P.Vretblad, Biochim. Biophys. Acta, 434, 169 (1976)

CROSS-LINKED ERYTHROCYTE MEMBRANE COLUMNS AS A TOOL FOR AFFINITY CHROMATOGRAPHY OF LECTINS

Francois Lawny, Marie-Hélène Bot, Edouard Lentwojt and Emile Segard

Institut de Technologie des Surfaces Actives, 60206 Compiègne, France

INTRODUCTION

The increasing use of lectins in biological and biochemical research has promted the development of many methods for their purification. The most widely used method is affinity chromatography, using the appropriate saccharide as a ligand according to the lectin's specificity. The support used can be either a polysaccharide be (Sephadex (1), Sepharose (2)), or polysaccharide itself (5,6), or glycoproteins (7), or synthetically prepared glycosides (8).
Most of these procedures are time-consuming and delicate to carry out, and sometimes the final support exhibits a low affinity and a low binding capacity. The purification of lectins would be easier if a general class of natural insoluble affinity supports were available. Since lectins are generally identified by their ability to agglutinate erythrocytes, we have tried to prepare a convenient affinity support with red cell membranes.
Recently, formaldehyde treated erythrocytes (9) were used as affinity adsorbants for lectins (10), while Avrameas et al. (11) used stromata entrapped in an insoluble gel of bovine serum alb ine derivative (B.S.A. co-crosslinked with glutaraldehyde). Total erythrocytes were used too after glutaraldehyde treatment with polyacrylamide beds to the isolation of lectins from Ricinus and different fungi (12).
The present paper deals with a method to obtain cross-linked stromata, packed in a column available as an insoluble affinity support, so that many different types of lectins are easily and quickly prepared, for example from peanut (P.N.A.), soybean (S.B.A.), Ulex europeus (U.E.A.), Sophora japonica (S.J.A.), Dolichos biflorus (D.B.A.) or wheat germ (W.G.A.).
Moreover, a packed bed of stromata prepared by this way is stable for long time and suitable for more than a hundred chromatographic runs. It is a good tool for a qualitative and precisely quantitative study of the binding site of lectins, the competitive elutions, the inhibitory power of their hapten sugar and also the structure of the membrane's antigens.

METHODS

Human erythrocytes were washed with 0.9 % saline phosphate buffer (S.P.B.) and a 3 % (v/v) suspension of human erythrocytes was made in the same buffer. Each 50 µl of a twofold serial dilution of a lectin solution and an equal volume of the erythrocyte suspension was incubated for 1 h at room temperature and the highest dilution that gave agglutination determined the titer.

Extraction from seeds

All seeds (100 g) were ground and defatted with ligroin (500 ml) in a Soxhlet apparatus.
The lectin was then extracted in a saline solution (NaCl 0.9 %) for 3 hours and precipitated by $(NH_4)_2\ SO_4$ at 4° C (470 g/l). After centrifugation the insoluble fraction was suspended in S.P.B. and stirred at room temperature for 3 hours. The suspension is then clarified by centrifugation at 15.000 g for 20 min and the agglutinating activity of the supernatant is controlled.

Preparation of the affinity columns - General method

As indicated by Dr. Faure (CNTS - Orsay, France), concentrated erythrocytes are dialysed in an artificial kidney against a S.P.B. with stepwise decreasing osmolarity (120 mosm, 60 mosm, 30 mosm, 0 mosm). If necessary, the stromata can be enzymatically treated, (by Neuraminidase, for instance) and finally centrifuged at 40.000 g for 30 minutes to obtain a dense white paste.
The stromata are pooled in a beaker, and 0.7 ml of 25 % aqueous glutaraldehyde solution is mixed with each 10 ml fraction of paste. They are then allowed to stand for 4 h at -30°C according to the method described by Gellf et al. (13) for the immobilization of enzymes. After thawing, the resulting sponge-like block is ground with a cell grinder and filtered on a Bücher funnel. The stromata are washed successively with 2 M NaCl, water, saturated glycine solution (twice), water, a concentrated solution of the appropriate hapten sugar and finally with S.P.B.

Adsorption-Elution

The adsorption of the lectin can be carried out either at 20°C or at 4°C at a flow rate allowing for a 20 min retention time of the lectin solution in the column. The column is washed with S.P.B. (as much as 50 times the volume of the injected sample can be applied).

RESULTS

The purity of lectins was checked by electrophoresis on polyacrylamide gels with and without S.D.S. The molecular weight measurements

LECTIN	Measured Molecular Weights	Expected Molecular Weights	Eluting Sugar	Enzymatic Treatment
P.N.A.	130.000	120.000 (17)	D- Galactose 0.1 M	Neuraminidase
S.B.A.	125.000	110.000 (18)	D- Galactose 0.1 M	Trypsin + $Mn^{++} 10^{-4}$M
U.E.A.	160.000	170.000 (18)	L- Fucose 0.05 M	-
S.J.A.	150.000	-	D - Galactose 0.1 M	-
D.B.A.	140.000	140.000 (19)	DGal NAc 0.005 M	-
W.G.A.	35.000	26.000 (18)	$(GlucNAc)_2$ 0.05 M	-

were made by analytical ultracentrifugation (centriscan M.S.E.). The characteristics of our lectins were found to be very similar to the expected values, as can be seen in the table.
All the lectins were easily bound by the support, and the adsorption was checked by controlling the lack of hemagglutinating activity at the outlet of the column. The amount of adsorbed lectin depends markedly on the preliminary enzymatic treatment which can be necessary of helpful (for example, neuraminidase treatment for P.N.A. adsorption), in relation to the number of receptors on the red cell membrane. Starting with 6 l of concentrated red cells it was possible to adsorb 150 mg of P.N.A., 150 mg of S.B.A., 100 mg of W.G.A., 50 mg of D.B.A. and U.E.A. and 40 mg of S.J.A., respectively.
The support itself can be used repeatedly during 6 months without loosing its binding capacity. Curiously bacterial contamination does not occur within this time, but the usual antiseptics can be used without damage anyhow.

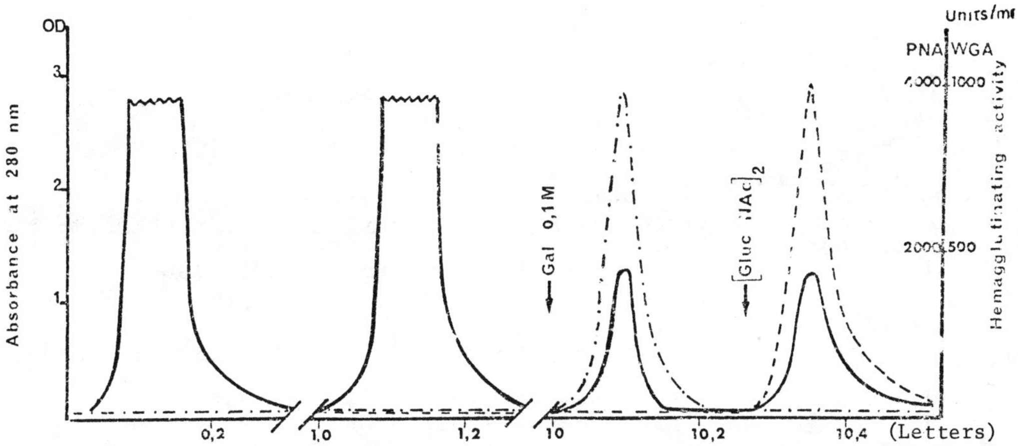

The competitive elution is performed by injecting the hapten sugar. The concentration of the saccharide can be pre-determined; the concentration used must be able to completely inhibit the agglutination of the erythrocytes at a lectin concentration of 1 mg/ml. The major part of the adsorbed activity is eluted in a single peak. The yield of the elution is only 10 % for the WGA when N-Acetyl.β. D.glucosamine is used. The elution is complete with di-N-Acetylchitobiose, however.

DISCUSSION

It is possible to purify many different lectins by using a single column: two blood group specific lectins (D.B.A., U.E.A.), two lectins known to require preliminary enzymatic treatment (P.N.A., S.B.A.) and two lectins that do not require either a specific blood group or an enzymatic treatment of erythrocytes (W.G.A., S.J.A.) for instance. Binding of S.B.A. occurs even with non-trypsinated erythrocytes; this is in accordance with results of Sharon and Lis

(15). Glutaraldehyde is perhaps able to modify the structure or the accessibility of some sites on the erythrocyte membranes. Blood group typing of glutaraldehyde treated erythrocytes has shown that only the Rhesus antigen is modified. A, B, O, M and N antigens are not affected by the treatment according to the expected action of the glutaraldehyde (4). Although the specificity of that type of support is very low, we can see that it is possible to separate a mixture of two lectins by specific elution with two different hapten sugars. The elution order is directly related to the affinity constant of the hapten as shown for W.G.A. with N-Acetylchitobiose.
However, many other applications of the support as natural immunoadsorbant are very interesting (purification of natural Anti A, Anti B, Anti Rh etc.) and will be studied in further experiments.

Aknowledgments: We thank Dr. Faure, Centre National de Transfusion Sanguine Orsay, France, for his precious help.

REFERENCES

(1) B.B.L. Agrawal, I.J. Goldstein, Biochem. Biophys. Acta 147, 262 (1967).
(2) J.A. Gordon, S. Blumberg, H. Lis and N. Sharon, F.E.B.S. Lett. 24. 193 (1973).
(4) P. Mosan, G. Buzo, H. Mazarguil, in "Anal. and Control of Immob. Enzym. Systems", p. 275, by D. Thomas and J.P. Kernevez, North Holland Amsterdam Publishers (1975).
(5) T. Terao, T. Trimura, T. Osawa, Hoppe-Seyler's Z. Physiol. Chem. 356, 1685 (1975).
(6) M.O.J. Olson, I.E. Liener, Biochemistry 6, 105 (1967).
(7) D. Le Vine, M.J. Kaplan, P.J. Greenoway, Biochem. J. 129, 847 (1972).
(8) M.E. Ravestin, A. Obrenovitch and M. Monsigny, F.E.B.S. Lett. 40, 62 (1974).
(9) B. Fauconnier, Ann. Inst. Past. 96, 110 (1959).
(10) R.W. Reitherman, S.D. Rosen, S.H. Barondes, Nature 248, 599 (1974).
(11) S. Avrameas, B. Guilbert, Biochimie 53, 603 (1971).
(12) G. Betail, M. Coulet, J. Guillot, Scandaria To, C.R. Soc. Biol. (Clermont Ferrand), 1304 (1971).
(13) G. Gellf, D. Thomas, G. Broun, Biotech. and Bioeng. 16, 315 (1974).
(17) R. Lotan, E. Stutelsky, D. Danon, N. Sharon, J. of Biol. Chem. 250, 8518 (1975).
(18) N. Sharon, H. Lis, Science, 4035, 949 (1972).
(19) M.E. Etzler and E. Kabat, Biochemistry 9, 869 (1970).

AFFINITY CHROMATOGRAPHY OF MEMBRANE VESICLES

G. Brunner*, E. Ferber*, K. Resch** and J. Golecki***

*Max-Planck-Institut für Immunbiologie, D-78 Freiburg, GFR
**Institut für Immunologie der Universität, D-69, Heidelberg, GFR
***Institut für Mikrobiologie der Universität, D-78 Freiburg, GFR

ABSTRACT

Affinity chromatography on ConA-Sepharose was used in the subfractionation of a thymocyte membrane fraction consisting of vesicles. In contrast to results obtained with other systems, the corresponding inhibitor alone was unable to elute the membrane vesicles; shearing forces were found to be necessary.

INTRODUCTION

We postulated (1,2) that for correct signal transformation and signal transduction - after the binding of a signal molecule of the environment to the cell - not only the receptor molecule in the cell membrane, but also the molecules neighbouring the receptor molecule (the whole receptor area), respond to the signal. To study this problem we tried to subfractionate plasma membrane vesicles of thymocytes using affinity chromatography (3,4) on ConA-Sepharose.
This paper, a continuation of the work presented in (3,4), discusses some methodological aspects. Several publications (6,7,2) will be compared critically. Concanavalin A (ConA) works as a mitogen on thymocytes by binding to sugar residues of membrane components. This binding can be competitively inhibited by α-methylmannoside (αMM).

METHODS

A microsomal membrane fraction of thymocytes was obtained as described (5). This fraction consists of membrane vesicles (70-150 nm in diameter) and up to 30% of the total protein is ribosomal protein. In some experiments, the microsomal fraction was further purified by a discontinuous sucrose gradient. The sucrose was removed by washing the vesicles twice for $1 \times 10^7 g_{av}$ x min at 4°C in the separation buffer (0.14 M KCl, 10 mM HEPES (2,4- (hydroxylethyl)-1-piperazinylethan sulfonic acid), 1 mM $MgCl_2$, 1 mM $CaCl_2$, 1 mM $MnCl_2$). This buffer was used for the following separation procedure which was performed at 24°C as described (4).

RESULTS

It is possible to subfractionate a microsomal or a plasma membrane

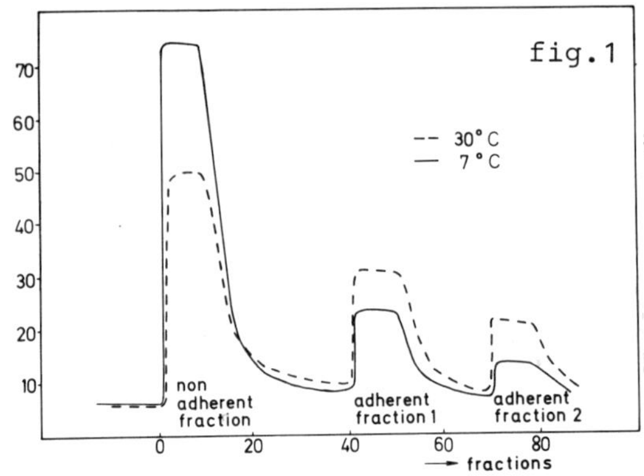

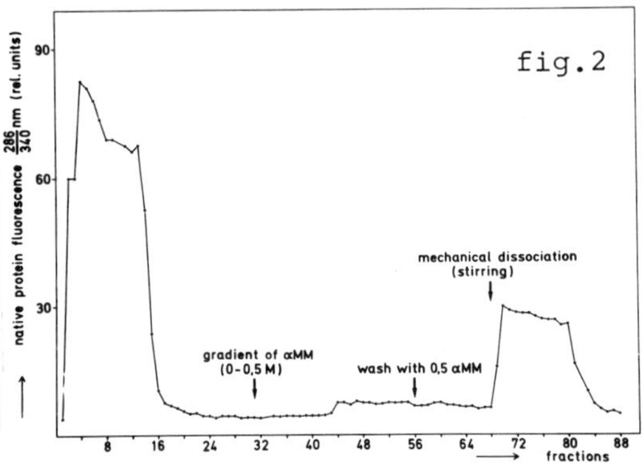

Fig. 1. The effect of temperature on the binding of membrane vesicles. The elution profile was measured by the O.D. at 280 nm and the separation was performed as described in (4).

Fig. 2. Dissociation of bound membrane vesicles (24°C). 5 mg microsomal protein were added to 60 ml ConA-Sepharose (50% (w/w) suspension). After elution of the non-bound membrane fraction as described (4), the ConA-Sepharose beads were rinsed with a linear gradient of αMM, ranging from 0 to 0.5 M without stirring. After that the beads were stirred.

Fig. 3. Effect of stirring on the binding of membrane vesicles (24°C). A, normal dissociation procedure (4) B, stirring during all procedure.

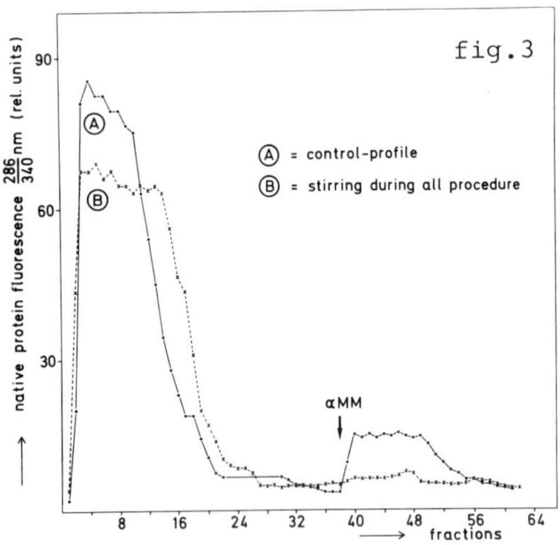

fraction into one fraction (60-80% of the total membrane protein) which does not bind and into fractions which bind to ConA-Sepharose (15 to 25% of total protein being found in "adherent fraction 1" and 5-10% in "adherent fraction 2") (fig.1). If enough membrane material is applied, a third and a fourth adherent fraction can be obtained. Negatively stained electron micrographs show that by using the microsomal fraction, free ribosomes are eluted with the non-bound fraction and that non-bound membrane vesicles and bound vesicles have a similar morphology.

The most interesting aspect of the affinity chromatography of membrane vesicles is the dependence of the dissociation of the membrane vesicles on shearing forces. Neither a gradient nor high concentration of αMM are able to elute significant amounts of membrane material (fig.2). Disturbing the settled gel is sufficient for the dissociation of the membrane vesicles and the presence of the inhibitor αMM is not necessary. The binding of membrane vesicles to ConA-Sepharose is prevented if the mixture is stirred throughout the whole separation procedure (fig.3) as well as by the simultaneous addition of αMM and membrane vesicles (4). The affinity chromatography of membranes is extremely temperature dependent. As shown in fig. 1 much more membrane material is bound to ConA-Sepharose at 30°C than at 7°C. The fractions produced by this procedure show differences in their enzyme activity and their lipid composition, as well as in binding properties for ConA (2).

DISCUSSION

Separation principle : If one may generalize from the ConA model, a special feature of affinity chromatography of membrane vesicles, as a mobile multicomponent system is, that elution with the competitive inhibitor alone is not possible. This has also been shown for lymphocyte plasma membranes (6). Furthermore, shearing forces allow dissociation of the bound membrane material even in the absence of the competitive inhibitor. This was also found in the case of membranes of plasmacytoma cells (7) and recently confirmed for postsynaptic membranes (8). These results suggest that different interactions seem possible in such a mobile multicomponent system:

(i) the affinity between receptor and ligand, the typical biospecific interaction. This is shown by the fact that the successively eluted fractions possess an increasing affinity constant for ConA (2) and by the fact that the binding could be inhibited by αMM.

(ii) interactions between the regions around the binding site of the membrane and the gel matrix. The inability of the inhibitor αMM to dissociate the membrane vesicles would support this. In contrast ConA-agglutinated Sepharose beads are quickly disaggregated by αMM (9).

(iii) multivalent interactions between one membrane vesicle and different Sepharose beads. This might explain the requirement for agitation of the gel (fig. 2).

(iv) the occurence of short range lateral movement (micropatching) of the receptor molecules in the membrane vesicles, a mobile multicomponent system, is suggested by the fact that continous stirring prevents the binding of the membranes (fig.3). A short time is necessary for binding to occur. The temperature dependence (fig.1) also reinforces this idea: at low temepratures the lipid phase of the membrane is more rigid and the lateral mobility of receptor is decreased.

(v) the effect of stirring on the elution of vesicles is probably not caused by the loss of receptor molecules from the membrane. This is indicated by successful rechromatography (4). Furthermore, the ConA receptor molecules seem to be integral proteins and gentle stirring should be not strong enough to remove the receptor molecules from the membrane.

The significance of this fractionation is at present not fully clear. In the case of lymphocytes, ConA binds to plasma membranes (9) as well as nuclear membranes (10) and the traditional problems in membranology complicate the interpretation of the results of the fractionation procedure: heterogeneity in the starting cell material, contamination with membranes of other cell organelles and alteration of membranes during storage. The fractionation of membrane vesicles is currently being discussed from two viewpoints:

i) The fractionation reflects a separation of inside out from right side out plasma membranes (6,7) as it is generally accepted that glycoresidues, to which ConA binds, are located on the outside of the plasma membrane. Therefore the non-bound fraction should consist of inside out vesicles. This is supported by immunological experiments and by labelling experiments using ConA-Ferritin (6). However, we have found (2) that the non-bound fraction also possesses ConA binding sites, in fact more that the bound fraction. These different results could be caused by unseald inside out vesicles in which the ConA binds to the inside of the vesicles.

ii) The fractionation reflects a subfractionation of plasma membranes. This is supported, by the fact, that the bound fraction can be further subfractionated in two or more fractions depending on the intensity of the used stirring forces. This finding indicates if this is not due to contamination, that a subfractionation of the right side out vesicles (adherent fraction) occurs, reflecting the heterogeneous structure of the plasma membrane. Further support comes from the different molecular composition (2) of the separated fractions.

REFERENCES

(1) G. Brunner, Membrane impression and gene expression, towards a theory of cytodifferentiation. Differentiation 8, 123 (1977).

(2) G. Brunner, E.Ferber, and K.Resch, Initial stages of differentiation of thymocytes in the plasma membrane, Differentiation 5, 161 (1976).

(3) G. Brunner, E.Ferber, K.Hansen, H.Knüfermann, and K.Resch, Isolation of plasma membrane fragments bearing ConA receptors, 9th FEBS Meeting, Budapest, Abstracts (1974).

(4) G. Brunner, E.Ferber, and K.Resch, Fractionation of membrane vesicles. I. A separation method for different populations of membrane vesicles of thymocytes by affinity chromatography on ConA-Sepharose, Anal. Biochem. 80, 420 (1977).

(5) G. Brunner, H.G.Heidrich, J.R.Golecki, H.C.Bauer, D.Suter, P.Plückhahn, and E.Ferber, Fractionation of membrane vesicles. II. A method for separation of membrane vesicles bearing different enzymes by free flow electrophoresis, Biochim. Biophys. Acta (1977), in press.

(6) F.S.Walsh, B.H.Barber, and M.J. Crumpton, Preparation of inside out vesicles of pig lymphocyte plasma membrane, Biochemistry 15, 3557 (1976).

(7) A. Zachowski, and A. Paraf, Use of a Concanavalin A polymer to isolate right side out vesicles of purified plasma membranes from eucariotic cells, Biochem. Biophys. Res. Commun. 57, 787 (1974).

(8) Bittinger, H., and H.P. Schnebli (1976) Concanavalin A as a Tool, John Wiley & Sons, London.

(9) U. Rutishauser, and L.Sachs, Receptor mobility and the mechanism of cell-cell binding induced by Concanavalin A, Proc. Nat. Acad. Sci. USA 71, 2456 (1974).

(10) A. Monneron, and D. Segretain, Extensive binding of Concanavalin A to the nuclear membrane, FEBS Letters 42, 209 (1974).

ON THE MECHANISM OF ADSORPTION OF ERYTHROCYTES TO HYDROCARBON-COATED AGAROSES

Gideon Halperin and Shmuel Shaltiel

Department of Chemical Immunology, The Weizmann Institute of Science, Rehovot, Israel

ABSTRACT

Hydrophobic interactions play an important role in the adsorption of erythrocytes onto hydrocarbon-coated agaroses, in spite of the fact that at neutral pH (where the adsorption is carried out) these cells have a negatively charged surface and the column materials (prepared by the CNBr procedure) are positively charged. This suggestion is supported by the following experimental evidence: (a) Columns with equal or even decreasing net charge exhibit an increased capacity for erythrocytes with increasing hydrocarbon chain length. (b) Even after acetylation of the columns (which was reported to cancel their positive charge) there is a chain-length dependent gradation in their capacity for guinea-pig or mouse erythrocytes. (c) Enzymatic removal of sialic acid residues from the surface of guinea-pig erythrocytes, with concomitant decrease in the negative charge of the cell surface, does not significantly affect the adsorption profile of these cells.

INTRODUCTION

Implantation of hydrocarbon chains on beaded agarose results in column materials suitable for resolution, purification and probing of proteins (for a review see ref. 1). We have recently shown (2) that erythrocytes can be efficiently adsorbed onto and desorbed from homologous series of hydrocarbon-coated agaroses (Seph-C_n*; n = 1, 2, 3, ...) under physiological, isotonic conditions. This reversible adsorption (Fig. 1) can be carried out with essentially no physical entrapment or lysis of the cells, as indicated by the fact that over 95% of the cells applied on the columns can be recovered. The erythrocytes eluted from such columns and those yet to be applied were found to be morphologically indistinguishable and to exhibit an identical osmotic fragility profile (2), suggesting that they are not significantly damaged by the adsorption-desorption procedure. This finding, together with the fact that erythrocytes from different sources exhibit different adsorption profiles on such column series, illustrates their potential use in discriminating between closely related cells and in probing cell surfaces.

---

* Seph-C_n represents Sepharose 4B activated with CNBr and reacted with an α-amino-alkane n-carbon-atoms long.

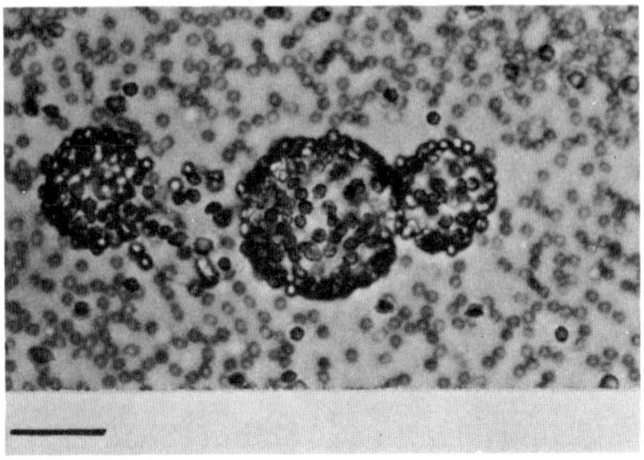

Fig. 1. Adsorption of guinea-pig erythrocytes on Seph-C_{10} as viewed under the microscope. Calibration, 50 μm.

MATERIALS AND METHODS

Homologous series of alkylagaroses (Seph-C_n ; n = 1-12) were prepared as described elsewhere (3), and their charge density was determined by potentiometric titrations (2). Acetylation of the columns was carried out following the procedure of Wilchek and Miron (4). The procedures used for the preparation and handling of erythrocytes, the determination of adsorption profiles of the cells on the column series and the determination of column capacities were described in detail previously (2). Removal of sialic acid residues from the surface of guinea-pig erythrocytes was carried out with *Vibrio cholera* neuraminidase and monitored by agglutination rate measurements as described by Marikovsky *et al.* (5).

RESULTS AND DISCUSSION

In view of the fact that erythrocytes are known to have a negatively charged surface at neutral pH (6,7) and since it is known that activation of agarose with CNBr and coupling with amines results in positive charges in the agarose derivative (8) we attempted to assess the contribution of ionic and of hydrophobic interactions to the binding of erythrocytes to alkylagaroses.

Figure 2 shows the lack of correlation between the charge density of the columns and their capacity to bind erythrocytes. It can also be seen that sets of columns which have an identical charge density (such as Seph-C_7, Seph-C_8 and Seph-C_9 or Seph-C_1 and Seph-C_{10}) differ markedly in their capacity for erythrocytes (Fig. 2, broken lines), the capacity increasing in all cases with the number of carbon atoms in the hydrocarbon chains bound to the agarose. Furthermore, it can be seen in Fig. 2 that while Seph-C_9 has a charge density ∿ 2,4-fold smaller than Seph-C_5, Seph-C_4 or Seph-C_2, its capacity to bind erythrocytes is ∿ 2.1-fold higher than Seph-C_5, ∿ 2.8-fold higher than Seph-C_4 and ∿ 6.5-fold higher than Seph-C_2. These results suggest that hydrophobic interactions most likely play a very important role in the adsorption of erythrocytes to alkylagarose columns.

In a recent publication, Wilchek and Miron (4) have shown that acetylation of alkyl agaroses with acetic anhydride abolishes their positive charge at neutral pH. In an attempt to evaluate the contribution of these positively charged groups to the binding of erythrocytes on Seph-C_n columns, we compared the adsorption profiles of red blood cells (from guinea-pigs, GPRC, and from mice, MRC) on homologous series of such columns before and after acetylation. As seen in Fig. 3, acetylation of Seph-C_n columns somewhat lowers their capacity to bind erythrocytes, especially

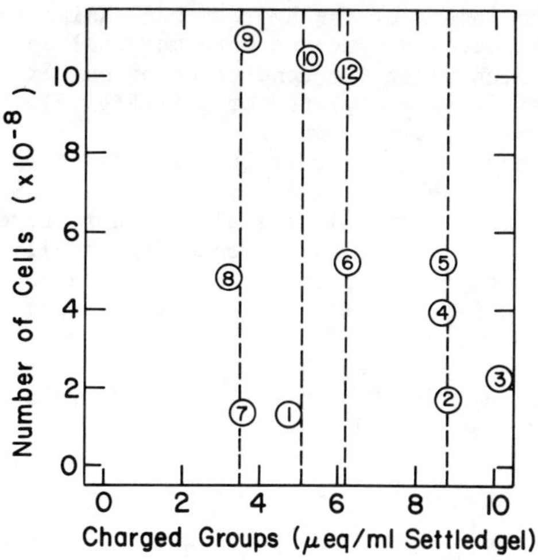

Fig. 2. Lack of correlation between the charge density of Seph-C_n columns and their capacity to bind erythrocytes. The charge density is given by the number of positively charged groups (μeq/ml of settled gel) at pH 7.0, as measured by potentiometric titrations. The capacity of the columns is expressed by the number of cells adsorbed by 1 ml of settled gel. Numbers within circles indicate the number of carbon atoms in each hydrocarbon chain of the column material (equivalent to "n" in Seph-C_n).

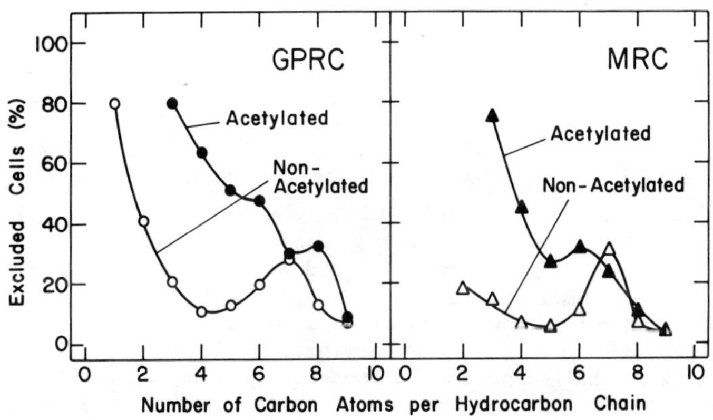

Fig. 3. Adsorption profile of erythrocytes (from guinea-pigs, GPRC, and from mice, MRC) on homologous series of Seph-C_n columns, before and after acetylation with acetic anhydride. An aliquot (0.15 ml) of a cell suspension (10^9 cells/ml buffer I (2)) were applied on 0.85 ml column material and the number of excluded cells was determined as described previously (2).

with the columns having shorter hydrocarbon chains. However, even after acetylation, the resulting homologous series exhibits a chainlength-dependent adsorption of erythrocytes which definitely increases with increasing chainlength.

These results suggest that while ionic interactions may contribute to the binding of erythrocytes onto Seph-C_n columns prepared by the CNBr procedure, this contribution does not seem to be a prerequisite for adsorption to occur. Furthermore, upon increasing the length of the hydrocarbon chains the relative contribution of the ionic interactions becomes more and more marginal and it practically vanishes with Seph-C_9, which under the conditions of our experiments binds all the GPRC or MRC cells whether it is acetylated or not (Fig. 3).

Another way we used for assessing the relative importance of the ionic and the hydrophobic interactions in the binding of erythrocytes to Seph-C_n columns made use of an enzymatic method for the removal of sialic acid residues from the cell surface. This modification is known to considerably reduce the available negative charge of the cell surface of erythrocytes, a process which can be monitored by agglutination rate measurements using poly-L-lysine (5). When guinea-pig erythrocytes were treated with *Vibrio cholera* neuraminidase and lost their available negative charge (as indicated by the poly-L-lysine test) they still bound to Seph-C_n columns with an adsorption profile which indicates that their binding is similar (if not tighter) than that of untreated cells (Fig. 4).

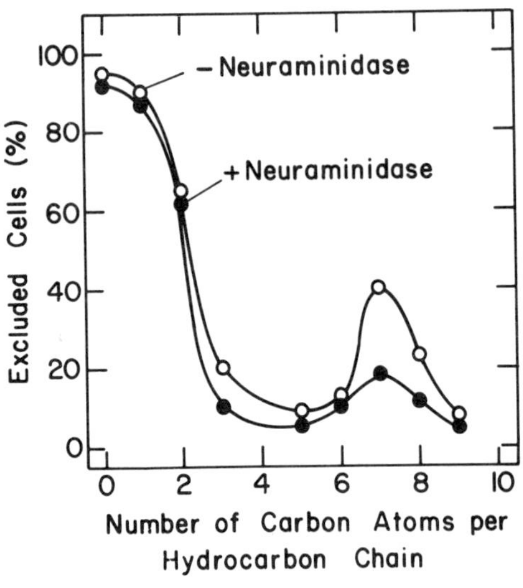

Fig. 4. Adsorption profile of intact (-o-) and of neuraminidase treated (-●-) guinea-pig erythrocytes. Two samples of 10^9 cells were incubated (60 min at 37°) either in 1 ml of buffered saline alone or in 1 ml buffered saline which contained also 20 units of neuraminidase. At the termination of the incubation period the cells were washed twice and suspended in isotonic phosphate buffered saline and the adsorption profile of the treated as well as the non-treated cells was determined as described previously (4).

In conclusion, the above experiments support the hypothesis that hydrophobic interactions are a major factor in the adsorption of erythrocytes onto alkylagaroses (4). The relative contribution of these interactions becomes, of course, more and more important as the length of the hydrocarbon chains increases. The possibility of gradually adjusting these hydrophobic interactions seems to be a very important feature of this approach, which provides an additional criterion for the resolution of cells and for probing their surface.

ACKNOWLEDGEMENT

This work was supported by a grant from the Gesellschaft fur Strahlen und Umweltforschung mbH, Munchen, Germany, and the National Council for Research and Development, Israel.

REFERENCES

1. S. Shaltiel, Hydrophobic chromatography. Use in the resolution, purification and probing of proteins, FEBS Proc. 40, 117 (1975).

2. G. Halperin and S. Shaltiel, Homologous series of alkylagaroses discriminate between erythrocytes of different sources, Biochem. Biophys. Res. Commun. 72, 1497 (1976).

3. S. Shaltiel, Hydrophobic chromatography, Meth. Enzymol. 34, 126 (1974).

4. M. Wilchek and T. Miron, The mode of adsorption of proteins to "hydrophobic columns", Biochem. Biophys. Res. Commun. 72, 108 (1976).

5. Y. Marikovsky, D. Danon and A. Katchalski, Agglutination by polylysine of young and old red blood cells, Biochim. Biophys. Acta, 124, 154 (1966).

6. H. A. Abramson, L. S. Moyer and M. H. Gorin (1942) Surface Chemistry of Cells pp. 303-328, Reinhold, New York.

7. R. F. Furchgott and E. Ponder, Electrophoretic studies on human red blood cells, J. Gen. Physiol. 24, 447 (1941).

8. J. Porath, J. C. Janson and T. Låås, Agar derivatives for chromatography, electrophoresis and gel-bound enzymes. I. Desulfated and reduced cross linked agar and agarose in spherical bead form, J. Chromatogr., 60, 167 (1971).

ENRICHMENT OF AVIAN ANTIGEN BINDING CELLS (ABC) BY AFFINITY CHROMATOGRAPHY

Hugo Wolf, Claus Gattringer and Georg Wick

Institute for General and Experimental Pathology, University of Innsbruck, Austria

SUMMARY

The present article deals with attempts for the fractionation or enrichment, respectively, of specific antigen binding cells (ABC) using bovine serum albumin (BSA) immunized chickens as an experimental model. In contrast to the situation in mammalian systems no well established methods for the separation of avian lymphocyte subpopulations have been devised so far. In this work it was shown by comparison of the following three principally different separation methods:

(1) Density centrifugation of cells forming rosettes with BSA-coated erythrocytes,
(2) Adherence of ABC onto antigen-derivatized surfaces (gelatin and polystyrene),
(3) Affinity chromatography using antigen-coated polysaccharide or polyacrylamide beads,

that the latter procedure is the most efficient method using polyacrylamide beads as a column filling material.

INTRODUCTION

For the study of various aspects of autoimmune disease the OBESE STRAIN (OS) of White Leghorn (WL) chickens represents an exceptionally convenient experimental animal model. These animals develop a hereditary, spontaneously occurring autoimmune thyroiditis as reviewed elsewhere (1). Early in life the thyroid glands of OS chickens become infiltrated by lymphoid cells, mainly B-cells, and most of these birds show clinical symptoms of hypothyroidism. In the serum of about 70 - 80 % of OS chickens autoantibodies to thyroglobulin can be demonstrated. Furthermore, a significant number of thyroglobulin-binding cells has been found in the blood and peripheral lymphoid organs of chickens of this strain, the appearance of which clearly preceeds the maximum degree of thyroiditis and the peak of autoantibody formation (2,3). It was deemed of interest to investigate whether these ABC differentiate into B-effector cells which finally mediate thyroid destruction. For functional analyses of ABC appropriate methods for the enrichment in an avian system had first to be established. Instead of using OS chickens as cell donors and thyroglobulin as an antigen we choose a BSA/anti-BSA system for this pilot methodological investigations. It appeared that for our special purpose, namely the yield of large numbers of specific ABC for future *in vivo* transfer experiments, affinity chromatography is the method of choice (4).

MATERIAL AND METHODS

Random bred NWL chickens were immunized intravenously with 4 mg BSA in 2 ml phosphate buffered saline (PBS, pH 7,2) and boosted weekly in the same manner until sacrifice at different times after the last immunization. Peripheral blood lymphocytes (PBL) and spleen cells from the same animal were obtained as described earlier (2), mixed and the resulting suspension of lymphocytes adjusted to a concentration of 1 - 4 x 10^8 cells/ml PBS containing 5 % normal chicken serum (NCS, inactivated at 56° C for 30 min, ultracentrifuged at 56.000 xG for 90 min, absorbed 1/5 v/v with pooled chicken red blood cells). For affinity chromatography (5) the efficiency of the following types of beads was tested:

(1) Agarose (Sepharose 6B, Pharmacia, Uppsala, Sweden; Biogel A5m, Biorad Laboratories, Richmond, Cal.)
(2) Polyacrylamide (Biogel P2 and Biogel P6, Biorad Labs.)
(3) Cross linked dextran (Sephadex G-25 medium, Pharmacia).

Coupling of antigen (4 mg/ml packed bead volume, PBV) to the agarose beads was done by means of cyanogen bromide as described by Cuatrecasas (6) and to polyacrylamide according to the procedure of Inman and Dintzis (7), using acylazide as an activated intermediate. Sephadex G-25 medium beads were coated with antigen via a long linear spacer as described for the so called fluoroimmunocytoadherence (FICA) technique by Richter and Wick (3). In each case the residual active groups of the beads were deactivated with 1 M glycine in 0,05 M bicarbonate buffer, pH 8.3. The used columns (K9, Pharmacia) were filled with 1,3 - 6,3 ml of ligand coated beads and 1 - 4 x 10^8 lymphocytes were applied in 1 ml portions. Fractions of 5 - 8 ml each depending on the initial volume of the cell suspension applied were collected on ice as long as cells appeared in the effluent (elution buffer:PBS, pH 7.2, 5 % NCS, 100 IU/ml penicillin/streptomycin). Finally, the ABC were eluted both in a specific (0.5 % BSA) and in an unspecific manner (vigorous mechanical agitation). The visualisation of ABC was done by a rosette assay using BSA-coated sheep red blood cells as indicator cells (2) and by indirect immunofluorescence (IIF) using BSA and a fluoresceinisothiocyanate-labeled rabbit anti-BSA 7S Ig conjugate. IIF proved to be about ten times more sensitive than rosette formation.

RESULTS AND DISCUSSION

As mentioned above a comparison of different enrichment techniques led to the conclusion that affinity chromatography is - in our hands - the best method for obtaining larger numbers of ABC (4), see Table 1.

TABLE 1 Enrichment of Avian Antigen Binding Cells by Affinity Chromatography on Different Columns

Matrix	No. of experiments	Average degree of enrichment tested by	
		[1]RFC	[2]MeIF
Biogel A5m	3	4,3	4,3
Sepharose 6B	12	3,5	6,2
Biogel P2	4	4,3	4,8
Biogel P6	4	5,8	7,8
	2	n.d.	8,5
Sephadex G-25 with spacer	2	2,5	2,5

1) Rosette forming cells
2) Indirect Membrane Immunofluorescence

As there seemed to be no temperature dependence for the degree of enrichment all final experiments were carried out at room temperature. Overlaying the columns with 1 - 4 x 10^8 cells/ml satisfying flow properties were obtained with all matrices, polyacrylamide, however, being superior to the other materials. For the improvement of the viability of the cells 5 % NCS was added to the medium and the effluent was kept on ice.

Chromatography on Sepharose 6B without specific elution showed a delay rather than a stable binding of ABC to the antigen-coated matrix as an about 5-fold enrichment occurred in the last fractions. This delay seems to be specific as proven by the use of a control column showed no ascending, but rather a horizontal elution profile. Moreover, lymphocytes sensitized against another, unrelated antigen (chicken thyroglobulin) were not enriched in the BSA-column, which is a further indication for the specificity of this delay in contrast to an unspecific enrichment due to a preferential retention of B-cells.

The highest degree of enrichment was reached by specific elution with 0,5 % BSA followed by mechanical agitation of the gel. Comparing the various matrices (see Materials and Methods) Biogel P6 gave the best results (Fig. 1).

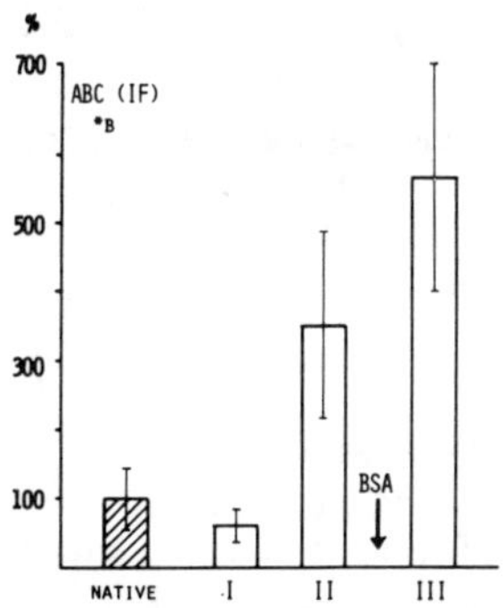

Pooled Fractions

+B IIF with rabbit anti-BSA conjugate

Fraction I: Early Effluent
Fraction II: Late Effluent
Fraction III: BSA Eluted Cells

Fig. 1. Evaluation of avian antigen binding cells by membrane immunofluorescence

The reason for this finding might be the completely uniform covalent binding of antigen to polyacrylamide beads, but not to agarose. Thus, by means of hemagglutination inhibition tests of the column effluent before specific elution it could be demonstrated that the passage of lymphoid cells entails a detachment of the antigen from agarose, but not from polyacrylamide beads. This held true even though low BSA concentrations, just sufficient for saturation of activated beads, were used for coating.

To achieve a better accessibility of the antigen for lymphocyte receptors BSA was coupled to Sephadex G-25 in another series of experiments via a long linear spacer. The use of this type of column resulted in a nearly complete depletion of ABC from the cell suspension due to a more stable binding to the antigen presented on this spacer. Conventional methods for elution of adherent cells were, however, insufficient in this case, and thus elution was attempted using different ionic strength salt solutions (0,2 - 2,0 M NaCl), low pH buffers (pH 6,0 - 3,0), Tween 20 (0,1 %), distilled water or 5 mM EDTA. All of these methods - except EDTA - were unsuccessful. Flow properties deteriorated rapidly with changing ionic strength, while Tween 20 and low pH heavily damaged the cells. However, in a direct comparison between the methods using spacer armed Sephadex G-25 and EDTA-elution or Biogel P6 with specific elution the latter enrichment procedure was still superior.

ACKNOWLEDGEMENTS

This work was supported by a grant of the Austrian Research Council (project no. 3120).

The supply of media by LS-Labor-Service GmbH & Co, Munich, FRG, is gratefully acknowledged.

REFERENCES

(1) G. Wick, R.S. Sundick, and B. Albini, A review: The Obese strain (OS) of chickens: An animal model with spontaneous autoimmune thyroiditis, Clin. Immunol. Immunopathol. 3, 272 (1974).

(2) E. Richter, and G. Wick, Thyroglobulin-binding lymphoid cells in Obese strain (OS) chickens, J. Immunol. 114, 757 (1975).

(3) E. Richter, and G. Wick, Fluoroimmunocytoadherence (FICA): A new method for the identification, enumeration and isolation of antigen-binding cells, Z. Immun. Forsch./Immunobiology 152, 351 (1977).

(4) C. Gattringer, H. Wolf,and G. Wick, Evaluation and improvement of different methods for the enrichment of antigen-binding cells of chickens, Folia biologica (1977), in print.

(5) H. Wigzell, and B. Andersson, Cell separation on antigen-coated columns. Elimination of high rate antibody-forming cells and immunological memory cells, J. Exp. Med. 129, 23 (1969).

(6) P. Cuatrecasas, Protein Purification by Affinity Chromatography. Derivatization of cross-linked polyacrylamide beads. Controlled induction of functional groups for the preparation of special-purpose, biochemical adsorbents, Biochemistry 8, 4074 (1969).

PART VI

AFFINITY CHROMATOGRAPHY OF HORMONE RECEPTORS

ISOLATION OF HORMONE RECEPTORS BY AFFINITY CHROMATOGRAPHY

Indu Parikh

Wellcome Research Laboratories, Burroughs Wellcome Co., Research Triangle Park, N.C. 27709

One of the most amazing properties of living cells is their ability to receive and respond to biochemical signals. Examples of reception and response to biochemical signals are observed all the way from the phenomenon of chemotaxis in the simplest form of living cells such as amoeba to the drug induced alteration in human behavior. As early as 1909 Paul Ehrlich introduced the idea of reception and resulting response of hormonal signals through cell surface receptors. A.J. Clarke in the 1930's introduced and developed the theory of drug and hormone action through their specific receptors. Since then progress has been made in identification and isolation of various hormone receptors.

Since the advent of affinity chromatography and the availability of a wide variety of chemical coupling reactions, together with various hydrophilic neutral solid supports, the isolation and purification of many of the hormone receptors has become experimentally feasible. However, molecules such as drug and hormone receptors are usually present in such minute quantities in biological tissues that conventional purification techniques such as gel filtration or separation based on electrophoretic mobilities may be inadequate for their purification (1-4). Detailed analysis of the methodology and applications of this powerful technique has been presented by various speakers of this symposium.

INSULIN RECEPTOR

Structures capable of specifically binding insulin, and having properties consistent with those expected for specific receptor interactions, have been identified in many cell types and in isolated membrane preparations (4). The receptors for this as well as other peptide hormones appear to be localized exclusively to the surface membranes of cells.

The ability to extract and solubilize the binding proteins in intact form from fat- and liver-cell membranes with nonionic detergents, and the availability of a rapid, quantitative assay for the soluble receptors, provide the necessary basis for studies designed to purify these membrane proteins (4-6). The extraordinarily small quantity of insulin receptors in mammalian-cell membranes and the absolute requirement for the presence of detergents to maintain their solubility pose special problems in purification studies. For example, the difficulties associated with purification of this receptor can in part be appreciated by considering that despite its large size (about 300,000) it represents only about 2 x 10^{-4}% of the protein of the liver homogenate and about 4 x 10^{-3}% of the membrane protein. Thus, it is estimated that to achieve complete purity would require a 400,000-fold purification of the crude liver homogenate. The formidable nature of this purification can be contrasted with the purification of another membrane-localized receptor, that for acetylcholine, which requires only 300- to 1000-fold purification in order to attain a homogeneous protein (7,8).

The ability of insoluble insulin-agarose derivatives to simulate the biological effects of native insulin in a variety of cells (4) and the direct interaction of fat cells with insulin agarose beads (9) suggest that such insoluble hormone derivatives may interact effectively with membrane structures and are, thus, potentially useful for purification of the receptors by affinity chromatography. The selective adsorption of liver-membrane fragments that contain glucagon receptors to glucagon-agarose columns (4) also points toward the promise of these procedures in receptor purification.

It has been shown that conventional procedures can be used for purification of the solubilized insulin receptors, and that the receptor proteins can be adsorbed to, and eluted from, affinity columns with an overall purification that probably exceeds 2 x 10^5-fold (10).

TABLE 1

Purification of the Insulin Receptor of Liver-Cell Membranes[a]

Procedure	Insulin-binding activity (pmoles/mg protein)	Purification
Crude liver homogenate	0.008	0
Liver membranes	0.15	20[b]
Triton extract of membranes	0.26	1.7[c]
$(NH_4)_2SO_4$, fraction 20-40%	0.75	3[d]
DEAE-cellulose chromatography	14	60[d,e]
Affinity chromatography[f]	About 2000	About 8000[d]
		About 250,000[b]

[a]The Triton extract was used for $(NH_4)_2SO_4$ fractionation, DEAE-cellulose chromatography, and affinity chromatography. Data from P. Cuatrecasas, Proc. Nat. Acad. Sci. U.S. 69, 318 (1972).

[b]Compared to crude liver homogenate. [c]Compared to liver membranes. [d]Compared to Triton extract of liver membranes. [e]Dialysis of Triton extract results in a 3-fold purification; DEAE-chromatography results in a further purification of about 20-fold. [f]These are tentative figures because of the difficulty in accurately determining the small amounts of protein obtained by these procedures.

This membrane bound receptor is conveniently solubilized by 0.1% Triton X-100. Another non-ionic detergent, Lubrol-PX, is nearly as effective as Triton X-100 in solubilizing the insulin receptor from liver of fat cell membranes. The presence of this detergent in the incubation medium, however, interferes with formation of insulin-receptor complex as determined by polyethylene glycol binding assay. Attempts to obtain soluble insulin receptor proteins by extraction of membranes with sodium dodecyl sulfate, dimethylsulfoxide, dimethylformamide, hexafluoroisopropanol, and pyridine are not successful. Solubilization of insulin receptor by Triton X-100 appears to be completely reversible (5).

Some of the more useful insulin-agarose derivatives are depicted in Fig. 1. Derivatives A and B are prepared by attaching insulin by the N-terminal residue of the B chain (B1phe) or by the single lysyl residue (B29) directly to the polymer backbone, by the cyanogen bromide procedure. C, one of the best derivatives, is prepared by the use of active carboxyl esters of agarose. The N-hydroxysuccinimide ester (11) of 3,3'-diaminodipropylaminosuccinyl agarose is reacted at 4° with porcine insulin (5 mg/ml) in 0.1 M sodium phosphate at pH 6.4 containing 6 M urea. Such derivatives contain 0.3-0.5 mg of insulin per milliliter of agarose. Derivative D is similarly prepared except that the buffer used for coupling is 0.1 M $NaHCO_3$ at pH 9.2. Derivative E is prepared by reacting the N-hydroxysuccinimide ester of diaminodipropyl aminosuccinyl-agarose with a large excess of

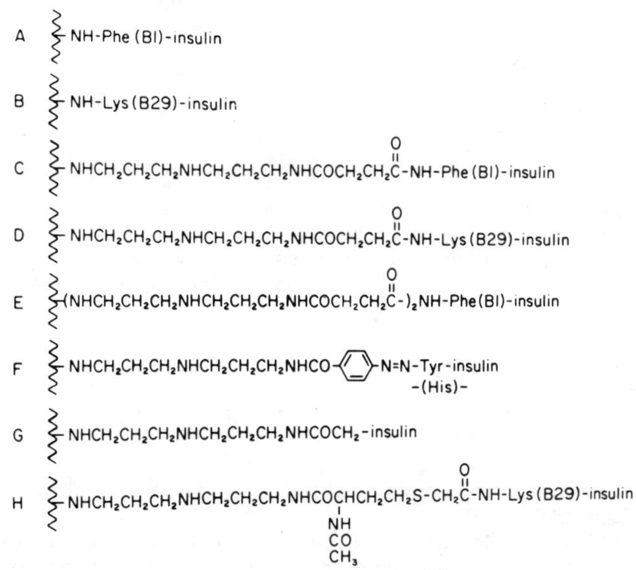

Fig. 1. Various insulin-agarose derivatives for affinity chromatography of insulin receptor (From Ref. 6)

3,3'-diaminodipropylamine, succinylating with succinic anhydride, (2) forming the corresponding N-hydroxysuccinimide ester, and coupling with insulin as described for D. Derivatives described by F are prepared by procedures similar to those used for preparation of azoglucagon-agarose (4). At pH values near 5.5, the predominant azo linkage is with histidyl residues, whereas at pH values near 7.5-8, a mixture of azohistidyl and azotyrosyl linkages is present. Derivative G, obtained by reacting the bromoacetyl agarose derivative (2) with insulin in 0.1 M $NaHCO_3$ buffer (pH 9.0) for 2 days at 24°, contains a mixture of linkage forms (amino, histidyl, and tyrosyl). Derivative H is obtained by reacting a sulfhydryl agarose (2) with an insulin derivative having both N-terminal residues blocked with acetyl groups.

Insulin coupled through a macromolecular spacer arm (see Fig. 2) to agarose are found to be most efficient in extracting the receptor from the Triton extracts of fat- and liver-cell membranes (3,6). The elution of the receptor in all cases is achieved with buffers, pH 5.5, containing 4 M urea. Urea dissociates the insulin-receptor complex, and this effect is completely reversed by reducing the urea concentration by dilution or by dialysis.

Since the solubilized insulin receptor appears to be a glycoprotein which is capable of interacting with certain plant lectins, adsorbents containing such lectins may be used to purify insulin receptor structures (12). Concanavalin A and wheat germ agglutinin bind quite well to the insulin-binding protein in the presence of 0.2% Triton X-100 which is used in the chromatography buffers. Fortunately, most of the lectin-binding glycoproteins of the membrane bind very weakly in the presence of this detergent. Thus, very small quantities of protein adsorb to these affinity columns, and it is possible to achieve substantial purification (3000-fold) of the insulin-binding protein by these procedures. Some advantages of these adsorbents, compared to insulin-agarose derivatives, include the ease of elution (solutions of the lectin specific sugars), the high capacity for the binding protein, and the avoidance of possible contamination by insulin. Recently Jacobs *et al*. (13) used a two-step purification method in which the eluate from insulin-agarose column was further purified to homogeniety on a Con A-agarose affinity column.

ESTROGEN RECEPTORS

Estrogen receptor proteins exist in at least three different molecular forms (8.6 S, 5.3 S, and 4.5 S) and are present in the cytoplasm of target tissues in very low amounts. Attempts to purify these relatively labile proteins by conventional methods have resulted only in modest purification and recovery has been poor (14).

The reversible nature of the estradiol-receptor interaction together with its stereochemical specificity and high affinity, i.e., a dissociation constant about 10^{-9} M at 4°, suggests that the receptor proteins may be ideally suited for purification by affinity chromatography.

The extremely small quantities and the relatively labile nature of the estrogen binding proteins from different target tissues present special practical difficulties in the handling and isolation of these receptors. It has been estimated that complete purification of the receptor proteins of calf uterine preparations would require 20,000- to 100,000-fold purification (15,16). Virtually all polymers used as solid supports in affinity chromatography exhibit marked adsorptive properties for free estradiol. Desorption of free estradiol during chromatography of samples containing estrogen receptors can result in "inactivation" or *apparent* removal of the estrogen-binding activity from the sample. Although free estradiol may not be present when the derivatized gel is washed with simple buffers, application of protein-containing solutions, e.g., albumin or uterine cytosol, alter the partitioning properties of estradiol between the gel and the aqueous medium and can thus markedly enhance the release of adsorbed hormone from adsorbents washed only with simple buffers. This serious problem can generally be avoided by exhaustive washing of the adsorbent with organic solvents (15).

The ester and azo bonds used in coupling the ligand to agarose or to other polymers are relatively unstable and may slowly release the ligand during chromatography. The lability of the bonds may depend on the pH and ionic strength of the buffer used, and the presence of reducing substances. Similarly, nucleophiles in the tissues extract containing the estrogen receptors may catalyze the hydrolysis or cleavage of such bonds.

The release of some free estradiol from the gel during chromatography of receptor-containing samples does not necessarily mean that the adsorbent will be ineffective in selectively extracting the receptor from the sample. If the affinity of the free hormone released from the gel is not very different from that of gel-bound hormone, the small amount of free hormone will not compete effectively with the much greater amount of immobilized hormone, and the receptor will thus preferentially bind to the solid support. If, however, the free hormone has a much greater affinity, e.g., by a factor of 1000, a very small fraction, 0.1%, of the total hormone will interfere with selective adsorption if it is present in free form. For this reason the release of adsorbed hormone (see above) is generally a more serious problem. Instability of the cyanogen bromide bonds formed is less likely to lead to the release of hormone in a form that effectively competes with the matrix-bound material.

Since the release of adsorbed estradiol during chromatography of the sample may lead to the erroneous conclusion that the column is removing the receptor from the sample, it is essential that steps be taken to examine the chromatographed samples for free estradiol as well as for the presence of estradiol-receptor complexes. Estradiol-agarose adsorbents containing macromolecular spacer arms (see Fig. 2), increase the chemical stability of the agarose bound estrogen due to their multipoint attachments to the matrix and provide other advantages in affinity chromatography (3,15,17).

The estradiol derivatives which have proved (15) to be most useful in the purification of receptors are those in which the hormone is attached to agarose through position 17 of the estradiol molecule. Although such estradiol derivatives demonstrate lower affinity than native estradiol for the receptor, they retain sufficiently high affinity to be useful in affinity chromatography. The affinity of 17-β-estradiol-17-hemisuccinate, the estradiol derivative which has proved to be the most useful, is only 300 times lower than that of 17-β-estradiol. Immobilization of these derivatives on certain of the agarose gels does not cause a further, serious decrease in the affinity for the receptor. For example, attachment of 17-β-estradiol-17-hemisuccinate to diaminodipropylamine agarose results in a 13-fold fall in affinity (K_i of about 10^{-5} M), whereas substitution on albumin-agarose only leads to a 2-fold fall in affinity (K_i of about 10^{-6} M) (13).

The specific estrogen derivative used has important implication for the subsequent steps of receptor elution from the gel. Because the estradiol derivatives which are used have substantially lower affinity than 17-β-estradiol for the receptor, it is not necessary that the amount of free estradiol which is added to the eluting medium be in great excess compared to the gel-bound ligand. Since the "excess" of free estradiol required for effective competitive exchange is related to both the concentration and the affinity of the particular gel derivative used, the specific ratio used in most work is adapted to reflect the particular conditions of the experiment. It is thus possible to use very low concentrations of estradiol of very high specific activity in the exchange reaction used in elution. This permits the use of the same radioactive estradiol for the subsequent assay of the binding activity present in the eluted sample, and it avoids contamination with large amounts of free, native estradiol.

Most gel derivatives which contain estradiol linked to agarose by bonding to the A-ring of estradiol possess very low affinity for the receptor (15). This is indicated by the total inability of the very stable 3-O-ether derivatives to bind estradiol receptors. This low affinity, coupled with the chemical instability of some of the other A-ring derivatives (which release free estradiol) probably

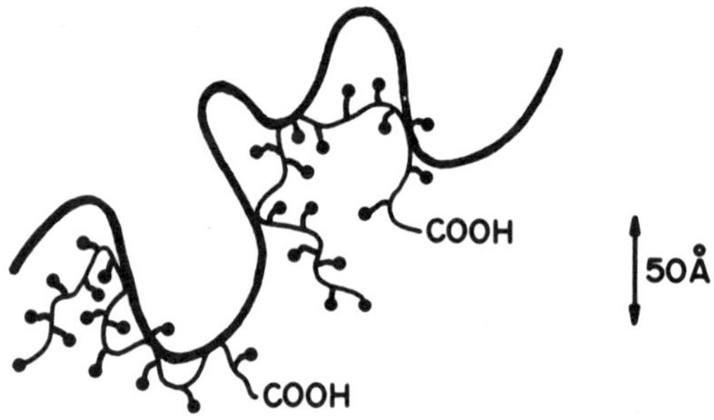

DENATURED ALBUMIN-AGAROSE

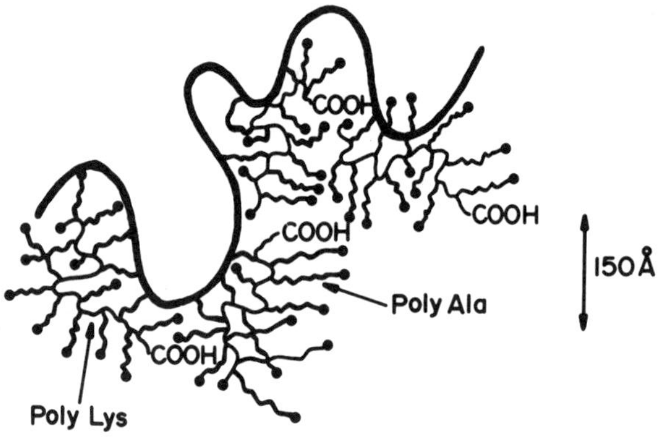

POLY-L-LYSINE-POLY-DL-ALANINE-AGAROSE

Fig. 2. Schematic presentation of macromolecular arms with multipoint attachments.

explains the failure of these adsorbents.

Molinari et al. (18) recently described a novel approach for isolation of estrogen receptor by affinity chromatography. Heparin attached covalently to agarose beads binds the native form of the estrogen receptor. The nuclear estrogen receptor as well as the "derived" cytosolic receptor do not interact with heparin-agarose. The native estrogen receptor is purified to homogeneity by affinity chromatography on heparin-agarose followed by gel filtration chromatography. The purified receptor protein is determined to be a single polypeptide chain of 69,000 MW (18).

TABLE 2

Hormone	Ligand	Source	Reference
1. Prolactin	Growth hormone	Mammary Gland	19
2. HCG	LH-HCG	Testis	20
3. Glucagon	Glucagon	Liver	21,22
4. Serotonin	5-HT	Blood Platelets	23
5. Cytokinin	Benzyladenine	Tobacco Leaves	24,25
6. Acetylcholine	Bungarotoxin	E. Electricus	7,26
7. Progesterone	Progesterone	Chick oviduct	27

During the past few years considerable progress has been made in the study, identification, isolation, and purification of a variety of membrane-localized receptors. Selected examples of application of affinity chromatography purification of hormone receptors are listed in Table II. The general approach has been to study in detail the binding interaction between a labeled hormone and a tissue or organelle that contains the putative receptor structures. The binding must show absolute specificity, the affinity must be consistent with the biological activity of the ligand, and the number of binding sites must be consistent with the physiological mechanisms operative in the intact system. As more information is gained concerning the nature of membrane receptors and their interaction with the natural regulators (hormones), greater focus will be placed on the nature of the molecular events which occur in the membrane following formation of the initial complex.

REFERENCES

(1) P. Cuatrecasas and C.B. Anfinsen, Affinity Chromatography, Ann. Rev. Biochem. 40, 259 (1971).

(2) P. Cuatrecasas, Protein Purification by Affinity Chromatography, J. Biol.Chem. 245, 3059 (1970).

(3) I. Parikh, S. March and P. Cuatrecasas, Topics in the Methodology of Substitution Reactions with Agarose, Meth. Enzymol. 34, 77 (1974).

(4) P. Cuatrecasas, Membrane Receptors, Ann. Rev. Biochem. 43, 169 (1974).

(5) P. Cuatrecasas, Isolation of Insulin Receptor of Liver and Fat-Cell Membranes, Proc. Nat. Acad. Sc. USA 69, 318 (1972).

(6) P. Cuatrecasas and I. Parikh, Insulin Receptors, Meth. Enzymol. 34, 653 (1974).

(7) E. Karlsson, E. Heilbronn and L. Widlund, Isolation of the Nicotinic Acetylcholine Receptor by Biospecific Chromatography on Insolubilized Naja Naja Neurotoxin, FEBS Lett. 28, 107 (1972).

(8) R.W. Olsen, J.C. Meunier and J.-P. Changeux, Progress in the Purification of the Cholinergic Receptor Protein from Electrophorus electricus by Affinity Chromatography, FEBS Lett. 28, 96 (1972).

(9) D.D. Soderman, J. Germershausen and H.M. Katzen, Affinity Binding of Intact Fat Cells and Their Ghosts to Immobilized Insulin, Proc. Nat. Acad. Sc. USA 70, 792 (1973).

(10) P. Cuatrecasas, Affinity Chromatography and Purification of the Insulin Receptor of Liver Cell Membranes, Proc. Nat. Acad. Sc. USA 69, 1277 (1972).

(11) P. Cuatrecasas, and I. Parikh, Adsorbents for Affinity Chromatography. Use of N-Hydroxysuccinimide Esters of Agarose, Biochemístry 11, 2291 (1972).

(12) P. Cuatrecasas and P.G.E. Tell, Insulin-like Activity of Con A and Wheat-germ agglutinin-Direct Interaction with Insulin Receptors, Proc. Nat. Acad. Sc. USA 70, 485 (1973).

(13) S. Jacobs, Y. Shechter, K. Bissell and P. Cuatrecasas, Purification and Properties of Insulin Receptors from Rat Liver Membranes, Biochem. Biophys. Res. Comm. 77, 981 (1977).

(14) G.A. Puca, E. Nola, V. Sica and F. Bresciani, Isolation and Characterization of Estrogen Binding Proteins of Calf Uterus, Advances in the Biosciences 7, 97 (1971).

(15) V. Sica, I. Parikh, E. Nola, G.A. Puca and P. Cuatrecasas, Affinity Chromatography and the Purification of Estrogen Receptors, J. Biol. Chem. 248, 6543 (1973).

(16) V. Sica, E. Nola, I. Parikh, G.A. Puca and P. Cuatrecasas, Purification of Oestradiol Receptors by Affinity Chromatography, Nature New Biol. 244, 36, (1973).

(17) I. Parikh, V. Sica, E. Nola, G.A. Puca and P. Cuatrecasas, Estrogen Receptors Meth. Enzymol. 34, 670 (1974).

(18) A.M. Molinari, N. Medici, B. Moncharmont and G.A. Puca, Estradiol Receptor of Calf Uterus: Interaction with Heparin-agarose and Purification, Proc. Nat. Acad. Sc. USA in press.

(19) R.P.C. Shiu and H.G. Friesen, Solubilization and Purification of a Prolactin Receptor from the Rabbit Mammary Gland, J. Biol. Chem. 249, 7902 (1974).

(20) M.L. Dufau, D.W. Ryan, A.J. Baukal and K.J. Katt, Gonadotropin Receptors, J. Biol. Chem. 250, 4822 (1975).

(21) F. Krug, B. Desbuquois and P. Cuatrecasas, Glucagon Affinity Adsorbents: Selective binding of Receptors of Liver Cell Membranes, Nature New Biol. 234, 268 (1971).

(22) C.B. Johnson, M. Blecher and N.A. Giorgio, Jr., Hormone Receptors-Activation of Rat Liver Plasma Membrane Adenyl Cyclase and Fat Cell Lipolysis by Agarose-Glucagon, Biochem. Biophys. Rec. Comm. 46, 1035 (1972).

(23) P.F. Pignatti and L.L. Cavalli-Sforza, Serotonin Binding Proteins from Human Blood Platelets, Neurobiology 5, 65 (1975).

(24) K. Yoshida and T. Takegami, Isolation of Cytokinin Binding Protein from Tobacco Leaves by Bioaffinity Chromatography and its Partial Characterization, J. Biochem. 81, 791 (1977).

(25) M.A. Venis, Stimulation of RNA Transcription from Pea and Corn DNA by Protein Retained on Sepharose Coupled to 2,4-Dichlorophenoxyacetic acid, Proc. Nat. Acad. Sc. USA 68, 1824 (1971).

(26) J. Schmidt and M.A. Raffery, Purification of Acetylcholine Receptors from Torpedo Californica Electroplax by Affinity Chromatography, Biochemistry 12, 852 (1973).

(27) R.W. Kuhn, W.T. Schrader, R.G. Smith and B.W. O'Malley, Progesterone Binding Components of Chick Oviduct, J. Biol. Chem. 250, 4220 (1975).

SOLUBLE BIOSPECIFIC MACROMOLECULE FOR PURIFICATION OF *ESTROGEN RECEPTOR*

J. Mešter, P. Hubert and E. E. Baulieu

Lab Hormones, 94270, Bicêtre, France

Biospecific adsorbents, i.e. insoluble materials containing covalently linked radicals capable of binding specific macromolecules have been extensively used for the purification of the latter (1,2). However, in the case of the intracellular *steroid hormone receptors*, certain difficulties were encountered due to the choice of insoluble support of the ligand. In particular such material has a relatively high ion-exchange capacity resulting in an important degree of non-specific retention of proteins along with the receptor and hence the danger of adulteration of the preparation at elution. Moreover, the local concentration of the ligand residues on the solid support used for affinity chromatography does not only depend on their overall concentration in the column or in the suspension, but also on the manner in which the synthesis was performed. In fact, due to the geometry of distribution of these residues within the adsorbent, only a fraction of them may be accessible for binding of the receptor molecules. From this follows that the conditions of binding and release of the receptor retained by the affinity adsorbent have to be established empirically (3,5).

In the present work we describe an alternative principle to the affinity chromatography, in which the usual solid support is replaced by a large water-soluble macromolecule *(dextran* $\overline{MW}$ *∿ 500,00).* A *7α-derivative of* E_2 *(estradiol)* attached to the dextran molecule via an 8-atom spacer chain was selected since this kind of derivative proved to be the most appropriate for solid-support affinity chromatography estrogen receptor purification (4,5). (More details concerning the synthesis of this compound will be published elsewhere (6)).

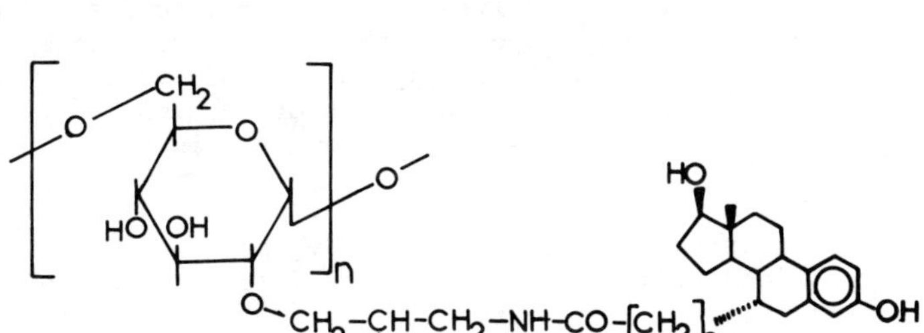

The *dextran-E_2 conjugate* contained approximately 3.10^{-5} equivalents of steroid per gram, as determined by U.V. (285 nm) absorption (corresponding to about one estrogen residue per 185 glucose units). The *dextran-linked E_2* was finally purified by extensive washing with ethyl acetate in a Soxhlet type liquid-liquid extraction apparatus in order to remove any traces of non-polymer bound steroid. The dextran-linked E_2 retained its capacity to compete for the estrogen binding sites of the *calf uterine cytosol receptor;* its K_i was calculated from the double-reciprocal plot to be 0.3 μM at 0°C; K_d for E_2 under these conditions was 0.2 nM. The competition was apparently due to a reversible binding of the polymer derivative to the receptor, as was demonstrated by the fact that, when an excess of ^3H-E_2 was added to the complex dextran-E_2-receptor, a rapid exchange of the polymer for the radioactive E_2 occurred at 25°C; the exchange was complete within 1 h. (In these and the following experiments, the separation of unbound from the receptor-bound ^3H-E_2 was carried out by the *charcoal adsorption technique* (10 min at 30°C) described previously (7)).

The crude *4S-trypsin estrogen receptor* was prepared according to Fox *et al.* (8) and separated from the remaining aggregates by gel filtration on an *Ultrogel (LKB) Aca-34* column. The 4S fraction of the receptor was then concentrated by ammonium sulphate precipitation and redissolved in 10 ml of Tris (50 mM)-EDTA (1.5 mM) buffer (pH 7.5). At this stage, the receptor is purified approx. 20 times (recovery approx. 30%) in respect to the initial cytosol preparation (6). We have verified that the dextran-E_2-receptor complex can be separated from the bulk of proteins present in the 4S-trypsin preparation: following incubation with a concentration corresponding to 250 K_i of the dextran-E_2 conjugate, the totality of receptor was removed from the 4S region of the elution pattern and migrated presumably as complex with dextran-E_2 in the void volume. On the contrary, when the same amount of non-estrogen-containing modified dextran preparation (identical to the dextran-E_2 conjugate except for omission of the last step of synthesis, i.e. fixation of the E_2 residues) was used, the totality of receptor eluted as the 4S form. In both cases, the majority of total proteins were eluted in the 4S region.

To purify the estrogen receptor, we pooled the void volume fractions containing the dextran-E_2-receptor complex (prepared from 400 g of calf uteri). This solution was then incubated with 1 mCi of ^3H-E_2 (58 Ci/mmol) for 1 h at 25°C in order to achieve the release of the receptor molecules from the dextran-E_2 conjugate; the estrogen binding sites of the receptor became occupied by the H-E_2 present in excess (approx. 1.5 μM final concentration corresponding to approx. 5 000 K_D). The solution was then cooled in ice and subsequently fractionated on an Aca-34 column (Fig. 1). The elution pattern showed three radioactivity peaks firstly, in the void volume, the radioactivity associated with the dextran-E_2 conjugate in a presumably non-specific manner, and/or with proteins bound to the polymer as well as that corresponding to any fraction of the receptor that became aggregated during the ammonium sulphate precipitation step; the second peak corresponded to the position of the 4S-trypsin receptor; and finally the unbound ^3H-E_2 activity. The optical density profile at 280 nm showed only one peak, namely in the void volume; this peak apparently reflects the presence of a certain amount of proteins initially bound to the macromolecular E_2 derivative and not released during the incubation with excess ^3H-E_2 (Fig. 1a).

When a portion of the dextran-E_2-receptor containing solution was incubated with ^3H-E_2 under identical conditions, except for an additional presence of a 20-fold excess (30 μM) of unlabelled E_2 (Fig. 1b), the radioactivity profile showed only two major peaks, namely in the void volume and in the free E_2 position. The remaining radioactivity in the "4S" region did not exceed 5% of that found in the absence of competing non-radioactive E_2, i.e. the amount corresponding to the isotope dilution effect. It may be concluded therefore that virtually all the binding sites of the protein isolated in the "4S" region were saturated at H-E_2 1.5 μM concentration indicating their high affinity, as is expected for the estrogen receptor.

The "4S" region peak was concentrated, freeze-dried and redissolved in 0.3 ml of 1% SDS. Electrophoresis in SDS-containing buffer (Fig. 1c) showed a single band of MW, approx. 63 000. Its specific activity was ∿ 13.000 pmoles of the bound ^3H-E_2/mg protein compared with about 8 pmoles/mg protein in the crude "4S" material; the degree of purification was therefore ∿ 1 600 fold in the biospecific step. The specific activity indicates a purity of ∿ 80% of the estrogen receptor.

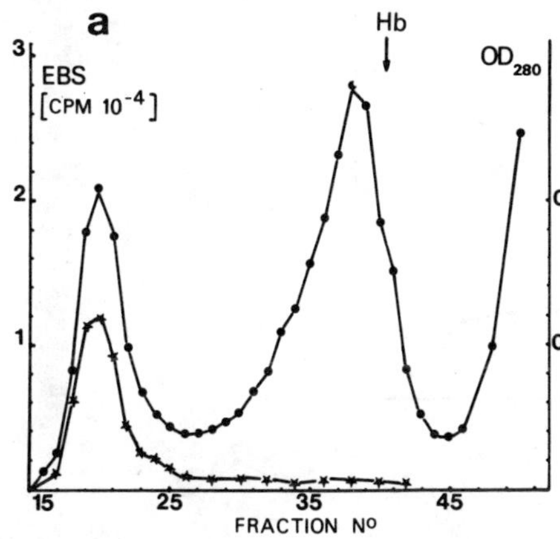

Fig. 1: Isolation of purified receptor.

The dextran-E_2 receptor complex was obtained in the void volume after gel filtration. The solution was divided into two parts.

a Four-fifths were incubated with 0.8 mCi of 6,7-^3H-E_2 (1.5 μM) for 1 h at 25°C, cooled and fractionated on an Ultrogel Aca-34 column. Optical density (x) and estradiol binding sites (EBS •) expressed as radioactivity of 0.1 ml portions are shown.

b One-fifth of the dextran-E_2 receptor solution was incubated with 0.2 mCi of the ^3H-E_2 in the presence of 30 μM unlabelled $^2E_2$ for 1 h at 25°C, cooled and fractionated on an Ultrogel Aca-34 column.

c SDS (1%)-polyacrylamide gel (7.5%) electrophoresis of the concentrated void volume (left) and the 4S region (right) of the eluted fractions shown as in a.

This result was obtained by relating the amount of the receptor measured in terms of ^3H-E_2 bound in the 4S peak assuming $\overline{MW}$ of 63,000 daltons and one binding site per molecule) to the quantity of the total protein as measured by the *amidoblack technique* (9). The purified receptor was unstable at the dilution encountered under our experimental conditions (∿ 0.5 μg) receptor protein/ml): for this reason it was not possible to carry out studies such as determination of the binding constants or electrophoresis in a non-denaturing system.

The quantity of material that can be purified by the technique described here is limited only by the size of columns used for gel filtration. In our conditions the maximum initial volume of cytosol was 1 l (400 g of calf uteri) leading to 10 to 20 μg of purified receptor (depending on the quality of tissue). The yield of purified receptor was 50-75% with respect to the amount of receptor in the crude 4S-trypsin (aggregate-free) preparation.

Fig. 1 (*contd.*)

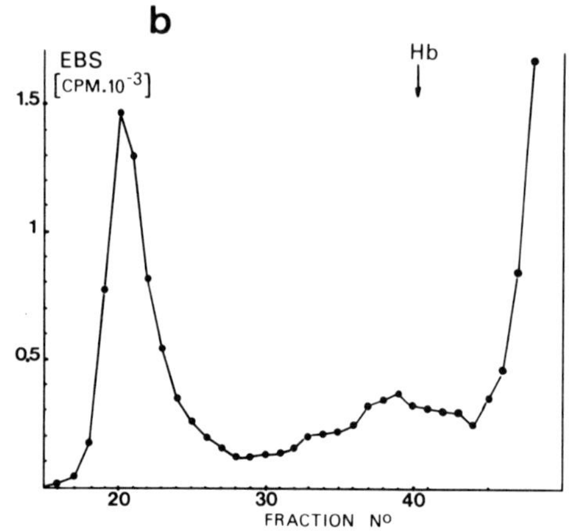

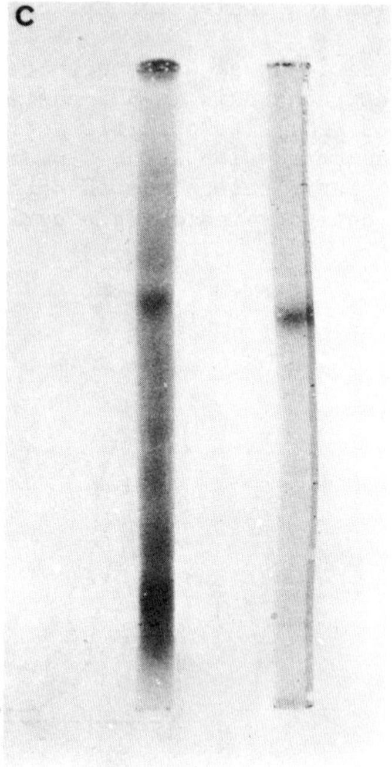

These experiments demonstrate the applicability of a soluble, high molecular weight estrogen derivative for the purification of estrogen receptor. In comparison with the conventional affinity procedure, this approach permits a stricter control of conditions of the binding as well as of the release of the receptor due to the fact that the affinity of the receptor for the biospecific polymer is determined under the conditions applying during the purification itself.

This relatively easy technique of gel filtration for isolation of complexes of specific proteins reversibly bound to high-molecular weight derivatives should prove applicable to purification of various biological macromolecules. Moreover, the principle of the method could be altered and the biospecific derivative chosen may possess another distinctive feature (e.g. polarity, charge, affinity for another material) instead of high molecular weight, that would facilitate its isolation from contaminating proteins. A large variety of proteins may thus become accessible to purification.

References

(1) C. R. Lowe and P. D. E. Dean, (1974) *Affinity chromatography*, Wiley, New York.

(2) W. D. Jakoby and M. Wilchek, (1974) *Methods in Enzymology, affinity techniques*, Academic Press, New York.

(3) V. Sica, E. Nola, G. A. A. Puca, P. Cuatrecasas and I. Parikh, (1976) *Receptors and mechanism of action of steroid hormones, part I*, Dekker, New York p. 85-108.

(4) H. Truong and E. E. Baulieu, Parameters influencing the purification of calf uterus estrogen receptor by affinity chromatography, *Febs Letters*, 56, 567 (1974).

(5) H. Richard-Foy, Thèse de Doctorat d'Etat ès Sciences, Univ. Paris VII (1977).

(6) P. Hubert, E. Dellacherie, J. Nell, J. Meŝter and E. E. Baulieu, Soluble biospecific macromolecule for purification of estrogen receptor. In preparation.

(7) J. Meŝter, D. M. Robertson, P. Feherty and A. E. Kellie, Determination of high-affinity oestrogen receptor sites in uterine supernatant preparations, *Biochem J.*, 120, 831 (1970).

(8) L. L. Fox, G. Redeuilh, P. Baskevitch, E. E. Baulieu and H. Richard-Foy, Production and detection and antibodies against the estrogen, receptor from calf uterine cytosol, *Febs Letters*, 63, 71 (1976).
(9) W. Schaffner, C. Weissmann, A rapid, sensitive and specific method for the determination of protein in dilute solution, *Anal. Biochem.*, 56, 502 (1973).

PART VII

AFFINITY TECHNIQUES IN THE ISOLATION OF DNA, RNA AND POLYSOMES

AFFINITY TECHNIQUES IN THE ISOLATION OF SPECIFIC POLYSOMES AND mRNA: PURIFICATION OF RAT ALBUMIN mRNA

John M. Taylor, Warren S. L. Liao, Tenny P. H. Tse, George H. Keller, and Ardell Conn

Department of Biological Chemistry and The Specialized Cancer Research Center, The M.S. Hershey Medical Center, The Pennsylvania State University, Hershey, PA, 17033, USA

ABSTRACT

Rat liver albumin mRNA has been purified to homogeneity by means of polysome immunoprecipitation and poly(U)-Sepharose affinity chromatography. Albumin-synthesizing polysomes could be identified by the binding of ^{125}I-labeled albumin antibody to rat liver polysomes. The binding is highly specific and occurs through an immunological recognition of the nascent peptide chains on individual ribosome monomers. Addition of an anti-antibody to the polysome-antibody complex allowed isolation of a specific albumin polysome immunoprecipitate. The immunoprecipitate was dissolved in detergent and the polysomal RNA was separated from protein by sucrose gradient sedimentation. Albumin mRNA was then separated from ribosomal RNA by affinity chromatography through the binding of the polyadenylate 3'-terminus of the mRNA to poly(U)-Sepharose.

The isolated albumin mRNA migrated as a pure species during gel electrophoresis and sucrose gradient sedimentation. The total translation product in an mRNA-dependent cell-free protein-synthesizing system derived from wheat germ, upon addition of the purified mRNA, was identified as authentic albumin by means of gel electrophoresis and tryptic peptide chromatography. A complementary DNA to the albumin mRNA was prepared and its hybridization kinetics were examined.

INTRODUCTION

The isolation of specific messenger RNAs has greatly facilitated the study of regulatory mechanisms in eukaryotic gene expression. Purified mRNAs have been employed in structure-function studies; and they have been used for the preparation of complementary DNA hybridization probes for quantitative analyses and gene isolation.

Several different experimental approaches to mRNA purification have been employed. The unique physical and chemical properties of certain mRNA molecules have been exploited, as in the purification of the guanine-rich silk fibroin mRNA by means of density gradient centrifugation and size fractionation (1). Size fractionation on linear sucrose gradients has also been utilized for the purification of several other mRNAs. These mRNA species frequently represented a major fraction of their cellular mRNA population, which facilitated isolation. A different approach to the purification of individual mRNAs has involved antibody techniques, which achieve the immunoprecipitation of specific polysomes synthesizing a single protein.

These immunochemical methods have been employed in the isolation of the mRNAs for chicken ovalbumin (2-4), mouse immunoglobulin light and heavy peptide chains (5-10), chicken histone V (11), ewe casein (12), and rat albumin (13).

The immunoprecipitation of specific polysomes depends upon the recognition of nascent peptide chains by antibodies prepared against the completed native protein. The antibody binding must be highly specific and ribonuclease must be inhibited if a pure, intact mRNA is to be isolated. Several experimental approaches are possible; and many technical requirements have been developed and refined by Schimke and his associates in their isolation of ovalbumin-synthesizing polysomes (2,3,14, 15). Various technical considerations will be discussed in this review, particularly as they have applied to the isolation of specific albumin-synthesizing polysomes from rat liver. The subsequent purification of the mRNA from immunoprecipitated polysomes by affinity chromatography will also be described.

The isolation of albumin-synthesizing polysomes involved the incubation of liver polysomes with an antibody prepared against native albumin, which binds to the nascent albumin peptide chains on the ribosomes (16). This reaction was followed by the incubation of the polysome-antibody complex with a second antibody, which was prepared against the first antibody (an anti-antibody) (13). The polysome-antibody-antibody complex was then sedimented through a discontinuous sucrose gradient to remove unreacted polysomes and unreacted antibody. The immunoprecipitated pellet was then dissolved in detergent, the polysomal RNA isolated by gradient centrifugation, and the mRNA purified with poly(U)-Sepharose by affinity chromatography (13). A brief outline of the procedure for indirect polysome immunoprecipitation is shown in Fig. 1. The technical requirements of this procedure will be reviewed.

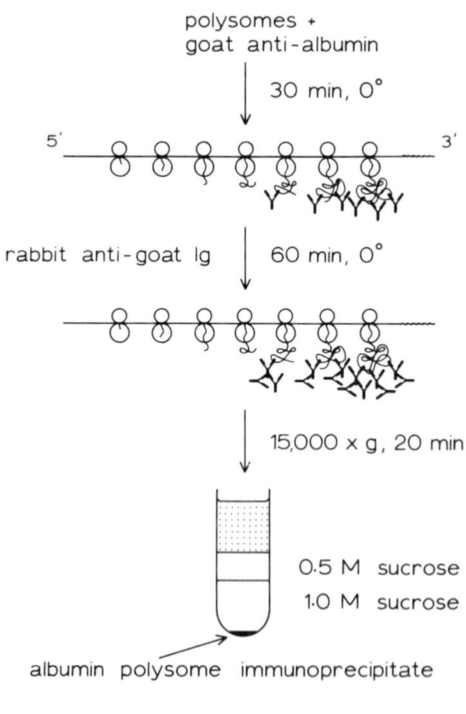

Fig. 1. Procedure for polysome immunoprecipitation

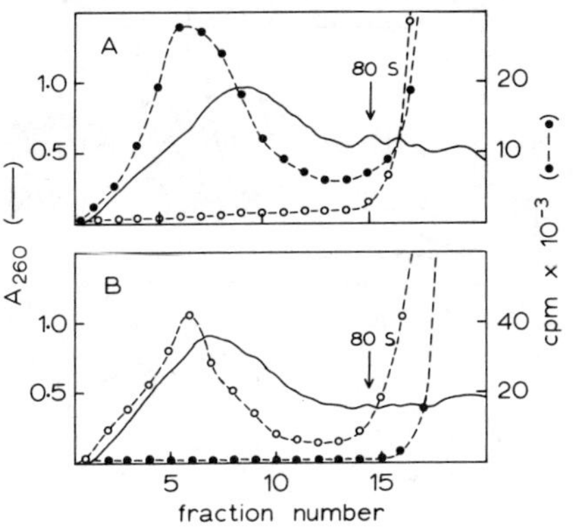

Fig. 2. Specific binding of ^{125}I-labeled antibodies to polysomes

Figure 2A shows rat liver polysomes (10 A_{260} units in 1 ml) incubated with either 10 μg of ^{125}I-anti-albumin (●---●) or 15 μg of ^{125}I-anti-ovalbumin (O---O) for 30 min at 0° (13). The polysomes were then sedimented in 15-50% linear sucrose gradients at 201,000 x g_{av} for 95 min at 4° in a Beckman SW41 rotor. Gradient fractions of 0.6 ml were collected and the radioactivity was determined. Figure 2B shows hen oviduct polysomes (10 A_{260} units in 1 ml) incubated with antibodies and examined as above.

PREPARATION OF SPECIFIC ANTIBODIES

Particular attention was directed towards antibody preparation and characterization (16). Antibodies against highly purified serum albumin were prepared in goats and the immunological specificity was examined. When reacted with an aliquot of [^{3}H]-leucine-labeled rat liver homogenate, the antibody precipitated only albumin as monitored by polyacrylamide gel electrophoresis in the presence of sodium dodecyl sulfate. The specificity of the antibody was also verified by immunoelectrophoresis against serum proteins. To minimize the size of polysome immunoprecipitates and decrease the extent of nonspecific adsorption, anti-albumin was immunopurified by affinity chromatography on columns of albumin-Sepharose. (Albumin was covalently linked to Cyanogen Bromide-Sepharose 4B, which was obtained from Pharmacia.) Antibodies against goat γ-globulin were prepared in rabbits; and they were partially purified by ammonium sulfate fractionation. Further purification of this material was not found to be helpful in polysome immunoprecipitation. Before their use in polysome immunoprecipitation, all antibodies were made ribonuclease-free by chromatography on columns of DEAE-cellulose and CM-cellulose (14, 16).

CHARACTERIZATION OF ANTIBODY BINDING

The specificity of the antibody for albumin-synthesizing polysomes was investigated by employing ^{125}I-labeled immunoglobulins in a variety of binding experiments. Under these conditions, albumin-synthesizing polysomes did not precipitate, and the binding of antibody did not cause significant changes in polysome sedimentation through sucrose gradients (Fig. 2). When rat liver polysomes were incubated with ^{125}I-anti-albumin, a significant peak of radioactivity was observed in the heavy polysome region, which was to be expected for albumin-synthesizing polysomes (Fig. 2). The radioactivity at the top of the gradient in Fig. 2 represents

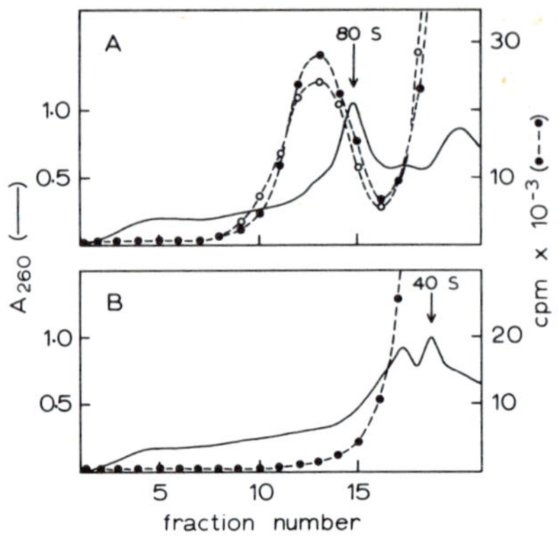

Fig. 3. Effect of pancreatic RNase and EDTA on the binding of ^{125}I-anti-albumin to rat liver polysomes

In Fig. 3A, polysomes (5 A_{260} units in 1 ml) were incubated with 5 μg of ^{125}I-anti-albumin for 30 min. Pancreatic RNase (8 μg) was added to the polysomes 30 min before addition of the antibody (O---O), or for 30 min after the antibody incubation period (●---●). Sucrose gradient analyses in both A and B were performed as in Fig. 2. In Fig. 3B, polysomes (5 A_{260} units in 1 ml) were incubated with 5 μg of ^{125}I-anti-albumin for 30 min. The mixture was then adjusted to 50 mM EDTA and incubated for an additional 30 min before sucrose gradient analysis.

unreacted immunoglobulins. Rat liver polysomes did not bind ^{125}I-anti-ovalbumin, indicating that nonspecific adsorption to ribosomal structural components did not occur. In a similar experiment, hen oviduct polysomes bound ^{125}I-anti-ovalbumin but not ^{125}I-anti-albumin. Preincubation of rat liver polysomes with unlabeled anti-albumin prevented the subsequent binding of added ^{125}I-anti-albumin; whereas preincubation with unlabeled anti-ovalbumin had no effect on the binding of ^{125}I-anti-albumin (16). The ^{125}I-anti-albumin did not bind to polysomes prepared from several other rat tissues.

The binding of ^{125}I-anti-albumin to nascent peptide chains on liver polysomes was demonstrated with the use of pancreatic RNase. In separate experiments, polysomes were incubated with RNase either before or after the binding reaction, and the results are shown in Fig. 3A. The enzymatic breakdown of polysomes to ribosomal monomers was the same in both cases. The binding of antibody to ribosomes is also the same for both reactions. This finding suggests that the antibody binding reaction is dependent on the presence of peptide nascent chains rather than polysome structure. The peak of radioactivity is primarily in the ribosome dimer region, indicating that an immunological cross-linking of nascent peptide chains can occur, but without resulting in large, heavy aggregates. That the cross-linking is to antigenic peptide chains and not to ribosomal structural components is suggested by Fig. 3B. In this experiment, polysomes were treated with EDTA after the binding reaction, with the result that the antigen-antibody radioactivity sedimented with the sucrose gradient distribution expected for free supernatant proteins. No radioactivity is observed in the 80 S or polysome regions. The antibody does not cause cross-linking between 40 S and 60 S ribosomal subunits.

The quantitative binding of ^{125}I-anti-albumin to albumin-synthesizing polysomes was investigated. Figure 4 shows that as increasing amounts of antibody were

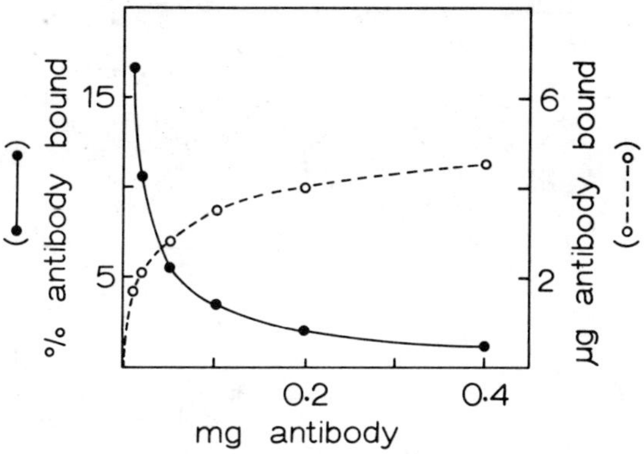

Fig. 4. Relative amount of ^{125}I-anti-albumin bound to rat liver polysomes

Figure 4 shows the binding of different amounts of ^{125}I-anti-albumin to 7 A_{260} units of rat liver polysomes in a constant volume of 1 ml. Following an incubation for 1 hr at 0°, the polysomes were examined on sucrose gradients as in Fig. 2. For each antibody concentration, the radioactivity in fractions 2 through 13 was added together and expressed as a percentage of the total radioactivity of the antibody added.

added to a constant amount of polysomes, the binding approached saturation. At low concentrations, a relatively high percentage of antibody was bound; whereas a relatively small fraction of antibody was bound at high concentrations.

The specific saturable binding of ^{125}I-labeled antibody to nascent peptide chains suggested that this reaction could be used to quantitate the relative amounts of specific polysomes. Antibody binding techniques developed by Palacios, Palmiter, and Schimke were employed to quantitate ovalbumin polysomes during hormone induction (14,15). We have employed similar methods to measure rat liver albumin polysomes under different physiological conditions (16,17). Specific ^{125}I-labeled antibodies have also been used to examine rat liver fibrinogen polysomes (18) and acetyl coenzyme A carboxylase polysomes (19). In all of these various studies, nonspecific adsorption of antibodies to polysomes was not observed. However, others have found that some antibody preparations bind nonspecifically to ribosomes, apparently through the Fc regions of the immunoglobulin heavy chains (5, 20). The use of Fab fragments minimized this problem (5,20).

POLYSOME IMMUNOPRECIPITATION

The immunoprecipitation of specific polysomes was found to require the addition of a second component to the polysome-antibody complex. Addition of antigen resulted in a direct precipitation of the complex (5,15). However, direct precipitation (with antigen) and related immunoadsorption techniques require large amounts of pure protein antigen which might be difficult to obtain in most cases. Alternatively, a second antibody which was prepared against the first antibody can be employed to form a precipitate polysome-antibody-antibody complex. The essential features of this indirect immunoprecipitation approach (with anti-antibody) were developed by Shapiro *et al.* (21) and Schechter (7), and the experimental conditions as required for albumin polysome precipitation are outlined

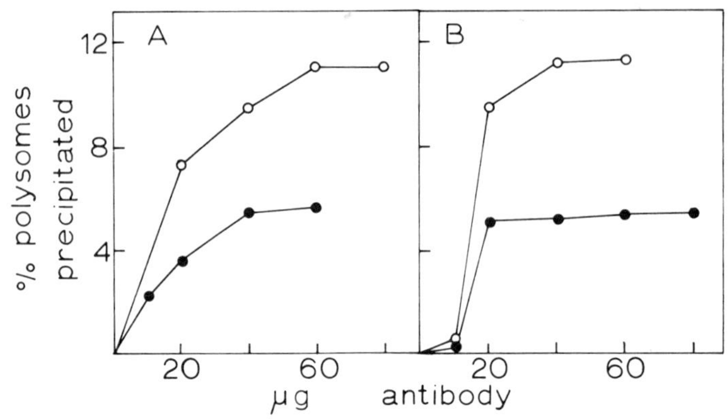

Fig. 5. Immunoprecipitation of albumin-synthesizing polysomes

Figure 5 shows indirect immunoprecipitation performed on 7 A_{260} units of [^{3}H]-leucine-labeled rat liver polysomes prepared from liver slices (5980 cpm or 7960 cpm per A_{260} unit) in a final volume of 1 ml; and the radioactivity was determined as described (13). Polysomes from normal rats (O——O) and from hypophysectomized rats (●——●) were assayed separately. In A, polysomes were incubated with varying amounts of goat anti-albumin in the first incubation, followed by 40 μg of rabbit anti-goat immunoglobulin per μg of goat anti-albumin in the second incubation. In B, polysomes were incubated with a constant amount (60 μg) of goat anti-albumin in the first incubation followed by varying quantities of rabbit anti-goat immunoglobulin per μg of goat anti-albumin in the second incubation.

in Fig. 1. It is also possible to covalently link the anti-antibody to an insoluble support such as agarose or p-aminobenzyl-cellulose (4) for use in immunoprecipitation. However, nonspecific adsorption of polysomes may be a problem for some matrix materials.

In considering the reaction conditions required for albumin polysome immunoprecipitation, the previous investigation of the binding of ^{125}I-anti-albumin to liver polysomes indicated that a concentration of 50-100 μg of goat anti-albumin per ml of polysomes (containing approximately 10 A_{260} units) approached saturating conditions. For the second antibody, it was determined that 1 mg of rabbit anti-goat γ-globulin precipitated approximately 50 μg of antigen (goat anti-albumin in this case) at the equivalence point (a 20-fold excess by weight).

The reaction conditions required for indirect immunoprecipitation were determined by taking into account the above antibody considerations and employing liver polysomes with [^{3}H]leucine-labeled nascent peptide chains (13). In the case of the first antibody, different concentrations of anti-albumin were used; and to complete the reaction, the ratio of second antibody to anti-albumin was held constant at a 40-fold excess by weight (twice that expected to be required for quantitative precipitation). Figure 5A shows that a maximum of about 11% of the postmitochondrial polysomes from normal liver are precipitated. In the hypophysectomized rat liver, albumin-synthesizing polysomes are reduced nearly 50% (17). Increasing the concentration of anti-albumin to 220 μg per ml (data not shown) did not result in a significant increase in the amount of immunoprecipitated polysomes, indicating that nonspecific adsorption of nonalbumin polysomes was not a significant problem.

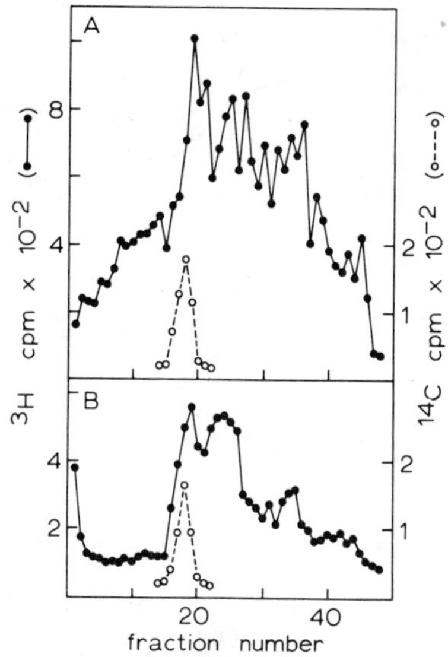

Fig. 6. Albumin-synthesizing polysome immunoprecipitate

Figure 6A shows the nascent peptide chains of (●——●) postmitochondrial liver polysomes which have been labeled for 5 min with [^{3}H]leucine (13) and examined by electrophoresis in polyacrylamide gels containing sodium dodecyl sulfate. In 6B, the albumin nascent chains (●——●) from the polysome immunoprecipitate are shown. Standard ^{14}C-labeled serum albumin (O---O) is shown for comparison. The direction of migration is from left to right.

To investigate the appropriate concentration of second antibody, the anti-albumin concentration was held constant at 70 µg per ml and different ratios by weight of the anti-antibody were used. Figure 5B shows that similar amounts of polysomes are immunoprecipitated at maximum levels. It was observed that the bulk amount of the immunoprecipitate was influenced mainly by the quantity of second antibody used. However, nonspecific adsorption or trapping could become a problem if the size of the immunoprecipitate is too large. RNA extraction would also be facilitated if the precipitate was small in size. Therefore, final reaction conditions were chosen to minimize the volume of the immunoprecipitate (70 µg of goat anti-albumin per ml and 40 µg of rabbit anti-goat γ-globulin per µg of first antibody). Substituting goat anti-chicken ovalbumin for the first antibody at 70 µg per ml resulted in less than 0.3% of the total radioactivity found in the immunoprecipitate.

CHARACTERIZATION OF THE POLYSOME IMMUNOPRECIPITATE

The specificity of polysome immunoprecipitation was investigated by examining the enrichment of albumin mRNA translational activity using a protein-synthesis assay in rabbit reticulocyte lysates (16). Albumin mRNA was found to be purified 9-fold in the immunoprecipitate compared to the initial liver polysome preparation (Table 1). This purification corresponds closely to the value that would be predicted based on the data presented in Fig. 5 and on the finding that albumin represents approximately 11% of the total protein synthesized in liver (17).

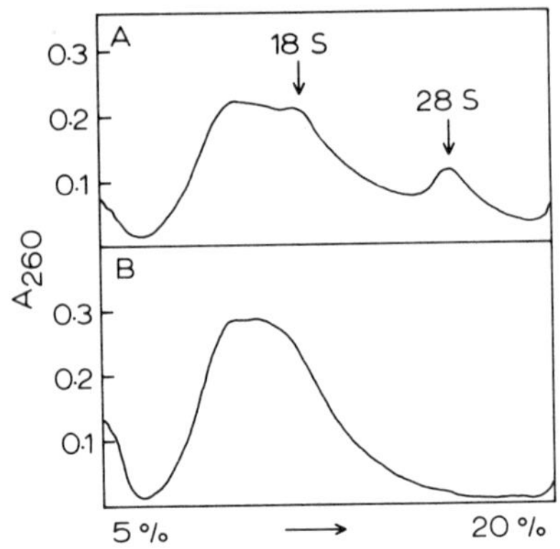

Fig. 7. Affinity chromatography of total liver mRNA on oligo(dT)-cellulose

Figure 7 shows the sedimentation profiles of the total liver mRNA eluates from oligo(dT)-cellulose affinity chromatography. Total liver RNA was prepared by phenol-chloroform extraction (17) and heat denatured as for poly(U)-Sepharose chromatography (13). The RNA was then adjusted to 0.50 M in sodium chloride and passed over oligo(dT)-cellulose in a buffer containing 0.4% sodium dodecyl sulfate, 20 mM HEPES, and 5 mM EDTA at pH 7.4. The column was washed with this high-salt binding buffer. The mRNA was eluted with 0.5% sodium dodecyl sulfate containing 5 mM EDTA at pH 7.4. This first mRNA eluate was then repassed over oligo(dT)-cellulose by the same procedure. The mRNA samples were sedimented in 5-20% linear sucrose gradients (13). Panel A shows the first mRNA eluate and panel B shows the repassed mRNA eluate.

The amount of nonspecific polysomal material in the immunoprecipitate was assessed by translational analysis (21). Albumin and ovalbumin mRNAs can be assayed in cell-free protein-synthesizing systems simultaneously without detectable cross-reaction. Thus, equal amounts of rat liver and chicken oviduct polysomes were mixed, the albumin polysomes were immunoprecipitated, and the extracted polysomal

TABLE 1 Purification of Albumin Polysomes by Immunoprecipitation

Fraction	RNA	Specific activity[a]	Total activity	Enrichment	Yield
	mg	cpm/µg RNA	cpm x 10^{-3}	-fold	%
Initial liver polysomes	17.3	563	9,740	1	100
Not precipitated	15.3	71	1,080	0.1	11.2
Albumin immunoprecipitate	1.6	5,070	8,110	9.0	83

[a]The assay for albumin mRNA translation activity was performed in the cell-free rabbit reticulocyte protein-synthesizing system as described (13).

RNAs were translated in rabbit reticulocyte lysates (21). Only 0.4% of the ovalbumin synthesizing activity was precipitated whereas albumin synthesizing activity was enhanced by the predicted 9-fold enrichment (ovalbumin polysomes comprised about 25% of the initial mixture).

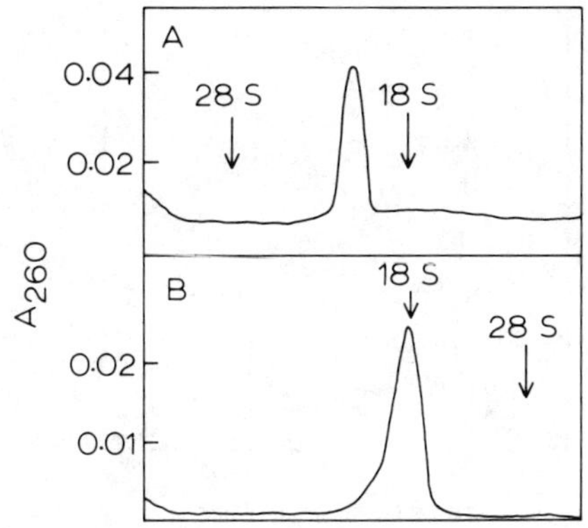

Fig. 8. Examination of purified albumin mRNA

Figure 8 shows 1.5 μg of albumin mRNA examined by gel electrophoresis, in A, on 2.5% polyacrylamide gels at pH 7.4 in 1% sodium dodecyl sulfate. In B, 5 μg of albumin mRNA were sedimented in 5-20% linear sucrose gradients containing 85% formamide (13). The direction of migration in A and B is from left to right.

The albumin polysome immunoprecipitate prepared from [^{3}H]leucine-labeled liver polysomes was examined by electrophoresis in polyacrylamide gels in the presence of sodium dodecyl sulfate (Fig. 6). No nascent peptide chains larger than native serum albumin were detected. In polysomes pre-treated briefly with RNase (15), immunoprecipitated material contained a range of nascent peptide chains decreasing to about 10,000 molecular weight. Since the antibodies employed have been prepared against native whole antigen, the RNase experiment suggests that the nascent peptides are apparently folding to yield their final mature conformation during synthesis.

In control studies (21), nonspecifically adsorbed nascent peptide chains have been shown to have an electrophoretic size distribution similar to that of the total nascent peptide chains on polysomes. The amount of nonspecifically adsorbed material was increased proportionally as the size of the immunoprecipitate was increased. These observations indicate that nonspecific precipitation represents random adsorption (or trapping) of polysomes. The relative amount of this nonspecific background dictates the lower limit for which polysome immunoprecipitation is a practical technique. Specific polysomes which represent less than about 1% of the starting material for indirect immunoprecipitation may be difficult to purify. Furthermore, large quantities of initial polysomes would be required.

PURIFICATION OF mRNA BY AFFINITY CHROMATOGRAPHY

Albumin mRNA was purified by poly(U)-Sepharose affinity chromatography (13). The albumin-synthesizing polysomes were dissolved in 2% sodium dodecyl sulfate and the protein was separated from the polysomal RNA by sedimentation through linear sucrose gradients (13). The RNA can also be prepared by phenol-chloroform extraction (17). The isolated polysomal RNA was then denatured by heating to dissociate any possible aggregates between ribosomal RNA and mRNA. Albumin polysomal RNA was next passed slowly over a column of poly(U)-Sepharose in a high salt-concentration buffer (containing 0.5 M sodium chloride). Under these conditions, the mRNA adsorbed to the bound poly(U) through the poly(A) segment at its 3' terminus (13).

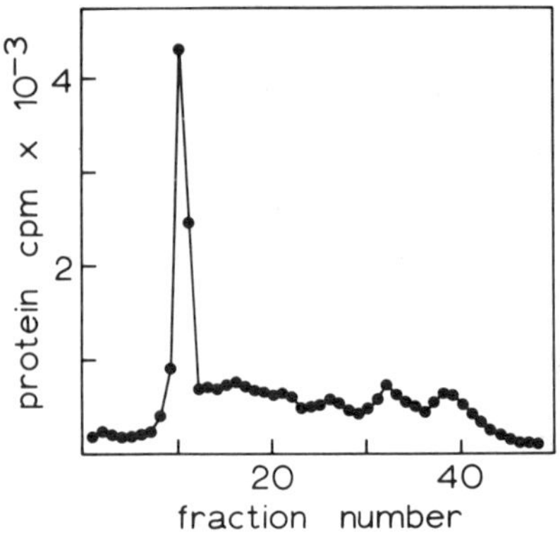

Fig. 9. Gel electrophoresis of the total albumin translation product from wheat germ extracts

Figure 9 shows an electrophoretic analysis of the albumin mRNA protein synthetic product (13). The isolated albumin mRNA was translated under optimum conditions in a cell-free protein-synthesizing system derived from wheat germ extracts (22). Following translation, the ribosomes were removed by centrifugation, and the total supernatant fluid was digested with RNase, adjusted to 1% in sodium dodecyl sulfate and 1% in dithiothreitol, dialyzed to remove free [^{3}H]leucine, and electrophoresis was performed on polyacrylamide gels. The gel was sliced and radioactivity was determined (22). The direction of migration is from left to right.

Nonspecifically adsorbed ribosomal RNA was removed by washing the gel with a low ionic strength buffer (0.05 M sodium ion). Finally, the albumin mRNA was eluted with 70% formamide. The formamide eluate contained approximately 76% of the initial albumin mRNA and was enriched 74-fold over the starting material (Table 2). None of the mRNA activity in the unbound RNA fraction would adsorb to the column when repassed twice more, suggesting that this mRNA contained either very short poly(A) segments or none at all. The low salt wash material, when examined on linear sucrose gradients, was found to be primarily ribosomal RNA. Assuming that the formamide eluate is pure mRNA and that the unbound mRNA is translated with equal efficiency, it can be estimated that the initial albumin polysomal RNA contains approximately 1.8% albumin mRNA.

TABLE 2 Purification of Albumin mRNA with Poly(U)-Sepharose

Fraction	RNA	Specific activity[a]	Total activity	Enrichment	Yield
	mg	cpm/µg RNA	cpm x 10^{-3}	-fold	%
Albumin polysomal RNA	3.86	2,460	9,510	1	100
Unbound	3.67	528	1,930	0.2	20
Low salt wash	0.01	8,150	82	3.3	1
70% Formamide eluate	0.04	181,000	7,240	74	76

[a]The assay for albumin mRNA translation activity was performed in the cell-free rabbit reticulocyte protein-synthesizing system as described (13).

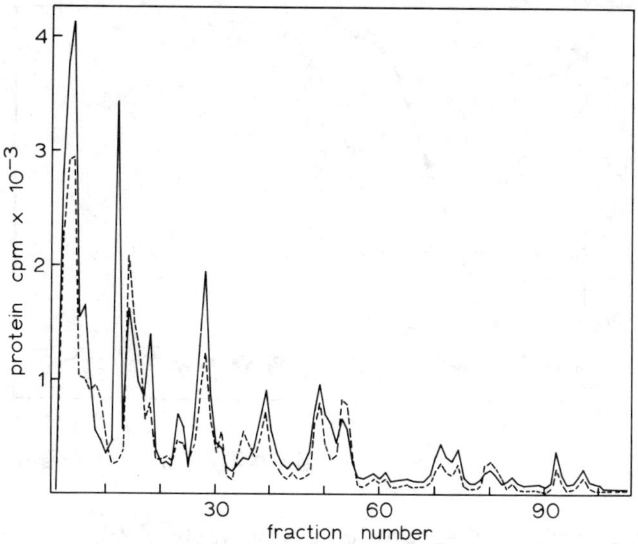

Fig. 10. Ion exchange chromatogram of peptides from the tryptic digestion of [^{3}H]albumin translation product and [^{14}C]albumin standard

In Fig. 10, purified albumin mRNA was translated in wheat germ extracts in the presence of [^{3}H]leucine (——), and the ribosomes were removed by centrifugation. The translation product was mixed with 5 mg of [^{14}C]albumin (---), prepared from rat serum, and the mixture was reduced, carboxamidomethylated, digested with trypsin, and examined by ion-exchange chromatography as previously described (13).

Oligo(dT)-cellulose has also been found to be satisfactory for isolating albumin mRNA from ribosomal RNA by affinity chromatography. However, the eluted mRNA material is usually contaminated by a significant amount of ribosomal RNA (Fig. 7A), whether or not a low salt washing step is included in the experimental protocol. Repassing the eluted mRNA over oligo(dT)-cellulose a second time removes the contaminating ribosomal RNA and allows a highly purified mRNA to be prepared (Fig. 7B). When a low salt washing step is included in the poly(U)-Sepharose procedure, very little ribosomal RNA is present in the eluted mRNA sample. This trace amount of contaminant material can be removed by recycling the mRNA through the affinity chromatography procedure. The chromatographic properties of both poly(U)-Sepharose and oligo(dT)-cellulose have been found to vary somewhat among the various batches obtained from commercial suppliers or from laboratory preparations, necessitating occasional rechecking of experimental protocols.

CHARACTERIZATION OF THE PURIFIED mRNA

Several criteria of homogeneity were applied to the purified albumin mRNA (13). The RNA in the formamide eluate from the poly(U)-Sepharose column was collected by sodium chloride-ethanol precipitation and examined directly by electrophoresis on polyacrylamide gels. The absorbance profile in Fig. 8A shows a single, symmetrical peak of mRNA migrating at about a 22 S position when compared to ribosomal RNA species. This observation is especially significant in view of the fact that

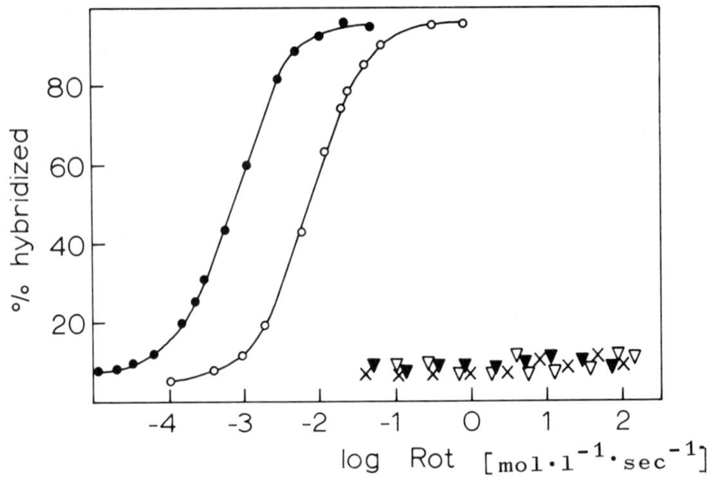

Fig. 11. Hybridization of albumin cDNA with various rat mRNA preparations

Figure 11 shows the hybridization of total poly(A)-containing RNA from rat liver (O——O), kidney (▼——▼), spleen (▽——▽), and brain (X——X), in up to 100,000-fold weight excess, to 50 pg of albumin cDNA. Total poly(A)-containing RNA was prepared by phenol-chloroform extraction (17) and poly(U)-Sepharose affinity chromatography (13). Excess purified albumin mRNA was also hybridized to the same albumin cDNA preparation (●——●).

no size-fractionation techniques of any kind have been used in the albumin mRNA purification. Therefore, gel electrophoresis becomes a meaningful criterion of homogeneity in this case. The mRNA gel also shows no evidence of contamination by 18S or 28 S ribosomal RNA, indicating the effectiveness of the poly(U)-Sepharose column for mRNA isolation.

The albumin mRNA preparation was also examined, under denaturing conditions, on linear gradients of 5 to 20% sucrose containing 85% formamide and low ionic strength buffers (Fig. 9B). Albumin mRNA sediments as an 18 S molecule (compared to the behavior of 18S ribosomal RNA on similar gradients) and again shows no evidence of ribosomal RNA contamination (i.e., no 28 S ribosomal RNA). The mRNA was recovered from these gradients and was translated efficiently in cell-free systems derived from reticulocyte lysates and wheat germ extracts. The mRNA molecule, then, is apparently intact and functionally active. Since the migration of an RNA species in polyacrylamide gels and sucrose gradients will be strongly influenced by nucleotide composition and secondary structure, only a range for molecular mass of 700,000-900,000 daltons (2060-2650 nucleotides) was estimated.

The unfractionated mRNA from the formamide eluate of the poly(U)-Sepharose column was translated in a cell-free protein-synthesizing system derived from wheat germ extracts (13,22). Following translation, the total translation product released from ribosomes was examined on polyacrylamide gels (Fig. 9). Analysis of the gels showed only one sharp peak of radioactivity located in the albumin position; and more than 90% of the total translation product reacted with specific albumin antibody. The small amount of low molecular weight peptides present was due to premature termination of translation with the release of peptidyl-tRNA from the ribosomes (22).

The nature of the albumin total translation product was examined further by tryptic peptide analysis (13). This material, which was labeled *in vitro* with [^{3}H]leucine, was combined with standard [^{14}C]albumin, then reduced, carboxamidomethylated, and digested by trypsin. The peptides were resolved by ion-exchange chromatography, and the radioactivity in each eluted column fraction was measured (Fig. 10). The data demonstrated that the isolated albumin mRNA directed the synthesis of authentic rat albumin in the wheat germ extracts. No peptide derived from the standard [^{14}C]albumin was without a corresponding ^{3}H-labeled peptide derived from the translation product. However, a ^{3}H-labeled tryptic peptide derived from the translation product (fraction 12) was identified without a corresponding ^{14}C fraction. This extra leucine-rich peptide was probably derived from the NH_2-terminal precursor region of the albumin product.

A radioactively-labeled complementary DNA copy (cDNA) of the purified albumin mRNA was prepared (23) and its kinetics of molecular hybridization were employed to gain further information regarding the mRNA. The reaction between pure albumin mRNA and its cDNA was nearly complete within a 100-fold range of Rot values and attained a level of 90% or more of the input cDNA (Fig. 11). These results indicated that the cDNA was transcribed from a homogeneous mRNA preparation. Hybridization with a cDNA of about 400 nucleotides in length gave a $Rot_{1/2}$ of 7.9 x 10^{-4} mol sec l^{-1}. Comparison of the albumin $Rot_{1/2}$ to the $Rot_{1/2}$ values obtained from the hybridization reactions of chicken ovalbumin cDNA-mRNA and rabbit globin cDNA-mRNA (correcting for the effects of cDNA length) indicated an average kinetic complexity of about 2800 nucleotides for albumin mRNA. This value is in reasonably good agreement with the predicted length of albumin mRNA. [Rat preproalbumin mRNA requires about 1800 nucleotides for the translated portion, contains a 3'-terminal poly(A) segment, and probably has an additional 200-400 nucleotides located in untranslated regions at the 5'-terminus and 3'-terminus (typical of eukaryotic mRNAs). The predicted length of albumin mRNA, therefore, would be approximately 2200-2400 nucleotides.]

Rat albumin cDNA was hybridized to the total poly(A)-containing RNAs from several rat tissues (Fig. 11). Hybridization of albumin cDNA with total liver mRNA gave a $Rot_{1/2}$ of 7.9 x 10^{-3} mol sec l^{-1}, indicating that albumin mRNA is 10% of the liver mRNA population. This finding is in close agreement with the relative amount of albumin-synthesizing polysomes (11%) in the liver and the relative level (11%) of albumin protein synthesis (17). There is a close correlation between the relative amounts of albumin mRNA, albumin-synthesizing polysomes, and the level of albumin protein production in the normal rat liver. Examination of the total poly(A)-containing RNA prepared from rat kidney, spleen, or brain by hybridization with albumin cDNA did not reveal the presence of albumin mRNA sequences in these tissues (Fig. 11). The various observations made with the albumin cDNA are all consistent with the preparation of a highly purified albumin mRNA.

DISCUSSION

Characterization of the albumin mRNA by cDNA hybridization reactions, translational analyses, and physical examination has demonstrated that a highly purified mRNA can be prepared by polysome immunoprecipitation and poly(U)-Sepharose affinity chromatography. These affinity techniques should be applicable for the isolation of a wide variety of mRNAs present in moderate to high abundance in various tissues. General technical requirements for polysome immunoprecipitation include antibody specificity, low nonspecific adsorption, and adequate quantities of undegraded polysome starting material. All isolation procedures require sterile technique (reagents are autoclaved to destroy ribonuclease), and ribonuclease inhibitors such as heparin or sodium dodecyl sulfate are included in the appropriate reagents.

A major objective in isolating a specific mRNA is to prepare a radioactively labeled cDNA for use as a hybridization probe in the quantitation of mRNA. The cDNA can also be employed to assess mRNA purity, and it can be utilized as a preparative reagent.

A specific cDNA which is covalently linked to cellulose can serve as an affinity chromatography medium for the detection and isolation of nuclear and cytoplasmic mRNA sequences, and for gene isolation. Employing oligo(dT)-cellulose as a primer for globin cDNA, Wood and Lingrel (24) prepared a highly specific cDNA-cellulose which could be used to prepare biologically active mRNA. Globin cDNA cellulose has also been employed for analytical purposes (25). In a similar manner, Anderson and Schimke (26) and Woo et al. (27) prepared ovalbumin cDNA-cellulose, and were able to partially purify ovalbumin anticodon gene sequences from fragmented chicken DNA. The coding strand of the ovalbumin gene has been partially purified by Woo et al. (28), who used immobilized ovalbumin mRNA on phosphocellulose as an affinity chromatography medium.

Affinity techniques have played a key role in the isolation of specific mRNAs. The current level of research activity suggests that this methodology will continue to evolve and have a major impact on the future development of the molecular biology of gene regulation.

ACKNOWLEDGEMENT

This investigation was supported by Grant Number CA16746, awarded by the National Cancer Institute, DHEW.

REFERENCES

1. Y. Suzuki, and D. D. Brown, Isolation and identification of the messenger RNA for silk fibroin from Bombyx mori, J. Mol. Biol. 63, 409 (1972).
2. R. Palacios, D. Sullivan, N. M. Summers, M. L. Kiely, and R. T. Schimke, Purification of ovalbumin messenger ribonucleic acid by specific immunoadsorption of ovalbumin-synthesizing polysomes and millipore partition of ribonucleic acid, J. Biol. Chem. 248, 540 (1973).
3. D. J. Shapiro, and R. T. Schimke, Immunochemical isolation and characterization of ovalbumin messenger ribonucleic acid, J. Biol. Chem. 250, 1759 (1975).
4. G. Schutz, S. Kieval, B. Groner, A. E. Sippel, D. T. Kurtz, and P. Feigelson, Isolation of specific messenger RNA by adsorption of polysomes to matrix-bound antibody, Nuc. Acids Res. 4, 71 (1977).
5. T. L. Delovitch, B. K. Davis, G. Holme, and A. H. Sehon, Isolation of messenger-like RNA from immunochemically separated polyribosomes, J. Mol. Biol. 69, 373 (1972).
6. T. L. Delovitch, S. L. Boyd, H. M. Tsay, G. Holme, and A. H. Sehon, The specific immunoprecipitation of polyribosomes synthesizing and immunoglobulin light chain, Biochim. Biophys. Acta 299, 621 (1973).
7. I. Schechter, Use of antibodies for the isolation of biologically pure messenger ribonucleic acid from fully functional eukaryotic cells, Biochemistry 13, 1875 (1974).
8. M. K. Legler, and E. P. Cohen, Estimation of the number of nucleotide sequences in mouse DNA complementary to messenger RNAs specifying a complete mouse immunoglobulin, Biochemistry 15, 4390 (1976).
9. M. Ono, T. Kondo, and M. Kawakami, Purification of immunoglobulin light chain messenger RNA by immunoprecipitation from mouse myeloma tumor MOPC-31C, J. Biochem. 81, 941 (1977).

10. M. Ono, T. Kondo, M. Kawakami, and T. Honjo, Purification of immunoglobulin heavy chain messenger RNA by immunoprecipitation from the mouse myeloma tumor MOPC-31C, J. Biochem. 81, 949 (1977).
11. A. C. Scott, and J. R. E. Wells, The isolation of chicken histone F2c (V) messenger RNA by immunoadsorption of F2c-synthesizing polysomes, Biochem. Biophys. Res. Commun. 64, 448 (1975).
12. L. M. Houdebine, and P. Gaye, Purification of the mRNA for ewe α-casein and β-casein by immunoprecipitation of polysomes, Eur. J. Biochem. 63, 9 (1976).
13. J. M. Taylor, and T. P. H. Tse, Isolation of rat liver albumin mRNA, J. Biol. Chem. 251, 7461 (1976).
14. R. Palacios, R. D. Palmiter, and R. T. Schimke, Identification and isolation of ovalbumin-synthesizing polysomes:specific binding of ^{125}I-anti-ovalbumin to polysomes, J. Biol. Chem. 247, 2316 (1972).
15. R. D. Palmiter, R. Palacios, and R. T. Schimke, Identification and isolation of ovalbumin-synthesizing polysomes: quantification and immunoprecipitation of polysomes, J. Biol. Chem. 247, 3296 (1972).
16. J. M. Taylor, and R. T. Schimke, Specific binding of albumin antibody to rat liver polysomes, J. Biol. Chem. 249, 3597 (1974).
17. G. H. Keller, and J. M. Taylor, Effect of hypophysectomy on the synthesis of rat liver albumin, J. Biol. Chem. 251, 3768 (1976).
18. H. Bouma, III, S. W. Kwan, and G. M. Fuller, Radioimmunological identification of polysomes synthesizing fibrinogen polypeptide chains, Biochemistry 14, 4787 (1975).
19. S. Nakanishi, T. Tanabi, S. Horikawa, and S. Numa, Dietary and hormonal regulation of the content of acetyl coenzyme A carboxylase-synthesizing polysomes in rat liver, Proc. Natl. Acad. Sci. USA 73, 2304 (1976).
20. Y. Ikehara, and H. C. Pitot, Localization of polysome-bound albumin and serine dehydratase in rat liver cell fractions, J. Cell Biol. 59, 28 (1973).
21. D. J. Shapiro, J. M. Taylor, G. S. McKnight, R. Palacios, C. Gonzalez, M. L. Kiely, and R. T. Schimke, Isolation of hen oviduct ovalbumin and rat liver albumin polysomes by indirect immunoprecipitation, J. Biol. Chem. 249, 3665 (1974).
22. T. P. H. Tse, and J. M. Taylor, Translation of albumin mRNA in a cell-free protein-synthesizing system derived from wheat germ, J. Biol. Chem. 252, 1272 (1977).
23. G. H. Keller, and J. M. Taylor, Synthesis of a complementary DNA to rat liver albumin mRNA, Biochem. Biophys. Res. Commun. 77, 328 (1977).
24. T. G. Wood, and J. B. Lingrel, Purification of biologically active globin mRNA using cDNA-cellulose affinity chromatography, J. Biol. Chem. 252, 457 (1977).
25. S. Levy, and H. Aviv, Quantitation of labeled globin mRNA by hybridization with excess cDNA covalently bound to cellulose, Biochemistry 15, 1844 (1976).
26. J. N. Anderson, and R. T. Schimke, Partial purification of the ovalbumin gene, Cell 7, 331 (1976).
27. S. L. C. Woo, J. J. Monahan, and B. W. O'Malley, The ovalbumin gene, purification of the anticoding strand, J. Biol. Chem. 252, 5789 (1977).
28. S. L. C. Woo, R. G. Smith, A. R. Means, and B. W. O'Malley, The ovalbumin gene, partial purification of the coding strand, J. Biol. Chem. 251, 3868 (1976).

SYNTHESIS OF A NOVEL MATERIAL FOR AFFINITY CHROMATOGRAPHY OF BIOPOLYMERS AND ITS APPLICATION FOR FRACTIONATION OF NUCLEIC ACIDS

Hans Bünemann and Werner Müller

Biologie III der Universität Bielefeld, Universitätsstraße, D-4800 Bielefeld 1, GFR

Commercially available agarose and acrylamide gels for affinity chromatography of biopolymers are too soft for their application on high viscous solutions of high molecular weight nucleic acids. A novel and more rigid support has been synthesized by the 'two step polymerization procedure' shown in Fig. 1.

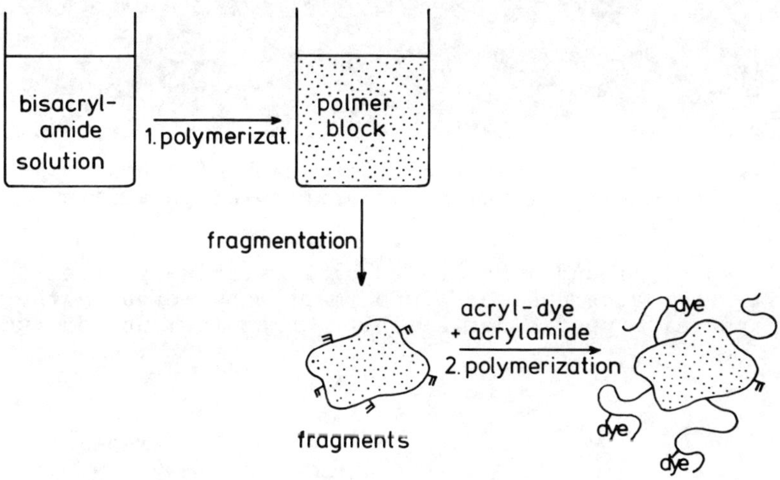

Fig. 1. Scheme for the 'two step polymerization procedure'

In the course of the first polymerization reaction a saturated solution of bisacrylamide in aqueous methanol is solidified by peroxidisulphate at 60°C. The polymer block of crosslinked bisacrylamide is easily broken and reduced to smaller particles by several passages through sieves with decreasing pore size diameter. The particles normally contain about 0,8 mmol of unreacted double bonds per gram dry weight, which are used for the coupling of suitable ligands in the course of the second polymerization reaction. For this coupling procedure an acrylamide derivative of the base pair specific (or biospecific) ligand is copolymerized with a large excess of acrylamide in a concentrated suspension of bisacrylamide particles in a suitable solvent. Under these conditions polymer acrylamide chains with a small number of ligands per chain are formed and immobilized immediately by incorporation of acrylamide residues on the bisacrylamide support. For a turnover of about 80% of monomers about 50% of the synthesized copolymer chains are covalently bound to the polymer matrix. The bisacrylamide particles show macroreticular pore size structure (Fig. 2) and enclose enzymes up to the dimensions of ferritin and helical nucleic acids up to molecular weights of about 500.000. They are chemically stable and don't change their volume substantially even under high flow rates and extreme salt concentrations such as 5 M sodium perchlorate.

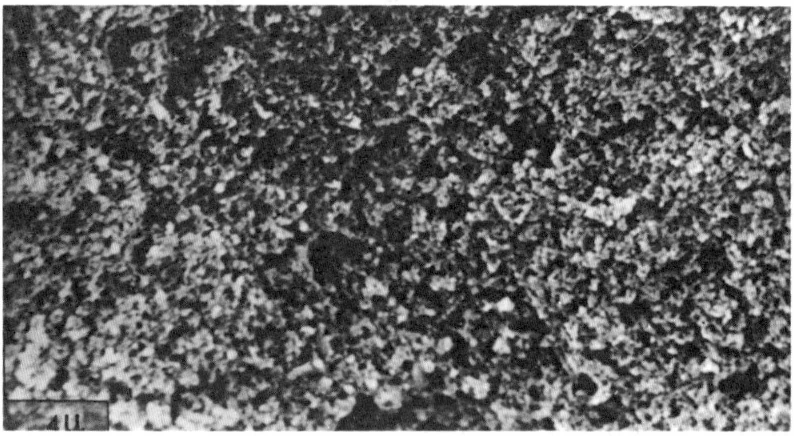

Fig. 2. Scanning electron micrograph of a bisacrylamid particle with macroreticular pore size structure.

The usefulness of ligand substituted bisacrylamide particles for the affinity chromatography of biopolymers was proved by the synthesis of special supports for affinity chromatography of nucleic acids.

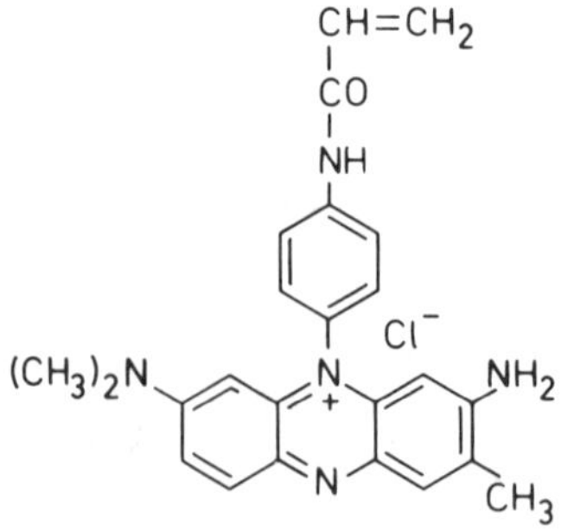

structure I

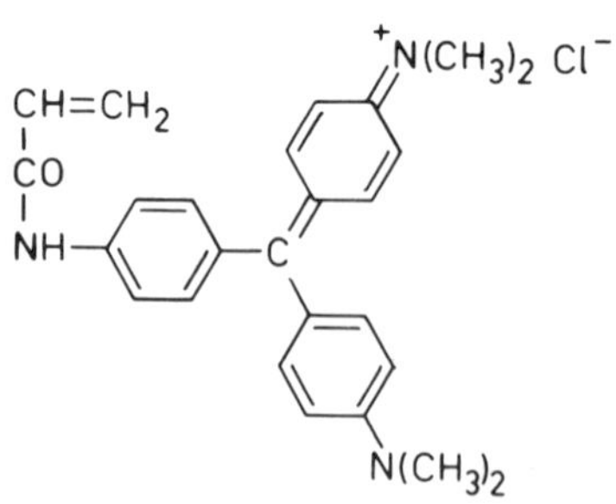

structure II

Fig. 3. Chemical structure of base specific dyes

For this purpose acrylamide derivatives of base specific dyes, such as guanine-cytosine specific N-acryloyl-4'-amino-phenylneutralred (APNR, structure I in Fig. 3) or adenine-thymine specific N-acryloyl-4-amino-malachitegreen (AMG, structure II in Fig. 3) were immobilized by copolymerization with acrylamide. The average chain length of the attached acrylamide polymers was determined to be about 150 acrylamide residues, substituted by a single dye per chain. AMG and APNR were chosen for their opposite base pair specificity as well as for the different structures of their complexes with double stranded DNAs.
AMG is assumed to bind similar as other adenine-thymine specific dyes in one of the two grooves of the DNA helix (1). Due to this type of binding the malachite green substituted support can be used successfully for separations of double stranded DNAs of different base composition on the one side (Fig. 4) and for separation of double stranded nucleic acids from single or partial single stranded nucleic acids on the other side (Figure not shown).

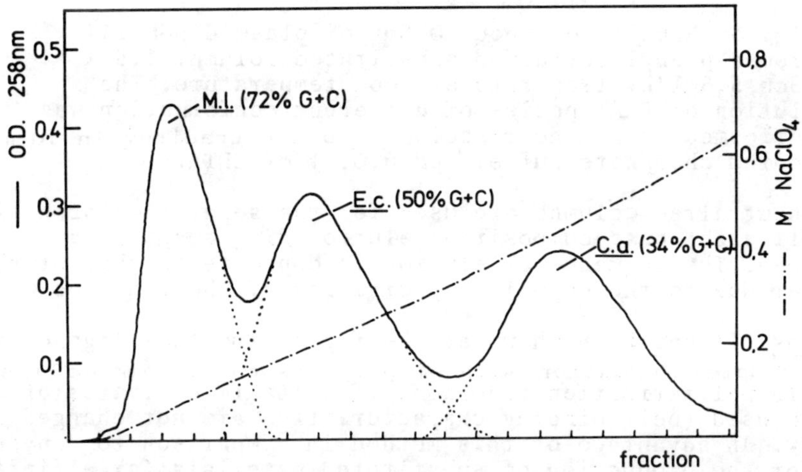

Fig. 4. Elution of 1mg of a mixture of 3 bacterial DNAs of an average molecular weight of about 700.000 from a malachite green substituted column: 1,5 cm x 16cm, 5,8ml/hr flow rate at room temperature. The DNA mixture contained DNAs of Clostridium acidiurici (34% G+C), Escherichia coli (50% G+C) and Micrococcus luteus in an optical density ratio of 1/1/1 at 258 nm. The column was equilibrated and operated with 10 mM sodium phosphate buffer, pH 6,0, 1mM EDTA. Elution of DNA was performed by a gradient of sodium perchlorate in the same buffer.

In contrast to AMG the guanine-cytosine specific APNR binds to DNA by intercalation (2). Its planar tricyclic chromophore intercalates preferentially between two neighbouring guanine-cytosine base pairs of double stranded DNA. The formation of such sandwich-like complexes of planar dyes with helical nucleic acids is facilitated energetically for supercoiled DNA species (3). Therefore the phenylneutralred substituted column can be applied for separation and isolation of circular DNA molecules of plasmids and viruses which often exist in the supercoiled conformation. As shown in Fig. 5 the supercoiled conformation of plasmid pGM 214 is eluted at higher salt concentration than its relaxed circular and linear analogues.

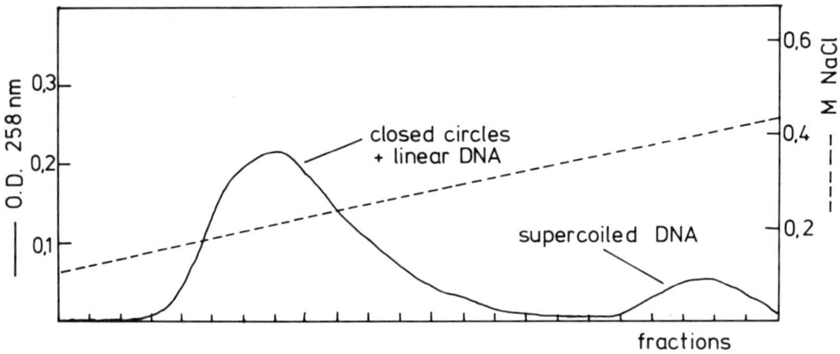

Fig. 5. Elution of about 0,5mg of plasmid pGM 214 from a phenylneutralred substituted column: 1,5 x 15cm, 5,5ml/hr flow rate at room temperature. The elution of DNA species of different conformation was performed by a linear sodium chloride gradient in 10 mM sodium phosphate buffer, pH 6,0, 1 mM EDTA.

If phenylneutralred columns are used for the separation of helical DNAs of different base composition elution diagrams, similar to Fig. 4 are obtained. The sequence of elution is opposite to that of the AMG polymer due to the opposite specificity of the two dyes.

Conclusions It has been shown at the first time that ligands for specific complex formation with biopolymers can be immobilized by a simple polymerization reaction. If suitable derivates of the ligand are used their binding characteristics are not changed substantially. An advantage of this method in comparison to conventional methods for the production of substituted materials for affinity chromatography is the fact that very long spacer arms of different length can be formed between the support and the ligand. Since the position of the dye substituent on the polymer spacer is variable, this material should give a versatile support for the isolation of groups of different enzymes which need spacer arms of different length for their reaction with the same kind of coenzyme.

Acknowledgements This investigation was supported by the Federal Government (BCT 105), GFR

REFERENCES

(1) W. Müller, H. Bünemann and N. Dattagupta Eur. J. Biochem. 54, 279 (1975)

(2) W. Müller and F. Gautier Eur. J. Biochem. 54, 385 (1975)

(3) W. Bauer and J. Vinograd J. Mol. Biol. 33, 141 (1968)

PREPARATION OF AFFINITY-ADSORBENTS CONTAINING NUCLEIC ACIDS IMMOBILIZED ONTO CARBOXYMETHYL-CELLULOSE

Hans Potuzak and Ulrike Wintersberger

Department of Biochemistry, University of Liverpool, England and Institute for Cancer Research, Vienna, Austria

INTRODUCTION

Affinity chromatography on columns containing nucleic acids dates back to 1968 when Rose Litman described the purification of micrococcal DNA polymerase on a column containing DNA immobilized to cellulose by irradiation with ultraviolet light (1), and Alberts et al. (2) as well as Gilham (3) independently suggested new methods for the preparation of celluloses containing polynucleotide ligands. Since that time chromatography on immobilized nucleic acids has established itself as a useful technique which has received wide application to both protein- and nucleic acid chemistry (4,5,6).

A simple and efficient procedure for the immobilization of nucleic acids to carboxymethyl-cellulose (CM-cellulose) which allows linkage of high amounts of polynucleotide in good yields is reported here. Affinity-adsorbents prepared by this technique have been successfully applied to the chromatography of proteins involved in cellular polynucleotide synthesis (7,8). They contain relatively high ligand-concentrations (up to 3.6 mg per ml column-volume) and are stable to elevated temperature, low ionic strength, and other desorbing conditions such as hydrophilic organic solvents. In addition to these advantages, columns can be reused repeatedly after washing with concentrated NaCl solutions *in situ* and allow high flow rates.

EXPERIMENTAL PROCEDURES

Pretreatment of the matrix

To obtain most efficient binding of nucleic acids, it is essential that the matrix is previously converted into the free acidic form. Especially, good results are obtained when pretreatment of the matrix is carried out as follows: 1g of CM-cellulose is suspended in 50 ml of 0.5 N NaOH and stirred for 30 min. After decantation from the supernatant the material is washed with water until the effluent reaches a pH of 8. The cellulose derivative is suspended in about 100 ml of water and the pH is adjusted to 3.5 by addition of 0.01 N HCl. After filtration under suction the material is washed three times with 5 ml ethanol/ether 1:1 (v/v) each, followed by 20 ml of ether and finally dried for 60 min at 40 °C.

Procedure for immobilization of nucleic acids

The required amount of the polynucleotide is dissolved in an adequate volume (10 ml per g CM-cellulose) of deionized water or 10 mM NaCl and added to the pre-treated matrix. Immobilization is achieved by slowly drying the mixture in an atmosphere of 40-50 % rel. humidity over a period of 60 h at 40 °C. The rate of drying which is critical in this procedure can be regulated by periodically covering the reaction vessel.

Desorption of nonreacted polynucleotide

As nucleic acids bind to celluloses by nonspecific interactions it is essential to remove all the nucleic acid adsorbed to the matrix by an efficient washing process. The polynucleotide-cellulose is therefore suspended in a desorbing solution containing hydrophilic organic solvents and gently agitated for two days at room temperature. Neutral solutions of low ionic strengths containing 50 % glycerol, ethylene glycol, or formamide are appropriate. To avoid denaturation of native double-stranded DNA, the solution contains 15 mM NaCl and 1.5 mM sodium citrate (0.1 SSC). After excessive washing with 0.1 SSC the material can be stored for long periods in neutral buffers containing 1 M NaCl.

Sucrose density gradient centrifugation

Depurination of both single- and double-stranded DNA under the reaction conditions described was investigated by determination of the decrease in molecular weight after hydrolysis of the column materials in 0.1 N NaOH at 70 °C for 10 min. Samples of 0.1 ml containing DNA concentrations of 1 mg/ml were applied to gradients of 5% to 20% sucrose in 0.1 N NaOH containing 0.9 M NaCl. Centrifugation was carried out in a Spinco SW 65 rotor at 2.1×10^6 g x h and fractions of 0.5 ml were collected and measured at 260 nm. The molecular weight of the undepurinated DNA was estimated by comparison with T4-DNA in neutral and alkaline sucrose gradients.

RESULTS AND DISCUSSION

A rapid and simple acid catalyzed reaction which allows immobilization of both single- and double-stranded DNA as well as RNA to CM-cellulose is described. The materials obtained exhibit stability to temperatures up to 80 °C and desorbing conditions such as low ionic strengths and hydrophilic organic solvents. As indicated in table 1, the yield of immobilization is especially high at input amounts of approximately 10 mg of polynucleotide per g CM-cellulose.

Binding of nucleic acids to insoluble matrices by chemical or physical methods often causes serious damage to the ligand due to denaturation and depurination. In order to estimate the extent of depurination of DNA under the reaction conditions described, we determined the decrease in its molecular weight after alkaline hydrolysis from the column. As under alkaline conditions the phosphodiester bonds of DNA on either side of a depurinated sugar residue are extremely labile to hydrolysis, sites of depurination can easily be detected by this method (9). In a native double -stranded DNA of a molecular weight of 5×10^6 treated under the conditions described, on the average two depurinated sites per strand were found. Single-stranded DNA was found slightly more sensitive to acidic conditions and we accordingly observed four sites of depurination in a single-strand of a molecular weight of 2.5×10^6. No depurination is expected to occur in the case of RNA, since as compared to DNA it exhibits much higher stability to acidic conditions.

To avoid denaturation of native double-stranded DNA under the conditions of the immobilization procedure, the reaction mixture should contain 10 mM NaCl. As estimated from the behaviour of double-stranded polynucleotides in solution, significant damage due to denaturation is not expected under those conditions.

In conclusion we describe a simple and efficient procedure for the preparation of affinity-adsorbents containing nucleic acids immobilized onto carboxymethyl - cellulose. The acid catalyzed reaction involved allows linkage of relatively high

TABLE 1 Dependence of nucleic acid bound to CM-cellulose on the input

Type of nucleic acid	input	immobilized	ligand concentration	yield
---	mg	mg	mg/ml	%
19 S denatured DNA	10	9.4	1.6	94
(calf thymus)	20	16.0	2.7	80
	30	18.2	3.0	61
	40	21.4	3.6	54
20 S native DNA	10	8.0	1.3	80
(calf thymus)	20	13.7	2.3	69
	30	14.9	2.5	50
	40	15.8	2.6	40
high mol.wt. RNA	10	5.7	1.0	57
(yeast)	20	10.8	1.8	54
	30	13.2	2.2	44
	40	13.6	2.3	34

amounts of polynucleotide leading to stable materials which seem to be particularly useful adsorbents for the chromatography of proteins involved in cellular polynucleotide synthesis. In addition to these advantages the reaction conditions are very mild causing only minimal damage to the ligand due to depurination.

REFERENCES

(1) Litman, R.M. (1968) J. Biol. Chem. 243, 6222-6233
(2) Alberts, B.M., Amodio, F.J., Jenkins, M., Gutmann, E.D., and Ferris, F.J. (1968) Cold Spring Harbor Symp. Quant. Biol. 33, 289-305
(3) Gilham, P.T. (1968) Biochemistry 7, 2809-2813
(4) Gilham, P.T. (1971) in: Methods in Enzymology (Grossman and Moldave, eds.) Vol. XXI pp. 191-197
(5) Alberts, B.M., and Herrick, G. (1971) in: Methods in Enzymology (Grossman and Moldave, eds.) Vol. XXI pp. 198-217
(6) Weissbach, A., and Poonian, M. (1974) in: Methods in Enzymology (Jakoby and Wilchek, eds.) Vol. XXXIV pp. 463-475
(7) Potuzak, H., and Wintersberger, U. (1976) FEBS-Lett. 63, 167-170
(8) Wintersberger, U., and Blutsch, H. (1976) Eur. J. Biochem. 68, 199-207
(9) Kuhnlein, U., Penhoet, E.E., and Linn, S. (1976) Proc. Nat. Acad. Sci. (U.S.A.) 73, 1169-1173

Correspondence and requests for reprints to:

Dr. Ulrike Wintersberger, Institute for Cancer Research, University of Vienna, Borschkegasse 8a, A-1095 Vienna, Austria

PURIFICATION OF ANTIBODIES SPECIFIC FOR PHOTOPRODUCTS OTHER THAN PYRIMIDINE DIMERS IN ULTRAVIOLET-LIGHT IRRADIATED DNA BY AFFINITY CHROMATOGRAPHY

Hanoch Slor

Department of Human Genetics, Sackler School of Medicine, Tel Aviv University, Israel

Abstract

Ultraviolet light (UV) irradiation of living cells induces the formation of various types of photoproducts in their DNA. These photoproducts are deleterious to the cell, interfere with m RNA production and DNA replication, and if not removed from the DNA by DNA-repair enzymes, the cells will be permanently damaged or die.

Immunization of Rabbits with UV-irradiated DNA produced in our laboratory a heterogenous population of antibodies specific for various types of photoproducts in irradiated DNA. In order to study the fate of such photoproducts in various cell types and understand their biological consequences, we have applied an immunological detection technique which is far more sensitive than previously employed biochemical methods. For the purpose of identification of the various types of photoproduct excision repair, we attempted to purify the various antibody types (e.g. thymine-dimer-specific and "minor"-photoproduct-specific). Using DNA-cellulose columns in which the DNA was either unirradiated, UV-irradiated, and UV-irradiated with consequent enzymatic photoreactivation, we were able to obtain purified antibodies specific for each type of those UV-photoproducts. Their nature and specificity was determined and their use in repair studies has revealed for the first time that in addition to DNA repair of thymine-dimers, mammalian cells repair "minor" photoproducts.

Introduction

We have recently obtained evidence that antibodies directed against photoproducts other than pyrimidine dimers ("minor" photoproducts) are present in serum of rabbits which were immunized with UV-irradiated DNA (1). In that study we have observed that following a complete monomerization of the pyrimidine dimers in UV-DNA by photoreactivating enzyme (yeast), there was a detectable reactivity with the anti-UV-DNA antisera. Complete photoreactivation was ascertained by the addition of ^{14}C-labeled supercoiled PM2 DNA to the reaction mixture, and periodically during the incubation, a sample was taken and incubated with a dimer-specific T4 UV-endonuclease. As long as there are dimers present on the PM2 DNA, the T4 enzyme will nick the DNA next to the dimer and the DNA molecule will be converted to the relaxed form which will be trapped on nitocellulose filter.

A further evidence for the presence of anti "minor" UV-photoproducts antibodies was obtained from an experiment in which DNA samples were UV-irradiated at various initial doses and were consequently subjected to a complete dimer monomerization by the photoreactivating enzyme. If there are anti "minor" photoproducts in the anti UV-DNA antibody population, it is expected that their reactivity with completely photoreactivated DNA will depend on the initial UV dose the DNA was irradiated with. Indeed, there was a linear relationship of antibody reactivity with the initial UV dose the DNA was irradiated prior to dimer monomerization.

The present communication describes our approach for the purification of both, anti pyrimidine dimers antibody and anti "minor" photoproducts antibody using affinity chromatography on DNA-cellulose columns. The specificity and nature of these antibodies was also determined.

Materials and Methods

Anti UV-DNA Antibodies: Equal proportions of calf thymus DNA and methylated bovine serum albumin (MBSA) at a concentration of 5mg/ml saline each, were complexed as described by Stollar et al (2). The complexed DNA-MBSA was then irradiated with UV-light (254 nm) for 60 min at a fluence of 25 ergs/mm^2/sec. Three rabbits were injected i.v in the ear with 0.5 ml of UV-DNA-MBSA on four consecutive days each week for a period of three weeks. One week after the last injection an additional 1 ml was injected as a booster. A week later rabbits were sacrified and their serum was collected and kept at -80°C.

Radioimmunoassay for anti DNA antibodies: ^{3}H-labeled DNA (E. coli, 30,000 cpm/μg), either unirradiated or UV-irradiated, was dissolved in buffer V of Lin a Riggs (3) consisting of 10 mM KCl, 3 mM $MgCl_2$, 0.1 mM Na_2EDTA, 5% DMSO, and 10 mM Tris-HCl, pH 7.45. Reaction mixture consisted of 0.1 ml DNA (3,000 cpm) and 0.1 ml diluted serum (in saline). Sera dilution was in the range of 1:40 to 1:10,000. After incubation for 60 min at room temperature the reaction mixture was slowly filtered through cellulose nitrate filters (millipore) which were boiled 5 min prior to use. Native DNA is not retained on the filter under these conditions whereas DNA bound to immunoglobulins is retained. Filters were rinsed with the same buffer, dried, and counted for radioactivity in a toluene-based scintilation fluid in a Packard Tricarb Spectrometer.

Photoreactivation of DNA: Yeast photoreactivating enzyme was purified according to the method of Minato a Werbin (4) and was free of DNase activity under the photoreactivation assay conditions used (4). The extent of photoreactivation was determined either by direct dimer content assay in hydrolized DNA according to Goldman a Friedberg (5) or Cook et al (6), or by following the conversion of UV-irradiated supercoiled PM2 DNA to the relaxed form upon incubation with T4 UV-endonuclease (1).

DNA-cellulose columns for affinity chromatography: Unirradiated DNA, UV-irradiated DNA (1000 ergs/mm^2), or photoreactivated UV-DNA, was bound to cellulose according to the method of Albert et al (7). A column (1.5 x 20 cm) was packed with a given type of DNA-cellulose and 250 ml of serum diluted 1:50 in saline was passed through, followed by 100 ml saline rinse. Antibodies retained on the column were eluted with acetic acid:0.5N NaCl, pH 3, and buffered immediately to pH 7.4 by 0.1 M Tris-HCl buffer.

Results

The scheme used for the purification of both, dimer-specific and non-dimer-specific anti UV-DNA antibodies is described in Fig.1.

Since the reactive antisera were also slightly reactive with unirradiated DNA (about 0.5% of the activity against UV-DNA), the first step in this procedure (column I) was designed to remove this activity by binding it to an unirradiated DNA-cellulose column. Indeed, the effluent of column I contained no activity against unirradiated DNA. The latter activity was retained on the column and could be eluted by acetic acid:NaCl. The anti unirradiated DNA antibodies were of the IgM type and were completely inhibited by anti IgM antiserum.

The effluent of column I which contained the antibodies against UV-DNA was applied to

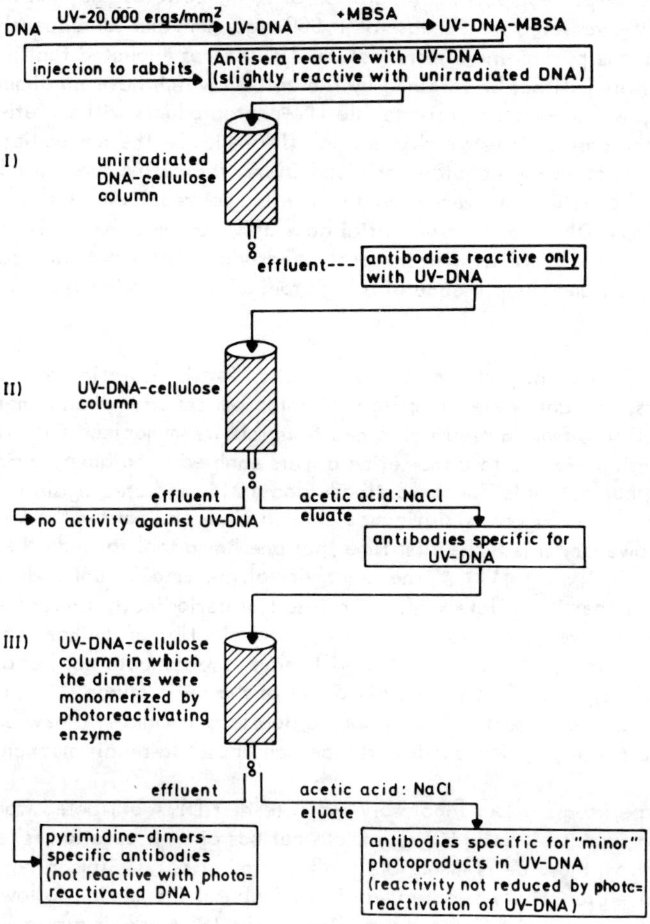

Fig. 1.

A SCHEME FOR THE PREPARATION AND PURIFICATION OF ANTIBODIES AGAINST PYRIMIDINE DIMERS AND ANTIBODIES AGAINST "MINOR" PHOTOPRODUCTS IN ULTRAVIOLET-LIGHT-IRRADIATED DNA a second column (column II) consisting of UV-DNA-Cellulose. Most of the non-reactive antibodies were not retained and are found in the effluent fraction. The anti UV-DNA antibodies were retained by the column and eluted with acetic acid: NaCl. Their activity was completely inhibited by anti IgG antiserum but not at all by anti IgM antiserum, indicating their IgG nature. The next step involved the preparation of DNA-cellulose column in which the DNA was previously UV-irradiated with a dose of 1,000 ergs/mm^2 and subsequently photoreactivated to monomerize all the pyrimidine dimers (column III). It was expected that antibodies specific for pyrimidine dimers will not be retained by the column which have no dimers, while the antibodies directed against non-photoreactivateable UV-photoproducts will be retained by the column. This indeed was the case as illustrated in section III of Fig. 1. The antibodies not retained by this column were not reactive with photoreactivated DNA, indicating their absolute specificity for dimers. The retained antibodies were eluted and tested for reactivity with both UV-DNA and photoreactivated UV-DNA (both at an initial dose of 1,000 ergs/mm^2). These antibodies were as reactive with the UV-DNA as with the photoreactivated UV-DNA, indicating a lack of specificity for dimers and the presence of a high specificity for other type(s) of photoproducts.

Discussion

We have described an affinity chromatography method for the isolation of antibodies specific for pyrimidine dimers, and antibodies specific for photoproducts other than dimers ("minor" photoproducts) in UV-DNA, from antisera obtained from rabbits immunized with UV-DNA. The use of photoreactivating enzyme to monomerize dimers enabled us to have immobilized antigenic sites (photoproducts) for the antibody population directed against "minor" photoproducts. The biological significance of such photoproducts in the DNA of the living cell is not known. However, this is the first time that one has a tool to study the fate of such photoproducts in the cell by testing the reactivity of the specific antibody with DNA of UV-irradiated cells, either immediately after the insult or periodically thereafter. No biochemical methods are available for accurate quantitative determination of "minor" photoproducts in DNA irradiated with "biological" UV dose (that will leave a reasonable number of survived cells, e.g. less than 500 ergs/mm^2). Thus our discovery of the anti "minor" UV-photoproducts, and our purification scheme presented in this communication, is making a new and powerfull tool for investigating the nature, biological importance, and possible repair mechanisms of "minor" photoproducts.

Preliminary experiments in our laboratory indicate that DNA extracted from UV-irradiated cells (Human lymphocytes, HeLa S3) at various periods after irradiation is less reactive than DNA extracted from these cells immidiately after irradiation. In human lymphocytes, for example, a 50% reduced reactivity is observed after 35 Min of incubation following irradiation of 500 ergs/mm^2. Such results indicate the existence of a DNA repair system which can excise "minor" UV-photoproducts from the damaged DNA. Details of the repair studies will be pubished elsewhere.

Acknowledgements

This research was supported in part by the US-Israel Binational Science Foundation (BSF), Jerusalem, Israel, and by the Israel Academy of Sciences.

References

1. Slor, H., Stela Nivi, Tama Lev-Sobe, and Errol C. Friedberg, in DNA Repair and Late Effects, H. Altman, edt. Rotzer-Druck, 1977 (in press)
2. Stollar, B.D., Sara Fucks, and Edna Mozes, J. Immunol. 73, 121 (1973).
3. Lin, S., and Riggs, A.D., J. Mol. Biol. 72, 671 (1972).
4. Minato, S., and Werbin, H., Biochemistry, 10, 4503 (1971).
5. Goldman Karin and Friedberg, E.C., Analyt. Biochem. 53, 124 (1973).
6. Cook, K.H., and E.C. Friedberg, Analyt. Biochem. 73, 411 (1976).
7. Alberts, B.M. et al, in Cold Spring Harb. Symp. Quant. Biol. 33, 289 (1968)

SUBJECT INDEX